EINSTEIN'S REVOLUTION

A STUDY IN HEURISTIC

EINSTEIN'S REVOLUTION

A STUDY IN HEURISTIC

ELIE ZAHAR

OPEN COURT
Chicago and La Salle, Illinois

To order books from Open Court, call toll-free 1-800-815-2280 or visit our website at www.opencourtbooks.com.

This book has been reproduced in a print-on-demand format from the 1989 Open Court printing.

Open Court Publishing Company is a division of Carus Publishing Company.

First printing 1989

Printed and bound in the United States of America.

Library of Congress Cataloging-in-Publication Data

Zahar, Elie.
Einstein's revolution.

Bibliography: p.
Includes index.
1. Relativity (Physics) 2. Physics—Philosophy.
3. Physics—Methodology. 4. Einstein, Albert, 1879–1955.
I. Title.
QC173.55.Z34 1988 530.1'1 88–9968
ISBN 0–8126–9066–4
ISBN 0–8126–9067–2 (pbk.)

PREFACE

This book can be regarded as an over-ambitious attempt to emulate Mach's celebrated *Science of Mechanics*. It is a textbook on relativity written from a historical-methodological point of view. It sets out to present the methodology of scientific research programmes (henceforth referred to as MSRP) as the most suitable framework for the theory of relativity. It should however be pointed out, right from the start, that the acceptance of MSRP is a precondition neither for an understanding nor even for an eventual appreciation of the present work.

Mach's classic offers two great advantages: it can, on the one hand, be used as a textbook even by those who do not share the author's philosophical views; from the *Science of Mechanics* one can, on the other hand, learn a lot about Machian philosophy without having to grasp all the technical details presented by the author.

Similarly, the present work sets out to achieve two seemingly incompatible aims: it is intended to be a fairly self-contained textbook on relativity, which could be used even by those who do not subscribe to MSRP; the only prerequisite for an understanding of the book—except possibly for the last chapter—is a knowledge of elementary algebra and elementary calculus. Moreover, a philosopher primarily interested in methodology and epistemology should be able to understand the philosophical theses put forward without having to wade through all the detailed technical arguments.

The structure of the book is, to a large extent, dictated by its dual aim. Both the introduction and chapter one are philosophical in character; the examples they contain are brief illustrations of the philosophical points made. Where the examples are drawn from relativity theory, forward references indicate the subsequent chapters where the physics is explained in greater detail and depth. As for chapters two to eight, I have tried, as far as possible, to follow one simple rule: namely to devote the early sections of each chapter to presenting, in a non-technical or in a semi-technical way, the philosophical thesis about to be defended. The intermediate sections constitute historical case studies which argue the philosophical thesis by establishing some important propositions of relativistic physics. In doing this, I often found it helpful, before embarking on history proper, to present modern, simplified versions of the proofs of certain theorems. Each chapter ends with a summing up, which is again largely non-technical.

The philosophical reader could, if he so wishes, skip the technical parts. He could read, first the introduction, then chapter one, then the opening and closing sections of each subsequent chapter. 'Could', but certainly not 'should', for methodologies prove their mettle through being confronted with, or rather through being applied to, actual history. It is my belief that seemingly plausible methodologies, like inductivism, falsificationism and conventionalism, break down when pitted against real, hence complex, history of science. To go back to the structure of the book: one can skip the technicalities but not, to my mind, without *some* philosophical loss. Conversely, as already mentioned, the philosophical sections can be omitted and the technical ones can be used as a textbook by those who do not want to bother with too much philosophy: in other words, by those who are happy to rely on the implicit methodology which MSRP claims to have articulated, i.e. characterised in general terms.

It would be impossible to mention all those who helped me towards writing this work. The list would be too long. I should however like explicitly to thank my colleagues Professor J. W. N. Watkins and Dr J. Worrall for their valuable suggestions and for their philosophical criticisms. I also gratefully acknowledge the help I received from Professor C. W. Kilmister, Professor M. Redhead and Professor S. Prokhovnik. There are finally two people to whom I owe more than I could possibly express: my late friend and teacher Imre Lakatos, and Professor J. Stachel. Needless to say, the errors contained in the book are to be attributed exclusively to the author.

TABLE OF CONTENTS

INTRODUCTION

§1. Basic Problems

As is well-known, special relativity is based on the following two axioms:

(P1) *(Relativity Principle)* True physical theories assume the same form in all allowable, i.e., in all inertial, frames of reference.

Einstein postulates an infinite set of inertial frames, any two of which move uniformly in a straight line with respect to each other. By definition, in every inertial frame a body not acted upon by external forces indefinitely keeps the same velocity.

Einstein's second axiom is:

(CL) *(Light-Principle, or Principle of the Constancy of the Speed of Light).* In empty space light has, independently of the state of motion of its source, the same constant velocity c in all directions (c = 300,000 km/sec).

Our starting point is the problem of accounting for Einstein's discovery of his second postulate (CL). (CL) sounds somewhat strange in that it refers to the velocity of light *without* specifying an underlying frame of reference, which alone lends meaning to the velocity concept. (Note that the 'empty space' mentioned in (CL) does not suffice to fix a frame.) In fact, (CL) can be properly grasped only in the light of postulate (P1) which tells us that, in order to express a law of nature, (CL) must necessarily hold in all inertial frames. This yields, as an immediate consequence of (P1) and (CL), the following proposition:

(P2) In all inertial frames light has, independently of the state of motion of its source, the same constant velocity c in all directions.

Note first that, in applying (P1) to (CL) and thus obtaining (P2), we have naturally treated c as an invariant; for otherwise, the light-principle would not have referred to the specific velocity c; it would simply have asserted that light travels with the same speed in all directions, thus leaving open the question whether this speed depends on the underlying frame.

One might also wonder why Einstein chose, as his second postulate, (CL) rather than the clearer and more meaningful proposition (P2). Note that (P2) incorporates the relativity postulate. Einstein may possibly have wanted to separate cleanly a universal relativity principle subsuming all true theories from any specific physical law like (CL). Such a separation was all the more desirable since, unlike (P2), (CL) holds in Maxwell's classical theory; and Einstein was intent upon emphasizing the continuity between classical and relativistic physics. In the opening paragraphs of his 'Electrodynamics of Moving Bodies' (EMB), he constructs special relativity by extending to the whole of physics, and more particularly to electrodynamics, a relativity principle which had hitherto applied to mechanics alone. This is why he might have opted for a principle like (CL) which holds in both classical and relativistic physics. However, classical electrodynamics implies that (CL) holds only in the ether frame, whereas Einstein based his synchronisation procedure on a (CL) understood as applying to every inertial frame, i.e. on (P2). This is why we shall, in what follows, make more frequent use of (P2) than of (CL). Unlike (CL) however, (P2) carries the following paradoxical implications.

To begin with, (P2) is extremely counter-intuitive in that it appears to be trivially falsified by everyday experience. If I move at a speed of 15 km/h, then a car moving at 30 km/h in the same direction will have a velocity of 15 km/h with respect to my frame of reference; or, at any rate, the car will seem to me to be moving more slowly than it 'actually' does. (P2) implies that, if the velocities were 150,000 km/sec and 300,000 km/sec instead of 15 km/h and 30 km/h respectively, then the car would still be moving, relatively to me, at a speed of 300,000 km/sec.

(P2) is a low-level hypothesis which looks like an empirical generalisation. It asserts that, if the speed of light is measured in any inertial frame, the result will be a constant c equal to 300,000 km/sec. This law does not seem to be on a par with the relativity postulate; the latter expresses such a high-level proposition that it is justifiably taken to be a meta-statement (see §8.1). The Relativity Principle is considered to be part, not of the blueprint of the universe, but of a constraint imposed on that blueprint.

Special relativity thus consists of two heterogeneous axioms. Prima facie, this violates a basic tenet of Einstein's methodology: the tenet that physical theories ought to be unified homogeneous wholes in which all primitive notions are closely interconnected.

Finally, Einstein uses (P2) in order to develop a new kinematics in which light has a maximal velocity. Kinematics provides a general framework for the whole of physics. Yet Einstein does not indicate why a property of a particular physical process, namely the constancy of the velocity of light, should be relevant to other physical phenomena, e.g. to the motion of a particle or the variation of energy with speed. In other words, Einstein does not tell us why the peculiar behaviour of light should have repercussions throughout physics.

For these reasons, it seemed obvious that (P2) was not the result of speculation undertaken on the basis of some plausibility argument but was intended to deal with a critical problem-situation. In fact, the question concerning the discovery of the invariance of c is part of a more complex problem consisting of the following three components:

(1) The (alleged) breakdown of classical physics (i.e. of Lorentz's Programme)
(2) Einstein's discovery of special relativity
(3) The acceptance of relativity theory

At first, I expected problems (1) and (3) to be trivially solvable. It seemed obvious that classical physics was abandoned because it clashed with certain observational results. As for relativity, it was accepted because it remained unrefuted and had moreover often been experimentally confirmed. Although, on further investigation, problems (1) and (3) turned out to be far from trivial, I still consider problem (2) to be the most interesting of the three.

§2. Methodology and History

The most usual textbook account of the discovery of relativity is the inductivist one. Since inductivism is a special kind of methodology, let me say a few words about the relation between methodologies on the one hand and historical explanations on the other.

It is a platitude that historical reconstructions are necessary because the historian does not know all the facts. There is in particular one type of fact to which the historian of science can, in principle, have no direct access; namely the intuitive system of appraisal which the scientist applies either in constructing his own theories or in choosing between available research programmes. In certain cases, scientists have written about their methodology and about their heuristics. However, the scientist's reports cannot always be taken at face value: he may either knowingly disguise the truth, or else he may be sincere but have a false consciousness of his own activity. To take an example: Newton claimed to have derived his theories from the phenomena; yet in Corollary V of the *Principia* he conceded that

> The motions of bodies included in a given space are the same among themselves, whether that space is at rest, or moves uniformly forward in a right line without any circular motion.

In this way, Newton gave expression to a classical principle of relativity which makes nonsense of his thesis that the Newtonian system, and in particular the absolute space hypothesis, were induced from the facts. Thus a scientist's explicitly professed methodology, as distinct from his singular and often tacit value-judgments, has by itself no methodological significance. When Newton claims that he has validly inferred his laws from observation, what the historian has to accept is the fact that Newton made such a claim, not that the claim is valid. Even the historian who decides to accept the scientist's own account of his heuristics is already reconstructing history; such an historian would in effect be subscribing to the very bold hypothesis that the scientist knows and says the truth about himself. In this context it is far safer to follow Einstein's advice that, in order to assess a scientist's work, one ought to concentrate on what he does and not on what he says.

Reconstructions are therefore not a luxury but a sheer necessity. Of course, a necessary condition for a reconstruction to be allowable is that it be consistent with the accepted facts. But how does the question of consistency between the facts on the one hand and some given methodology on the other arise in the first place? As is well known, methodologies largely consist either of prescriptive statements or of value-judgments constituted into systems of appraisal; hence, they can be neither refuted nor confirmed, let alone verified, by historical 'facts'. However, to each methodology there corresponds the historical thesis that great scientific breakthroughs took place in accordance with the tenets of the methodology in question. For example, to the inductivist prescription that one ought to 'observe' and then 'induce' theories from experimental results, there corresponds the thesis that great revolutions were achieved by scientists who actually made new observations and then derived theories from them. Naturally, an historian who attributes to a scientist a given methodology has to do so in a manner compatible with what is known about the scientist's discoveries and publications. Consistency with the facts may appear to be a stringent condition, but it is in effect very weak. Consider for example an historian who puts forward a methodology M to which a scientist S is supposed to adhere *intuitively* (i.e., in the majority of cases, S will be incapable of articulating M in general terms). Suppose it is found that, in some situation in the past, S violated M. The historian can always adduce certain external factors—psychological, sociological or even physiological factors—which 'explain' why S's actions did not conform to M. Mere consistency with the facts is therefore a weak constraint, which is why the notion of a rational reconstruction becomes all-important. An allowable reconstruction is said to be rational if it minimises the recourse to external factors for which there exists no independent evidence; where 'external' means 'not subsumed under the methodology in question'. Obviously, no reconstruction is entirely rational but one reconstruction can be more rational than another.

§3. The Michelson-Morley Experiment

Most of the answers to the question of why Einstein's programme superseded Lorentz's refer to the weaknesses of Lorentz's solution to the problem posed by Michelson's results. I shall argue (in chapter two) that these alleged weaknesses are illusory; but let me start by giving, in classical terms, a schematic description of the experiment which was first performed by Michelson in 1881, then repeated with increased precision in 1887 and after.

Michelson used an interferometer consisting of two perpendicular arms: BE = L and BD = ℓ. At B a half-silvered mirror makes an angle of 45° with BE. A light source A emits a beam which is divided at B into two rays: a reflected ray R_1 which travels along BD, is reflected at D, then goes back to B; and a ray R_2 which is transmitted along BE, falls perpendicularly on a mirror at E, then returns to B where it is partially reflected before interfering with R_1. Suppose BE lies in the direction of the earth's motion through the ether, and consider a frame of reference fixed with respect to the earth; then, on the classical account, (c − v) and (c + v) are the speeds of R_2 between B and E and between E and B respectively (where: v = velocity of the earth, c = speed of light). Hence, the time taken by R_2 to return to B is:

$$t_2 = (L/(c - v) + L/(c + v) = (2L/c)\gamma^2,$$

where $\gamma = (1 - v^2/c^2)^{-1/2}$. (Because of its central role in relativity theory, the coefficient $(1 - v^2/c^2)^{-1/2}$ is denoted by a special symbol.)

Let u be the speed of R_1 along BD. The velocity of R_1 in the ether is $(\vec{v} + \vec{u})$ from which it follows that $|\vec{v} + \vec{u}| = c$, i.e., $v^2 + u^2 = c^2$; therefore $u = (c^2 - v^2)^{1/2}$. Hence the time taken by R_1 to return to B is

$$t_1 = 2\ell/(c^2 - v^2)^{1/2} = (2\ell/c)\gamma.$$

If the arms of the interferometer are equal, i.e. if L = ℓ, then

$$t_2 - t_1 = \frac{2L}{c}\gamma^2 - \frac{2L}{c}\gamma = \frac{2L}{c}\gamma(\gamma - 1).$$

If the apparatus is rotated through 90°, the time taken by R_1 to come back to B is increased by $(2L/c)\gamma(\gamma - 1)$, while the time taken by R_2 is diminished by the same quantity. The total time difference is therefore

$$\frac{4L}{c}\gamma(\gamma - 1) = \frac{4L}{c}\left[\left(1 - \frac{v^2}{c^2}\right)^{-1} - \left(1 - \frac{v^2}{c^2}\right)^{-1/2}\right] \doteqdot \frac{2Lv^2}{c^3}$$

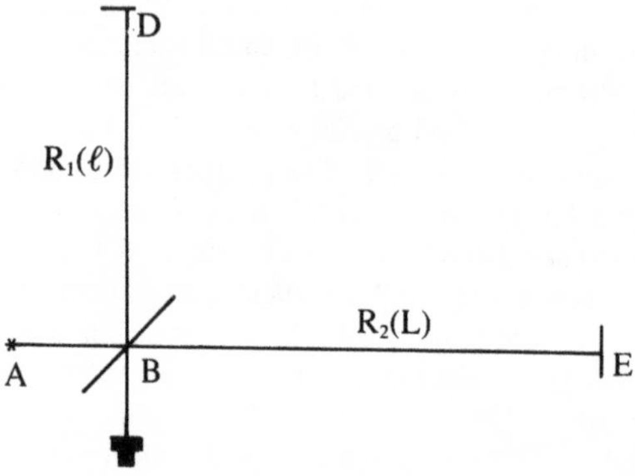

Figure 1.

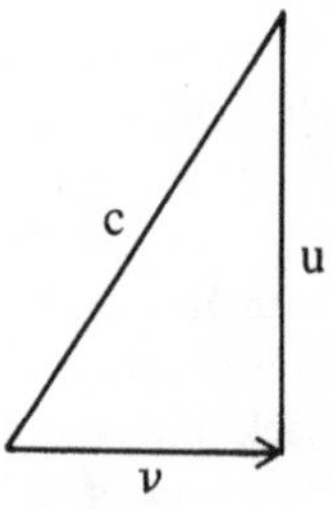

Figure 2.

This time difference should cause a certain shift of the interference fringes. No such shift was observed; so Michelson claimed to have refuted Fresnel's original hypothesis of an ether at rest and confirmed Stokes's theory of total ether drag. However, in 1881 Michelson had taken the speed of R_1 to be c in both directions, thus obtaining a time difference of $4Lv^2/c^3$, which is twice as large as the correct one. Lorentz pointed out this mistake in Michelson's calculations and showed that the real shift fell within the limits of observational error.

Michelson, together with Morley, repeated the experiment in 1887; they increased precision by making the light rays R_1 and R_2 travel several times between the mirrors. Still, no shift of the fringes was observed. Lorentz was now convinced that the Michelson-Morley result was a serious difficulty, not just for Fresnel's, but also for his own theory.

§4. The Inductivist Account of the Michelson-Morley Experiment

Let us now go back to the inductivist explanations of the genesis of special relativity. Such explanations are very satisfying in that they yield a total solution to all the problems: (1), (2) and (3), mentioned above.

Inductivists maintain that the proposition (P2) concerning the invariance of c is a valid generalisation of Michelson's result. According to Max Born:

> . . . the second statement, that of the constancy of the velocity of light, must be regarded as being experimentally established with certainty. (Born 1962, p. 225)

Reichenbach claims:

> . . . it would be mistaken to argue that Einstein's theory gives an explanation of Michelson's experiment, since it does not do so. Michelson's experiment is simply taken over as an axiom. (Reichenbach 1958, p. 201)

Inductivism makes a dual claim, namely that Einstein succeeded where Lorentz failed. This is underlined by Kompaneyets in his (otherwise excellent) textbook:

> A direct experiment was performed which showed that the velocity of light cannot be combined with any other velocity and, in all reference systems, it is equal to a universal constant c. This was the famous Michelson experiment. (Kompaneyets 1962, p. 191)

In other words, it is alleged that the experiment established the invariance of c, which entails the breakdown of the addition law of velocities; since the addition law follows from the Galilean transformation which forms part of Lorentz's system, the latter is refuted by the same experiment. Thus inductivism claims that the Michelson-Morley experiment simultaneously defeated Lorentz and established a fundamental principle of Einstein's system.

Concerning the counter-intuitive character of Principle (P2), it allegedly stems from the fact that classical kinematics was induced from the observed motions of bodies whose speeds are usually small compared with the velocity of light; these results were then recklessly extrapolated, their empirical origin was forgotten and, finally, people took them to be self-evidently true. However, it is Nature's privilege to choose to behave very differently in the case of velocities comparable with the speed of light; we simply have to observe Nature more carefully and obey the dictates of experience, however counter-intuitive they might appear to be. As for the acceptance of (P2) by the scientific community, it is alleged that, having been validly induced from experience, (P2) was simply recognised as true (presumably within the limits of observational error).

Unfortunately, in spite of its appealing character, the inductivist account breaks down both for logical and for historical reasons. From a logical point of view, it has by now become a platitude that observation reports neither establish nor even probabilify higher level theories. Michelson's 'protocol' sentences are not equivalent to the general proposition that in all inertial frames the speed of light is a universal constant independent of the velocity of the source. Moreover, scientists were aware that scientific theories cannot be inductively established. Einstein, who had been strongly influenced by Hume, knew that inductive inferences are invalid. Planck, when he joined the relativistic camp, had good reasons to believe, not that relativity theory was experimentally established, but on the contrary that it might already have been refuted by Kaufmann's experiment (see chapters six and seven). As far as history is concerned, Michael Polanyi pointed out that:

> Michelson's experiment had a negligible effect on the discovery of Relativity. (Polanyi 1958, p. 10)

Shankland's account supports Polanyi's claim:

> When I asked him (i.e., Einstein) how he had learned of the Michelson-Morley experiment he told me that he had become aware of it through the writings of H.A. Lorentz, but only after 1905 had it come to his attention. (Shankland 1963)

Thus, it is an accepted fact that Einstein read Lorentz's 1895 *Versuch einer Theorie der elektrischen und optischen Erscheinungen in bewegten Körpern*, where Michelson's experiment is discussed at length. It is also a fact that Einstein denied that Michelson's results played any direct inspirational role in the genesis of special relativity. An historian according to whom Einstein was an inductivist could say that Einstein was simply lying or that, in later life, he had simply forgotten about the true origins of special relativity. However Holton (1969) convincingly showed that, throughout his life, Einstein, though an admirer of Michelson's work, always stopped short of saying that the discovery of relativity had anything to do with Michelson's experiment. The inductivist is therefore left with the hypothesis that Einstein systematically lied, throughout his life, about the role of Michelson's experiment. This is a logically possible explanation but it is out of character for Einstein, who was known for his intellectual honesty and who, in different circumstances, freely admitted his debt to other scientists. So the inductivist account can be made consistent with the facts, but there exists no independent evidence for the psychological hypothesis that Einstein, though normally an honest man, always lied about the influence of Michelson's work on his own. There is of course another solution open to the inductivist: he could accept that Einstein was telling the truth and simultaneously contend that some external factors caused Einstein to act irrationally, i.e. to violate the canons of inductivism by basing special relativity on some speculative argument; but, while the inductivist would have to specify these external factors and find some independent evidence for their existence, it will be shown that an historian who subscribes to MSRP needs resort to no external considerations at all (see chapter three). Thus, in the absence of independent evidence, an inductivist reconstruction is less rational than one made in the light of MSRP.

§5. The Falsificationist Account

The basic prescription of Popper's falsificationist methodology is: aim at replacing an empirically refuted theory T by a new hypothesis T′, which explains the success of T, explains the refuting evidence and is moreover independently testable (i.e., T′ makes predictions different both from the predictions of T and from the evidence which refuted T). Corresponding to this prescription is the historical thesis that major scientific breakthroughs were achieved by theories being refuted then modified in a content-increasing way. Scientific revolutions are supposed to occur in response to experimental refutations which constitute important moments in this dynamic of change.

At first sight it looks as if this thesis should apply—as Popper himself at one time thought it did—to the scientific revolution ushered in by Einstein's special relativity theory. The falsificationist account runs as follows: Michelson's experiment, which was called "the greatest negative experiment in the history of science" (Lakatos & Musgrave 1970, p. 162, quoting Bernal), refuted Lorentz's classical system; in order to explain away Michelson's result, Lorentz allegedly resorted to an auxiliary assumption, the Lorentz-Fitzgerald contraction hypothesis *(LFC)*, which was however ad hoc. On his own admission Einstein was familiar with part of Lorentz's work, namely with the "Théorie électromagnétique de Maxwell" (*TEM* 1892) and with the *Versuch einer Theorie der elektrischen und optischen Erscheinungen in bewegten Koerpern* (*Versuch* 1895 cf. Born 1956). Finding the LFC ad hoc, Einstein proposed his special theory of relativity (*STR* 1905) as a better, i.e. non ad hoc, alternative. The STR was corroborated and thus came to be accepted by the scientific community. Note that this explanation, which is clearly superior to the inductivist one, *depends in an essential way on the notion of ad hocness.*

§6. Holton's Account

In his "Einstein, Michelson and the 'Crucial' Experiment" Holton gives further support to the claim that the LFC was ad hoc:

> This saving of Hülfshypothese [i.e., the LFC] is introduced completely ad hoc. No explicit comment is made which connects this assumed shrinkage with the Lorentz transformations in their still primitive form as published earlier in the book [i.e. Lorentz's 1895 Versuch]. (Holton 1969, p. 171)

Holton however attaches to ad hocness a meaning different from that given to it by Popper. For Holton the LFC is ad hoc because it is not integrated into the rest of Lorentz's system. For example, it is allegedly unconnected with the Lorentz transformation equations.

Before we can assess either Popper's or Holton's account of the origins of relativity, we therefore have to define more precisely what is meant for a theory to be ad hoc with respect to some fact. One such definition will be given in the course of dealing with the problems posed by the logic of scientific discovery (see chapter 1.12). It can however be said at this stage that, even if Popper's or Holton's approach proves correct, it might solve problems (1) and (3) mentioned above, but would not by itself explain how (P2) was discovered.

One advantage of inductivist methodology is that it yields a genuine logic of discovery: one discovers theories (or rather one ought to discover them) by inducing them from observation. Both Popper and Holton recognize that theories transcend facts. Scientific hypotheses must of course be such as to yield all accepted empirical results, but the latter do not suffice to determine the theories completely: there exists between scientific theories and factual statements a non-rational gap. Does this imply that we must give up the hope of finding an internal explanation of the discovery of scientific hypotheses? I do not think so. Of course, I do not wish to assert that there exist no 'creative' jumps in the process of discovery, but I maintain that such non-rational moments are much narrower in scope than the gap between observation statements and theories.

— 1 —

METHODOLOGICAL PRELIMINARIES

§1.1 Logic of Discovery or Psychology of Invention?

§1.1.1 Popper's Darwinian Account versus Lakatos's MSRP

Until recently, it was very much part of orthodoxy in philosophy of science that there is no room for rational heuristics. Popper, Reichenbach and other members of the Vienna Circle all agreed that there is a sharp distinction between the 'context of discovery' and the 'context of justification'. Only the latter lies within the domain of methodology whose proper task is to evaluate theories laid on the table, i.e., theories already constructed. As for the context of discovery, it belongs to the psychology of invention. Neither calling for, nor even being susceptible of, any rational reconstruction, the process of discovery allegedly rests on what Einstein called 'empathy with nature'.

Most philosophers have said little about how theories arrive 'on the table' in the first place. An exception was Popper in a book entitled *Die beiden Grundprobleme der Erkenntnistheorie,* written in 1930–31 but published only in 1979. Popper's account of the emergence of new theories is linked to an overall Darwinian world-view: if new theories turn out to explain certain facts, this is not the result of rational and concerted efforts on the part of the scientists; it is an effect of the superabundance of available theories. New hypotheses emerge like spontaneous mutations; most of these will be experimentally falsified and rejected; but, since there has been an overproduction of hypotheses, some of them will survive. In Popper's own words,

> there exists no *law-like dependence* between receptions, [i.e.] between new objective conditions, and the emergence of reactions (or rather, there is only *one* form of dependence, namely the *selective* one, which renders non-adaptive reactions worthless and forces the organism to come up with something new or else go under—a form of dependence which cannot possibly enable the unexplained aspect of this process to be clarified). (Popper 1979, p. 27; my translation)

To repeat: Popper offers a Darwinian account of the progress of knowledge. Progress is supposed to result negatively from the elimination, by natural selection, of defective alternatives. As for the positive adaptation of theories to facts, it is the effect of chance; there is no intended adequacy between theory and reality. Only overproduction makes it possible for certain theories both to escape falsification and to explain the past empirical success of their rivals. Hence, there is no genuine logic of discovery, only a psychology of invention juxtaposed to a methodology which appraises fully fledged theories.

Several recent philosophers have suggested that, despite Popper and the old orthodoxy, there is room for a rational heuristics in physical science. The most important is, in my opinion, Imre Lakatos, who explains the development of science in terms of what he calls the methodology of scientific research programmes (MSRP). Every research programme is characterised by a hard core and a positive heuristic giving rise to a sequence of theories. The hard core consists of certain propositions shared by all elements of the programme. The positive heuristic consists of certain directives, possibly deriving from a metaphysical principle, which guide research by indicating both the method by which new theories should be constructed and the manner in which the whole programme should deal with empirical refutations. Thus, the development of a programme is simultaneously governed by its hard core and by its positive heuristic. The core is protected by the methodological decision of never regarding it as falsified even in cases of empirical refutation; this immunisation policy constitutes the so-called negative heuristic of the programme, each of whose members is subject to two conditions: it should logically imply the hard core while conforming to the positive heuristic (henceforth referred to simply as the heuristic). Within a given programme, a theory represents progress if it satisfies three conditions: it should entail novel empirical predictions; some of these predictions ought eventually to be confirmed; and finally, the theory should be structured in accordance with the heuristic. For instance, no hypothesis should resort to factors judged to be occult in the light of the metaphysical system underpinning the programme. Closely connected with this definition of progress are three distinct notions of ad hocness. A theory is said to be ad hoc_1 if it makes no novel predictions. It is ad hoc_2 if none of its novel predictions have been actually verified: for one reason or another the relevant experiments may not have been carried out, or (much worse) an experiment devised to test a novel prediction may have yielded a negative result. Finally, the theory is said to be ad hoc_3 if it is obtained from its predecessors by a modification of the auxiliary hypotheses which violates the heuristic of the programme.

Though the hard core and the heuristic do not *uniquely* determine the sequence of theories to be constructed, they constitute a genuine logic of discovery; they provide positive hints as to what to do in a period of crisis. Lakatos regards this kind of heuristics as belonging to some sort of limbo which is rational and non-psychologistic (see Lakatos 1978).

However neither Lakatos nor any other philosopher has, to my knowledge, made this suggestion of a rational heuristics at all precise. In this chapter I try to remedy this defect by presenting a deductivist thesis which, though strictly inconsistent with Lakatos's position, is in the spirit of his programme while remaining compatible with Popper's anti-inductivist views about the progress of science.

Right from the beginning, we ought to distinguish sharply between two questions. The first concerns the possibility of rationally reconstructing the process of discovery; the second is whether such a reconstruction has any role to play in the appraisal of theories after their construction. We shall see in what sense Popper and Reichenbach are right in answering the second question in the negative. As for the first question, I intend to show that the process of discovery is much more rational than it appears at first sight; that it is neither inductive nor largely intuitive; that it does not belong to any kind of limbo, but largely rests on deductive arguments from principles which underlie not only science and deductive metaphysics but also everyday decisions. The choice of consistent sets of such principles constitutes the positive heuristics of research programmes. Thus I hope, on the one hand, to underline the continuity between science and common sense knowledge and, on the other, to show that the heuristics of programmes operate deductively (at the metalevel) and not in any mysterious way. Rather than try to provide a general a priori definition of heuristics, let me begin by examining some specific methods used in the construction of theories, with a view to extracting from them a more general structure. At the same time I shall pursue a policy of attrition, which consists in gradually reducing the scope and importance of the allegedly irrational aspects of the process of discovery.

§1.1.2 Ad Hocness and the Adaptation of Theories to Facts

Since, as was said above, 'ad hocness' depends in an essential way on the notion of *novelty of facts*, this notion has to be examined in some detail.

Before embarking on a general discussion let us consider a few concrete examples of novel facts. Lakatos mentions the return of Halley's comet as a new fact anticipated by the Newtonian programme and, of course, it has to be agreed that the discovery of any new (type of) fact is the discovery of a novel fact. The converse however does not hold; for, if we simply equate novelty with *temporal* novelty, we are driven into a paradoxical situation. We should, for example, have to give Einstein no credit for explaining the anomalous precession of Mercury's perihelion because this fact had been recorded long before general relativity was proposed. Similarly, we should have to say, contrary to informed opinion, that Michelson's results did not confirm special relativity and Galileo's experiments on projectiles did not support Newton's theory of gravitation. Lakatos, who did not easily dismiss the judgments of physicists, was aware of this difficulty and tried to avert it by shifting his original view and saying that, *in the light of a new theory,* some known facts may 'turn into' novel ones. Like Feyerabend (1975), Lakatos was thereby drawing a natural conclusion from Popper's thesis that all observations are theory-laden; so a change in theory may well result in a change of fact. For example, whereas Balmer merely 'observed' that the hydrogen lines obey a certain formula, Bohr connected these lines with the energy levels of the electron in the hydrogen atom.

However, this modified notion of 'novel facts' is open to the following fatal objection. Any theory is a set of propositions connecting different terms and concepts. We can always define the properties of a physical entity like 'mass' through the relations which 'mass' bears to other notions within a given theory. Consequently a new hypothesis will generally ascribe new meanings to old terms. For instance, any experimental consequence of relativity theory involving 'mass' say, would trivially become the expression of a novel fact: the fact that a steel ball rolling down a slope takes a certain time to reach the bottom would automatically become a novel fact, once the steel ball is regarded as possessing relativistic mass. This is obviously absurd.

Although Michelson's result contains a reference to 'length' which acquires a new meaning in special relativity, one must not claim that the result has thereby been changed into a novel fact. The Michelson result is indeed novel vis-à-vis special relativity, but its novelty does not rest on this interpretation of 'length'; for the 'crucial' experiment can be described in an observational language which remains unaffected by theory-change. 'The arms of the interferometer have equal lengths' can for instance be replaced by 'The extremities of the two arms can be made to coincide by placing the two arms alongside each other'.

When does a prediction lend—if experimentally corroborated—*genuine* support to a theory, and when is such support only spurious? Consider the following situation. We are given a set of facts and a theory $T[\lambda_1,\lambda_2,\ldots,\lambda_m]$ containing an appropriate number of parameters. Very often, the parameters can be adjusted so as to yield a theory T* which 'explains' the given facts; it may even happen, given sufficiently many degrees of freedom, that new dramatic relations between old facts can be exhibited, or rather fabricated. Suppose for example that

(1) $\vdash T[\lambda_1,\ldots,\lambda_m] \Rightarrow (I \Rightarrow P[\lambda_1,\ldots,\lambda_m])$,

where I describes some initial conditions and expresses auxiliary assumptions independent of the parameters, while $P[\lambda_1,\ldots,\lambda_m]$ is a prediction dependent on the values assigned to $\lambda_1,\ldots,\lambda_m$. Suppose further that a sequence of experiments is carried out, where the initial conditions described by I are realised and some outcome P* is actually observed. Thus $I \wedge P^*$ is ascertained. It might be possible to establish the existence of a set Q* of numerical m-tuples such that:

$$(2)\ \vdash (T[\lambda_1,\ldots,\lambda_m]\wedge I) \Rightarrow ((<\lambda_1,\ldots,\lambda_m>\in Q^*) \Rightarrow (P[\lambda_1,\ldots,\lambda_m] \Rightarrow P^*)),$$

or even

$$(2')\ \vdash (T[\lambda_1,\ldots,\lambda_m]\wedge I) \Rightarrow ((<\lambda_1,\ldots,\lambda_m>\in Q^*) \Leftrightarrow (P[\lambda_1,\ldots,\lambda_m] \Rightarrow P^*))$$

Hence, in view of (1):

$$(3)\ \vdash (T[\lambda_1,\ldots,\lambda_m]\wedge I) \Rightarrow ((<\lambda_1,\ldots,\lambda_m>\in Q^*) \Rightarrow P^*).$$

We may even be able to prove:

$$(4)\ I \Rightarrow ((<\lambda_1,\ldots,\lambda_m>\in Q^*) \Rightarrow (P[\lambda_1,\ldots,\lambda_m] \Rightarrow P^*)),$$

or even

$$(5)\ I \Rightarrow ((<\lambda_1,\ldots,\lambda_m>\in Q^*) \Leftrightarrow (P[\lambda_1,\ldots,\lambda_m] \Rightarrow P^*))$$

At any rate, if we choose $\lambda_1^*,\ldots,\lambda_m^*$ so that $<\lambda_1^*,\ldots,\lambda_m^*>\in Q^*$, and then take $T^* \equiv T[\lambda_1^*,\ldots,\lambda_m^*]$, we obtain

$$(6)\ \vdash T^* \Rightarrow (I \Rightarrow P^*).$$

For example, by expanding a function in terms of spherical harmonics and then adjusting the coefficients of the expansion, the precession of Mercury's perihelion can be accounted for within Newtonian physics (Adler, Bazin & Schiffer 1965, ch. 6, 6.6). In such a case we should certainly say that the facts provide little or no evidential support for the theory, since the theory was specifically designed to deal with the facts. Facts can therefore play a crucial positive role in the evaluation of numerical parameters; in certain cases a metaphysical proposition, or meta-proposition, determines a physical theory to within certain quantities; the latter are then fixed by a series of observations, which cannot be regarded as corroborating the theory precisely because they are implicated in its construction. Hence, the way in which a theory is constructed is relevant to the appraisal of its empirical merits. If we are given only the end product T* which explains some fact *a* (say), we shall in general be unable to determine whether *a* lends genuine support to T* or whether T* was cleverly engineered to yield *a*. This suggests the following redefinition of the notion of 'novel fact':

(*) A fact will be considered novel with respect to a given hypothesis if *it was not used in the construction of the hypothesis*.

Relations (1)–(5) explicate what is meant by the adjustment of parameters to facts and so illustrate a special, though widely applied, method of using experimental results in the construction of theories.

Of course, under definition (*), any temporally new (type of) experimental result *e* will be novel since no theory could have been proposed in the light of the evidence *e*. Temporal novelty is thus a sufficient, but by no means necessary, condition for novelty. A temporally new fact may have greater *psychological* impact than some known fact but this, on its own, is irrelevant to the objective empirical support which it lends to a hypothesis.

It follows from (*) that empirical non-ad hocness is a three-place relation N(H,T,e) between a heuristic H, a theory T and an observation statement e. This relation obtains iff T explains e and T was constructed by means of H, where H is independent of e. Thus: if e is empirically verified, then e supports, not T alone, but T and H jointly. It is conceivable that two research programmes P_1 and P_2, possessing two different heuristics H_1 and H_2, should give rise to the same theory T and that a fact e should provide support for (H_1,T) but not for (H_2,T). Thus, it is more appropriate to speak of support for a research programme P than for an isolated hypothesis T. This is because P possesses a heuristic H which governs the construction of T.

This new criterion of novelty of facts implies that the traditional methods of historical research are even more vital for evaluating experimental support than Lakatos had already suggested. The historian has to read the private correspondence of the scientist whose ideas he is studying; his purpose will not be to delve into the psyche of the scientist, but to disentangle the heuristic reasoning which the latter used in order to arrive at some new theory. In other words: the historian has to reconstruct the research programme within which the scientist was working. This reconstruction is necessary because only finished theories, but not the heuristic methods leading to them, are usually published.

It may of course look as though (*) constitutes a psychologistic, or person-relative, definition of evidential support. Relations (1)–(5), at any rate, show that the process of adjusting parameters to experimental results is an objective one. It may however be necessary to adapt more abstract mathematical entities, like matrices, operators and even logical functors, to known facts. This is why it is preferable to adopt the more general, though admittedly less precise, definition (*) of the novelty of a prediction. To repeat: this definition is to be given an objective meaning. Thus Popper and the Vienna Circle are, in a sense, right in distinguishing between the context of discovery and that of justification: in the methodological appraisal of theories, it is not the genesis of scientific hypotheses as such but certain logical relations which should play a significant role. In certain cases, it can be mathematically proved that the presence of sufficiently many open parameters guarantees that some finite set of facts will unfailingly be accommodated by a law of a given form; for there are at least as many unknowns as there are independent equations. Since a mathematical theorem tells us nothing about physical reality, it follows that the facts which help towards constructing a theory cannot simultaneously strongly support it.

§1.1.3 An Ad Hoc Derivation of the Contraction Hypothesis

Both in "The Relative Motion of the Earth and the Ether" (1892) and in the *Versuch* (1895) Lorentz pursued two distinct lines of reasoning, each leading to a contraction hypothesis which accounts for Michelson's result. One of Lorentz's derivations is ad hoc but the other is not. As an illustration of definition (*) above, let us reproduce Lorentz's ad hoc argument which issues, not in one, but in a whole spectrum of contraction hypotheses.

In deducing the prediction:

$$\Delta t = t_2 - t_1 = \left[\frac{2L}{c}\gamma^2 - \frac{2L}{c}\gamma\right] \quad (\gamma \underset{\text{Def}}{=} (1 - v^2/c^2)^{-1/2})$$

which conflicts with Michelson's result, two major concepts are used, namely 'length' and 'velocity.' Through tampering with the properties of length or of velocity or of both, one might hope to turn the deduction of $\Delta t = (2L/c)\gamma^2 - (2L/c)\gamma$ into a prediction of the null outcome $\Delta t = 0$, which Michelson 'in fact observed'.

Lorentz showed how he *could* explain Michelson's result by altering the properties of the length of a rigid rod: if we assume that the lengths of the two arms of the interferometer, the one moving transversally and the other horizontally through the ether, are shortened by the factors λ_1 and λ_2 respectively, then:

$$\Delta t = \frac{2L\lambda_2}{c}\gamma^2 - \frac{2L\lambda_1}{c}\gamma = \frac{2L\gamma}{c}(\lambda_2\gamma - \lambda_1).$$

Thus, if λ_1 and λ_2 are chosen in such a way that $\lambda_1/\lambda_2 = \gamma$, i.e. if $\lambda_2 = \lambda_1 (1 - v^2/c^2)^{1/2}$, then $\Delta t = 0$ and Michelson's result is accounted for. The equation $\lambda_2 = \lambda_1(1 - v^2/c^2)^{1/2}$ obviously possesses infinitely many solutions in λ_1 and λ_2. Referring back to relations (2) and (3) above, we can say that Q* consists of all ordered pairs $<\lambda_1,\lambda_2>$ such that $\lambda_2 =$

$\lambda_1(1 - v^2/c^2)^{1/2}$. Lorentz remarks that the particular solution: $\lambda_1 = 1, \lambda_2 = (1 - v^2/c^2)^{1/2}$, which constitutes the Lorentz-Fitzgerald Contraction Hypothesis (LFC), is the one already implied by his molecular forces hypothesis (MFH). The derivation of the MFH, which is in effect Lorentz's non ad hoc argument, will be discussed in some detail in chapter two.

Lorentz was aware that his two lines of reasoning were independent of each other, as is clearly indicated by the following quotation from "The Relative Motion of the Earth and the Ether" (1892):

> It may be worth mentioning that the result obtained in the case of electric forces yields, when applied to molecular forces, exactly the value given above for α [α is in fact $(1 - \lambda_1/\lambda_2)$]. (LCP vol. 4, p. 222)

Both heuristic paths were retraced in the *Versuch* of 1895: the non ad hoc path in section 23 and the ad hoc one in Section 90 (LCP vol. 5, pp. 1-138). Only Section 90 was included in the 'Relativitätsprinzip', which was later translated into English. Given the popularity of this book, it is hardly surprising that a myth arose about the ad hoc character of the Lorentz-Fitzgerald Contraction Hypothesis.

§1.1.4 The Correspondence Principle

Certain adequacy requirements which, according to Popper, ought to be met by scientific hypotheses turn out to be heuristic devices governing the construction of these hypotheses; no wonder therefore that the latter actually satisfy the requirements in question. Thus a rational account of heuristics dispels part of the mystery surrounding the miraculous 'progress' of science. Consider for example the condition that a new theory should explain the past empirical success of the rival it is meant to supplant. This requirement appears in the history of science as a heuristic tool, the so-called correspondence principle, which was explicitly enunciated as such by Niels Bohr: quantum laws ought to be so constructed as to tend to their classical counterparts under certain conditions (e.g. $h \to 0$ where the '$\to$' means 'tends towards'). Note that Bohr put forward this principle in 1923, i.e., before matrix or wave mechanics took on their final form. The correspondence principle, which proved very fruitful in the development of atomic physics, had of course been tacitly applied in other domains long before the birth of quantum theory.

Let $\varphi = 0$ be a law, which satisfactorily explains all the phenomena within some domain D but breaks down outside D. We may thus be led to construct a new equation $\Phi = 0$, which not only avoids the known refutations of $\varphi = 0$ by the adjustment of certain parameters, but also tends to $\varphi = 0$ when certain quantities, by tending to 0, ensure that we remain within D. I do not intend to go into the detail of how these parameters and quantities are arrived at; suffice it to say that they may be introduced, as in (1) above, in order to yield some known recalcitrant facts. (The Fitzgerald contraction factor $(1 - v^2/c^2)^{1/2}$, through which the parameter v/c was introduced, is a case in point.)

Let us note that, in all non-trivial cases, both $\Phi \rightarrow \varphi$ and $\Phi \neq \varphi$ hold; i.e. the law $\Phi = 0$ can both contradict and slide smoothly into $\varphi = 0$. Yet, in all this, we remain strictly within the boundaries of deductive logic. This is how quantum mechanics contradicts classical laws while corresponding to them; this is also how Newtonian gravitation, though strictly incompatible with Kepler's laws, is nonetheless based on them. In all these cases we have to do with a syntactic-mathematical type of continuity which, coupled with the semantic stability of our observational language, entails continuity at the *empirical* level. This is consistent with the occurrence of dramatic revolutions at the ontological level; that is, in case we still insist on providing theories with semantic interpretations. For example, there is no indication that the ontology of quantum mechanics, whatever it eventually turns out to be, will prove continuous with that of classical physics; yet quantum mechanics is governed by the correspondence principle. It is worth noting that, in order to prevent metaphysical upheavals from destabilising science, Duhem and Poincaré restricted themselves to the consideration of low-level observation coupled with a purely syntactical approach to theory construction.

As an important illustration of the correspondence principle, let us examine Planck's modification of Newton's second law of motion (for more on this see §7.1). Planck used the equation:

$$\vec{f} = \frac{d}{dt}(m\vec{v})$$

in order to derive

$$\vec{f} = \frac{d}{dt}\left(\frac{m\vec{v}}{\sqrt{(1 - v^2/c^2)}}\right),$$

which is in general incompatible with classical dynamics. Planck proceeded as follows: taking stock of Einstein's approach, he demanded that Newton's 2nd law

$$\vec{f} - m\vec{a} = \vec{f} - \frac{d}{dt}(m\vec{v}) = 0$$

be replaced by a Lorentz-covariant equation $\psi = 0$. Planck knew that the Lorentz transformation tends to the Galilean one when $(v/c) \to 0$. He consequently required that $\Phi \to (\vec{f} - m\vec{a})$, as $(v/c) \to 0$. It follows by continuity that $\Phi_0 = \vec{f} - m\vec{a}$, where Φ_0 denotes the value of Φ for $v = 0$. Then, he had to find a particular case where both the force and its transformation law are known independently of the motion caused by the force. Lorentz, Einstein and Poincaré had already determined the transformation rules for the electromagnetic field and, long before that, Lorentz had found the force acting on an electron to be equal to: $e(\vec{E} + (\vec{v}/c) \times \vec{H})$, where: e = charge of the electron, $\vec{v}$ = its velocity; $\vec{E}$ = electric field, $\vec{H}$ = magnetic field, and c = speed of light. Planck chose a moving inertial frame in which the electron is instantaneously at rest; he applied the law $\vec{f}' - m\vec{a}' = 0$ to the immobile electron, thus obtaining $e\vec{E}' - m\vec{a}' = 0$, where the dashes refer to the physical quantities in the moving frame; he finally applied an inverse Lorentz transformation, which took him back to the stationary frame and yielded the equation

$$e(\vec{E} + \frac{\vec{v}}{c} \times \vec{H}) = \frac{d}{dt}\left(\frac{m\vec{v}}{\sqrt{(1 - v^2/c^2)}}\right).$$

Since the left hand side is the so-called Lorentz-force, Planck equated the force with the vector:

$$\frac{d}{dt}\left(\frac{m\vec{v}}{\sqrt{(1 - v^2/c^2)}}\right).$$

Thus he overthrew Newton's law of motion

$$\vec{f} = \frac{d}{dt}(m\vec{v}),$$

which had hitherto been regarded as the cornerstone of physics, by a proof involving the correspondence principle applied to that same law. The relation of Kepler to Newton, which is more complicated than that of Newton to Planck, will be examined below (in §1.1.7). Through the correspondence principle and the method of directly adapting theories to certain facts, the quasi-inductive path of science loses its miraculous character; it becomes an intended effect.

It might be felt that there is nothing sacrosanct about the correspondence principle. After all, a new theory needs only explain the *empirical* success of its predecessor without *having* to be continuous with any previous hypothesis. The answer to this objection is that there exist general inductive reasons why continuity must govern the transition from one theory to the next. When scientists speak of past empirical successes, they have in mind *virtual* as well as *actually observed* facts. They think of smooth domains consisting both of presumed and of ascertained facts. They set out to account, not only for known isolated observations, but also for allegedly observed regularities. Such regularities are regarded as being approximately captured by existing laws. The scientists may of course err in their assessment, but this is irrelevant to the present argument: it is still the case that the only way of getting at these *presumed* regularities is through continuity either with past theories or with some of their low-level consequences. Hence the correspondence principle which, though it may not operate as a heuristic device, remains an essential adequacy requirement.

Let us give an example. Strictly speaking, Newton had to account, not for Kepler's laws, but only for Tycho's data. However, even if he had countenanced a completely ad hoc theory, Newton could hardly have picked off each of Tycho's countless data then built them, one by one, into his theory. More importantly: Newton took it for granted that Kepler's laws, which subsume infinitely many virtual facts, correctly interpolate Tycho's finite set of observations. ('Correctly' is here to be understood as 'approximately truly'.) So the presumed empirical success which Newton set out to explain went far beyond any *finite* set of numerical results.

Even quantum mechanics, though it violates Leibnizian continuity and thus breaks with a long tradition, obeys the correspondence principle and is, in this sense, continuous with classical mechanics. It therefore seems as if the correspondence principle is with us to stay, if only as a desideratum.

§1.1.5 The Metaphysical Character of the Hard Core

The principles of correspondence and of the adaptation of theories to facts, which leave a great latitude for the choice of new theories, are common to all research programmes; their main function is to preserve all empirically ascertained knowledge. As is shown by the examples above, the process of discovery brings into play other principles (for example the covariance requirement), which are specific to particular programmes (in this case the relativistic one). I now propose to examine the nature and origins of these supplementary principles.

In order for the hard core to be protected against experimental refutation, it should, when taken in isolation, prove unfalsifiable; it ought to be metaphysical in Popper's sense. This is a conclusion which Lakatos did not clearly draw. It is my claim that certain metaphysical propositions have prescriptive counterparts which can, in turn, be 'translated' into meta-statements about scientific hypotheses. An ontological thesis can obviously impose conditions on the form which scientific theories may assume; that is, *ontology may be taken to have prescriptive import*. It is essential that such prescriptions be translated into propositions, or into meta-propositions, if one wants the positive heuristic to operate *deductively* by providing premises for the logical derivations which determine theories. Thus, we have the following scheme:

Metaphysics (hard core) - - - - → Prescriptions - - - - → Meta-statements (positive heuristic)

In other words: the heuristic may reflect certain aspects of the hard core; the distinction between hard core and heuristic is not as absolute as Lakatos imagined. In this context, we should note that the correspondence principle can be given the form of a meta-statement. As above, let $\varphi = 0$ denote a known law which is to be modified; the correspondence principle could read as follows: 'The new law is of the form $\phi = 0$, where ϕ is a function of the quantity λ such that $\Phi \rightarrow \varphi$ as $\lambda \rightarrow 0$' (λ may be some known parameter like v/c). This proposition is obviously a meta-sentence.

Let me give two examples. Part of the ether programme's hard core is the proposition that physical reality consists of a universal medium possessing mechanical, or electro-mechanical, properties; hence the prescription that theories should ideally derive all known phenomena from the states of one ether. The corresponding meta-principle is that all laws of nature contain only the concepts of position, time, mass, density and (possibly) charge (see chapter two). Special relativity is based on the metaphysical proposition that no privileged inertial frame exists. This leads to the prescription that all theories should assume the same form in all inertial frames; the corresponding meta-statement is that all laws of nature are Lorentz-covariant (see chapter three).

§1.1.6 The Logic of Discovery

By means of examples taken from the history of science I propose to show that the logic of scientific discovery is not inductive; that it does not resemble artistic creation either, for it is largely deductive. Against Lakatos, I maintain that heuristics possesses *no* special status half-way between logic and psychology. Unlike Popper, I think that deduction constitutes the most important element in the process of invention. It has been said that deduction itself contains irrational elements in that every step within a proof is not uniquely determined by the previous steps. My answer to this objection is two-fold. First, the objection is consistent with my aim of breaching the alleged dichotomy between the irrational invention of hypotheses on the one hand and the deduction of empirically decidable consequences on the other;

if my claim holds, then the supposedly irrational character of the context of discovery would be no different from that of the context of justification. Secondly, I do not believe that mathematicians produce proofs by a process resembling a sequence of spontaneous mutations, but that they reason in advance about what kind of proof is likely to lead them to some desired result. They carry out something like proof-analysis, for example a piece of meta-reasoning based on an analogy with some available proof which is known to go through in some neighbouring domain. Mathematicians generally have an idea about the overall structure of a deduction before they fill in the tedious details.

The premisses on which the construction of new theories rests are precisely the metastatements described in (3) above; these naturally include the correspondence principle. In my opinion, these metastatements either are, or else flow from, largely non-technical philosophical principles used in everyday life. They may form part of commonsense knowledge, or they may be innate and thus possess a genetic basis. Their emergence could be explained, along Darwinian lines, in the way in which Popper accounts for the invention of new hypotheses: namely through mutations independent of all receptions. What is important for my thesis is that these principles should have been present either before, or else concurrently with, the birth of science proper. Their insertion into the evolutionary process, which leads to the constitution of scientific programmes, makes systematic research possible. Such principles operate heuristically either by generating new hypotheses, or through being applied to existing laws in order to create theories conforming more closely to the principles in question; the new theories are subsequently confronted with the facts and the same process is then repeated all over again. In other words: the emergence of consciousness and of its innate tendencies may, one day, be explained in Darwinian terms; but, once given, consciousness makes purposive action possible. Such purposive action manifests itself most brilliantly in the sciences, whose development from then on follows, not a Darwinian, but a Lamarckian path.

Let me formulate, as clearly as I can, some of the metaphysical propositions on which the logic of discovery rests. According to Emile Meyerson, the whole of science is informed by the identity principle (see §1.3) which consists in denying the apparent diversity of the phenomena; or, rather, in deriving this diversity from one fixed set of laws. This is the so-called legal version of the identity principle which all species need in order to survive. According to the causal version, nature consists of substances governed by strict conservation laws (see Meyerson 1908, ch. 1). The human mind has an irresistible tendency to hypostasize natural processes, thus turning them into things whose total quantity remains constant. This is an innate propensity, which already leads the child to a belief in the persistence of material objects. The identity principle, both in its legal and in its causal versions, thus clearly precedes science proper. Of course, it also permeates science itself. In the eighteenth and nineteenth centuries it led to the discovery of the conservation laws of momentum and energy. The *unity requirement,* which played a central role in Einstein's methodology, clearly derives from the legal version of the identity principle: all phenomena should be subsumed under one all-embracing theory, possibly under a unique geometry. Nature does not split into disjoint domains subject to different laws. Should such a schism appear within science, then one ought to try and deduce all the laws, possibly in some modified form, from a unique hypothesis. A first and simplest step towards such a unified solution consists in letting one of the domains annex the others; i.e., in extending the laws governing one domain to all the others.

We can also mention the principle of *the proportionality of effects to causes*. According to Meyerson, the human mind wishes to *identify every effect with its cause,* which is of course impossible. Without going that far, it can be argued that in everyday life we instinctively assume the magnitude of the effect to be an increasing function of the intensity of its cause; hence the local proportionality of cause to effect.

There is also the *principle of sufficient reason,* which was clearly ennunciated by Leibniz but had already been tacitly used by the pre-Socratics.

We finally encounter arguments based on *an intuitive notion of probability;* for example, the assumption that long sequences of coincidences do not occur in nature because they are unlikely. The rejection of 'conspiracy theories' belongs to the same order of ideas: i.e., of theories which postulate deep causes, or deep asymmetries, together with compensatory factors preventing such causes from ever surfacing at the level of phenomena. An exact and systematic compensation is considered implausible because highly improbable. In everyday life we would, for example, be suspicious of the statement that a certain person X was intrinsically good but that external circumstances, or various hidden psychological mechanisms, *always* made X do the morally wrong thing. It is also this intuitive notion of probability which leads us to consider a theory as strongly supported by the distant and unexpected facts which it entails; it seems implausible that this should be the effect of pure chance having nothing to do with the truthlikeness of the theory.

I have already emphasized the fact that the principles enumerated above are used in everyday life and that they preceded, or arose concurrently with, science proper. Admittedly, such principles are made more precise by the mathematical techniques currently used in the sciences; but this should not mask their continuity with commonsense knowledge. (Anyway the desire for precision, which mathematics embodies, may well be innate and hence precede the birth of science.) Let me also add that the deductions which start from the principles mentioned above and end with the construction of new theories are not always watertight; they may contain what Lakatos called 'hidden lemmas'. In the examples below I hope to show that such logical gaps are bridged by considerations of simplicity and that they are anyway far less important than some philosophers imagine. They certainly do not warrant the view that scientific discovery *consists in such gaps*, or in such 'intuitive jumps', rather than in deductive reasoning. Moreover, physicists often recognise that the simplicity requirements they invoke are expedients which science ought subsequently to shed.

§1.1.7 Examples from the History of Science

Let me start with the simplest examples illustrating the principles mentioned in §1.1.5 above.

In Ptolemaic astronomy each of the five planets performs a complex movement one of whose components simulates the revolution of the sun about the earth. This is a huge cosmic accident, or rather a series of coincidences which are accepted without explanation. It can be mathematically proved that, relatively to the sun, all five components would disappear at one stroke. This, according to Dreyer (1953, p. 312), was a main starting point of Copernican astronomy. Whatever the case may be, it is true that Copernican theory explains these coincidences by deriving them from a single source, namely from the wrong, but perfectly natural, choice of the earth as our primary frame of reference.

In classical mechanics inertia and weight are two fundamentally different qualities; the first is the resistance of a given body to impressed force, the second its receptiveness to any gravitational field. Yet, as if by accident, these two qualities are measured by the same scalar, namely the mass m. Thus: $m = m_i = m_g$, where: m_i = inertial mass and m_g = gravitational mass. Refusing to believe in a simple coincidence, Einstein proposed to construct a theory which would explain the identity of gravitational and inertial masses and thus unify gravitation and kinematics. In view of the correspondence principle, he had to take account of the empirical consequences of $m_i = m_g$, in particular of the law that in a gravitational field all bodies experience the same acceleration irrespective of their internal constitution. This law had been noted by Newton and had subsequently been severely tested and corroborated. Hence, at the observational level, the field is determined by the acceleration it generates. Since acceleration is a kinematical, i.e. a geometrical, quantity, the simplest solution consists in letting geometry annex gravity. I have already said, in (5) above, that annexation is the simplest form of unification; it is also clear that, if annexation is to take place at all, then kinematics should annex gravitation, not the other way around; for kinematics, i.e. our space-time system, is the most fundamental structure within physics. Einstein was also aware that, as shown by Newton, the field must depend on the distribution of masses in the universe. According to special relativity, every mass has an energy equivalent and vice-versa. According once again to the correspondence principle, special relativity ought to be a limiting case of the new hypothesis; so the required geometry had to be linked to the energy content of space. Einstein was thus led to geometrize gravitation, i.e. to construct general relativity (see §§8.1–3).

Consider now the reduction of optics to electromagnetism. From his equations Maxwell inferred that all electromagnetic disturbances are propagated *in vacuo* with the velocity of light. He concluded that light must be an electromagnetic wave. Was this a sudden intuitive insight, or did Maxwell simply commit a logical error by affirming the consequent? I find both of these explanations implausible, unless by Maxwell's *'intuitive insight'* we mean a compressed and possibly semi-conscious version of the following argument: two distinct ethers, the so-called luminiferous ether and the electromagnetic one, had been postulated; both carry waves; it is highly unlikely that two wave processes should belong to disjoint domains and yet possess numerically identical velocities; hence optics and electromagnetism ought to be connected in some fundamental way. I have already remarked that annexation is the simplest method of unification. However, one important question remains: should optics be annexed to electromagnetism or vice-versa? Maxwell had already consolidated electromagnetism into an integrated system possessing a wide spectrum of frequencies which could easily accommodate the visible interval. Electromagnetism was bound to become the dominant theory.

A good example of a conspiracy theory is provided by Lorentz's hypothesis (see chapters two and three). Lorentz postulated an absolute medium which, *in the light of his own hypothesis,* gradually became undetectable: compensatory factors affect our measuring instruments in such a way that every uniformly moving system behaves like the absolute frame. By establishing the first-order covariance of Maxwell's equations, Lorentz had shown that motion through the ether could give rise to no first-order effects. By deriving the contraction of material rods from his axioms, he also accounted for Michelson's null results. Lorentz's hypothesis was thus becoming the kind of conspiracy theory which would inevitably arouse Einstein's suspicions; but, as will be shown, more than mere suspicion was needed in order to lead Einstein to special relativity. It is nevertheless worthy of note that, as soon as Lorentz realised that his transformations form a group, so that perfect symmetry obtains between the ether and every other inertial frame, he was converted to relativity. Though he still hankered after the ether, he knew that Lorentz-covariance had become the operative principle; he too found it implausible that the ether should be totally concealed by compensatory factors.

The magnitude of the relativistic upheaval seems to exclude any attempt at explaining its origins in gradualist terms. The invariance of c violates commonsense to such an extent that only inductivism, or creative intuitionism, seems capable of accounting for its discovery. According to Reichenbach, Michelson established an experimental result which Einstein later generalised and elevated to the rank of a postulate (see Introduction §4). However, apart from the truism that every experimental result can be explained in infinitely many ways, it seems certain, from an historical point of view, that Michelson's experiment *played no direct part* in Einstein's thinking. Should one then conclude that relativity issued forth from an irrational flash of intuition? This kind of mysticism however explains nothing: it simply expresses one's ignorance. More specifically, it does not explain why Einstein chose one *particular* process, namely the propagation of light, as the cornerstone of his *general* kinematics.

I should like to present a gradualist explanation of the origins of relativity. Let me first make a general logical point: a slight and seemingly negligible alteration in the axioms of an existing theory may entail wildly divergent consequences lower down the logical scale. Thus, gradualism at the theoretical level is perfectly compatible with revolution at the lower level of empirical predictions. For example: the extension of the relativity principle from mechanics to electrodynamics, which was quite an intelligible move, implies that the rest mass of a particle, hitherto regarded as the amount of stuff or substance in a given body, could vary; this testable implication was totally counter-intuitive (see §7.3). Let us secondly look at history proper: as it turns out, the invariance of c, though presented as an axiom, is the end-point of a long chain of reasoning bringing into play the same philosophical arguments which had brought forth classical physics. Einstein tells us that, towards the end of the 19th century, physics displayed the kind of dualism which he found unacceptable. There was, on the one hand, a science of mechanics subject to a known relativity principle and expressed by ordinary second-order differential equations. On the other hand, Maxwell had put forward an electromagnetic theory in the form of first-order partial differential equations, which seemed to single out a unique frame and thus violate relativity. Nature appeared split into two domains governed by antithetical laws. It had moreover been noted that electromagnetic induction depends not on the absolute but on the relative velocities of moving bodies. Nature thus conspired to conceal from us the asymmetry which ought to exist between the ether and other inertial frames. Einstein found this situation wholly unsatisfactory. Relativity was meant both to eliminate all conspiracies and to restore the unity of physics; it was to attain these two objectives at the same time. The principle of relativity was extended to the whole of physics and in particular to electromagnetism (see §3.2). This immediately yielded the relativity of electromagnetic induction and the invariance of the constant c which occurs in Maxwell's equations. Thus the counter-intuitive 'postulate' of the invariance of c, namely (P2), follows from the plausible generalisation of the relativity principle. One wonders though why Einstein decided to keep Maxwell's theory unchanged, even at the cost of modifying mechanics. Once again, this was not the outcome of a sudden inspiration. Einstein knew that all attempts to explain electromagnetic phenomena in mechanistic terms had failed; this was why Lorentz had tried to carry out a reduction in the opposite direction. Thus if one of the two theories, Newton's or Maxwell's, was to give way to the other, it would have to be the former. Einstein consequently extended the relativity principle to the whole of physics, whilst taking Maxwell's equations as his fixed point. In this sense electromagnetism was annexing mechanics; annexation proved once more to be the simplest means of unification.

To sum up: the invariance of c was the last step in a long deductive argument involving philosophical principles, or rather meta-principles. The argument was *about the structure of theories* as well as about certain empirical facts; it is a second-order form of reasoning. I do not claim to have identified all the premisses used by Einstein. There are probably gaps in his argument which may have been bridged by considerations of simplicity or of convenience. For example: Einstein could have modified both Newton's and Maxwell's theories; adhering to one of them was however simpler. All the same: Einstein did not experience a flash of intuition but worked through a series of deductive steps, making use of the principle of unity and of the exclusion of all conspiracy theories.

I should finally like to consider one of the most controversial episodes in the history of science: Newton's discovery of universal gravitation.

This example is important in that it underlines the role which mathematics can play in physical discovery. It also shows that the transition from Kepler to Newton is somewhat similar to that from Newton to Einstein. This lends support to the thesis, implicit in the construction of MSRP, that the intuitive methodology used by great scientists was largely stable over time. Hence the discovery of relativity, though it started a great *scientific* revolution, marked *no new epistemological or methodological* departure. According to Max Born (1964) Newton deduced his theory from Kepler's laws. Duhem (1906) and Popper (1972, ch. 5) correctly pointed out that these laws contradict universal gravitation and hence cannot validly imply Newtonian theory. By describing Newton's path to discovery as comprising two distinct stages, I propose to show that Born's views can be reconciled with Popper's and Duhem's.

The first stage is deductive in the ordinary sense of the word. It consists, on the one hand, in deriving the second law of motion $\vec{f} = m\vec{a}$ from the Cartesian principle of inertia and, on the other, in extracting certain logical consequences from Kepler's theory. One of Newton's most important discoveries is in fact a theorem of kinematics, i.e. of pure mathematics, which naturally involves the differential calculus. Newton proved that, as a consequence of Kepler's laws, every planet is subject to a central acceleration inversely proportional to the square of the distance between the planet and the sun, where the proportionality factor is the same for all planets. Conversely: every mobile P, whose acceleration is directed towards some fixed point S and is inversely proportional to $\overline{SP}^2$, describes a conic section with one focus at S; moreover, the radius vector connecting S to P sweeps out equal areas in equal intervals of time. Gravitational theory is wholly contained in these two propositions, without being logically implied by them. Let us note that Newton formulates these two propositions in terms, not of central accelerations, but of centripetal forces. This is because, right from the start, he makes use of the equation $\vec{f} = m\vec{a}$, which I have so far not had to postulate. Mach (1893, ch. II) wondered why Newton put forward two laws, namely the principle of inertia and the equation $\vec{f} = m\vec{a}$, of which the first is a trival corollary of the second. Yet Newton had good reasons for proceeding as he did; he had borrowed the law of inertia from Descartes and then used the principle of the proportionality of cause to effect in order to derive his second law of motion (see Westfall 1971, p. 463). Physicists had always looked upon force as a dynamical cause; the problem was to determine its precise mechanical effect. Aristotle had directly connected force with velocity; but the law of inertia, which was used by Galileo and then given its classical formulation by Descartes, asserts that motion is a state of the mobile, hence requires no cause in order to persist. Force was thus needed only in order to alter this state. Since force is an instantaneous cause, such an alteration will be proportional to the cause acting throughout an interval of time, i.e. to the product of this interval by the force. By a convenient choice of units, the force will therefore be equal to the rate of change of the quantity of motion, i.e. to the derivative of the momentum. Thus:

$$\vec{f} = \frac{d}{dt}(m\vec{v}) = m\vec{a}.$$

This explains why Newton chose to formulate two laws of motion, namely the principle of inertia and the force law, of which the second is not *postulated*, but *derived* from the first. In this way, we obtain two schemes:

(α) [(Law of Inertia) Λ (Effect ∝ Cause)] $\Rightarrow [\vec{f} = m\vec{a}] \Rightarrow$ [(Kepler's Laws) $\Rightarrow$ (Centripetal force inversely proportional to the square of the distance from the sun and directly proportional to the mass of the planet)]; and

(β) **[(Law of Inertia) Λ (Effect ∝ Cause)]** ⇒ [$\vec{f} = m\vec{a}$] ⇒ [(Centripetal force inversely proportional to the square of the distance from a fixed point S) ⇒ ((Path is a conic section with one focus at S) & (Equal areas law holds relatively to S))]. At this point, we have to mention Newton's third law which says that every action provokes an equal opposite reaction (where, in this context, both 'action' and 'reaction' mean 'force'). This principle has a double origin, empirical and purely theoretical. On the one hand, Newton conceived of force in terms of pressure, tension, or muscular effort; i.e. in anthropomorphic terms. In statics, every pressure is neutralised by an equal counter-pressure. As for gravitational action, we know that Newton assimilated it to a tension in a rope connecting two material bodies; since the tension is the same all along the rope, the two bodies must attract each other with equal and opposite forces. Gravity thus conforms to Newton's third law. On the other hand, Newton was well aware that the third law flows from the Cartesian principle of the conservation of total momentum, i.e. from a generalised principle of inertia to which Newton had given a vectorial form (see Westfall 1971, chs. VII-VIII). Consider an isolated system consisting of two particles P_1 and P_2. Let m_1, m_2 and $\vec{v}_1$, $\vec{v}_2$ be the masses and the velocities of P_1, P_2 respectively. The total quantity of motion, which is expressed by $(m_1\vec{v}_1 + m_2\vec{v}_2)$, must be conserved; i.e. $m_1\vec{v}_1 + m_2\vec{v}_2 = \text{constant}$. Differentiating, we obtain:

$$\frac{d}{dt}(m_1\vec{v}_1) = -\frac{d}{dt}(m_2\vec{v}_2).$$

Thus, by virtue of Newton's Second Law: $\vec{f}_{12} = -\vec{f}_{21}$, where $\vec{f}_{ij}$ denotes the force exerted on P_i by P_j. In the particular case of gravitation, it follows that two bodies attract each other with equal and opposite forces.

Basing himself on this third law, Newton established two propositions; the first asserts that the centre of gravity of an isolated system remains either at rest or in uniform rectilinear motion in absolute space; according to the second proposition, two bodies alone in space describe, each relatively to the other, and each with respect to their common centre of gravity, paths which are all similar to one another (see Newton 1686, vol 1 pp. 19, 164–6). Let us denote by (γ) the conjunction of these two propositions.

We have now come to the end of the first stage in the process of discovery initiated by Newton. The second stage consists in a deductive *second-order* argument, or deductive meta-argument, which examines several propositions in order to reconcile some of their logical consequences. Newton remarks that, in view of his third law, the sun cannot be motionless (Newton 1686, vol. 1 p. 164). Note that this consequence already flows from the principle of sufficient reason: if a body A, by virtue of its material nature alone, attracts another body B, then there is a sufficient reason for B also to act on A. Kepler had already conjectured that, if A and B were alone in the universe, they would both spontaneously move towards their common centre of gravity. Moreover, in his "Astronomia Nova", he had attributed the variation in the distance between Mars and the sun to a magnetic force; but such a force must necessarily give rise to an *interaction* and thus cause a motion of the sun. It was precisely in order to avoid this undesirable consequence that, in his *Epitome,* Kepler replaced the magnetic interaction by a *unilateral* force exerted by the sun on the planets. As for Newton, having accepted the sun's motion, he had simultaneously to make sure that Kepler's laws were approximately obeyed; i.e., he had to make sure that the correspondence principle held good. This was already ensured by the second law of motion $\vec{f} = m\vec{a}$, which entails that the acceleration of a mobile is inversely proportional to its mass. Since the sun was assumed to be very massive in comparison with the planets, it could safely be regarded as motionless. Thus one of Kepler's important assumptions was approximately satisfied.

Making use of some consequences of Kepler's laws, Newton attacked the problems posed by the motion of systems of point masses. Note that, though Kepler's theory is strictly false, some of its logical implications could well be strictly true. Thus, Newton was in no way violating deductive logic. Let us recall that Kepler derived his laws from certain forces which the sun supposedly exerts on the planets. Since, for Kepler, there were no interactions between the various planets, their number played a role in the determination neither of their trajectories nor of the equal areas law. Hence, the simplest system to which Kepler's laws apply consists of the sun S and of a unique planet P. A corollary of Keplerian theory is that, relatively to S, P describes an ellipse with one focus at S. Newton took this corollary to be strictly true and then proceeded to exploit it. From (γ) above, he concluded that the centre of gravity G of S and P can be considered as immobile; also that P and S describe ellipses having a common focus at G; finally that the equal areas law applies to both S and P (Newton 1686, vol. 1 p. 167).

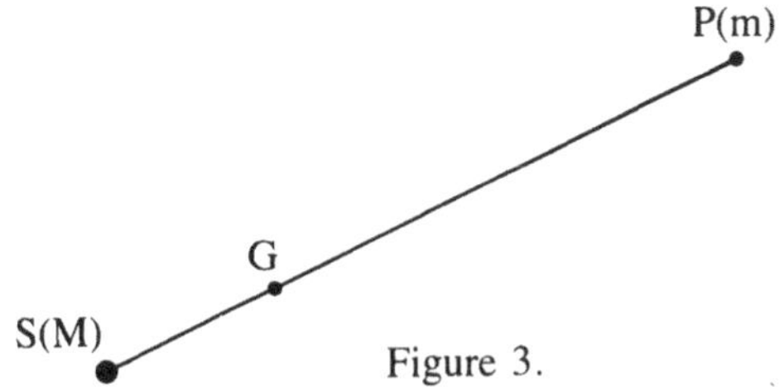

Figure 3.

From (α) above Newton inferred that the force acting on P is constantly directed towards G, hence also towards S, and that its magnitude is proportional to $1/\overline{GP}^2$. Since $\overrightarrow{GP} = (M/(m + M))\,\overrightarrow{SP}$, where M and m denote the masses of S and P respectively, the force to which P is subject is inversely proportional to $\overline{SP}^2$ as well as to $\overline{GP}^2$. Similar considerations apply to S because of the obvious symmetry between P and S.

If, in accordance with (α) above, we suppose that the force at P is proportional to m, then by symmetry, i.e. by the principle of sufficient reason, the force at S must be proportional to M; but, by virtue of Newton's third law, these two forces have equal magnitudes; hence each one of them must be proportional to $mM/\overline{SP}^2$. We have thus arrived at the law of universal gravitation.

To sum up: the path leading to universal gravitation consists of two stages of which the first is the more arduous and more important. In this first stage Newton derived his laws of motion from the Cartesian law of inertia together with the principle of the proportionality of cause to effect; he also *deduced* an inverse square law for central accelerations from Kepler's theory. The second stage involves a meta-argument which resorts to the correspondence principle and to the principle of sufficient reason in order to determine a new gravitational theory. Thus is solved the paradox, underlined by both Duhem and Popper, of how Newton could have derived his theory from laws contradicting it: Newton *reflected* on Kepler's laws without supposing them to be true; he extracted certain *logical* consequences from them and went through a chain of *deductive* reasoning aimed at determining a proposition X such that X conforms to Newton's third law and also tends to Kepler's theory in certain limits. This is obviously a meta-proof since the unknown X is a proposition and not a quantity. Putting it a little differently: Newton's meta-proof does not contain Kepler's laws as premisses but is *about* these laws; and, in a meta-argument, one can reason about contradictory propositions without running into contradictions. This is why I cannot agree with Popper when he writes:

> It is important to note that from Galileo's or Kepler's theories we do not obtain even the slightest hint of how these theories would have to be adjusted—what false premises would have to be dropped, or what conditions stipulated—should we try to proceed from these theories to another and more generally valid one such as Newton's. Only after we are in possession of Newton's theory can we find out whether, and in what sense, the older theories can be said to be approximations to it. (Popper 1972, p. 210)

It should be conceded that Newton could have arrived at his inverse square law by a different route from the one described in the *Principia*. It was not unusual, in the 17th century, to assimilate the intensity of gravity to that of light, which was known to obey an inverse square law. As was later 'proved' by Kant, one can use the principle of sufficient reason in order to show that a gravitational action which emanates from a unique body in space and is propagated without leaving any remnants behind must be uniformly distributed over the surface of a sphere; since the latter has an area equal to $4\pi R^2$, the inverse square law follows. However, the symmetry, which is essential if this argument is to go through, breaks down in the case of a two-body system; unless one assumes that each of the two bodies behaves as if the other were absent. Anyway, this second proof does not invalidate the one presented by Newton in the *Principia;* he may well have reached the same goal by two different routes.

§1.1.8 Conclusion

Science can adapt its theories to certain facts selected in advance. These facts may have refuted extant theories which scientists then *knowingly* set out to transcend. At the same time, by making use of the correspondence principle, scientists *preserve* empirically ascertained knowledge. But this strategy, which accounts for the quasi-inductive path of science, still leaves too great a freedom in the choice of possible hypotheses. This freedom is considerably narrowed by the intervention of philosophical principles and metaprinciples. Such principles are *stable* in the sense that they preceded the birth of science proper and have since remained largely constant; they could be innate and even possess a genetic basis; their emergence could thus be explained, in evolutionary terms, by mutations most of which would have been subsequently eliminated by natural selection. Moreover, such principles are mostly non-technical and often vague. If they were all made precise and then conjoined, they might well entail contradictions. (After all, they may have arisen from a confrontation with very different physical situations.) The heuristic of a research programme is determined by the *coherent choice* it operates among these principles and by the more or less sharp formulation it gives to each of them. In this effort towards increased precision, *mathematical idealisation can play a major role*. One can cite, as examples, the mathematical definitions of continuity, of differentiability, of analyticity, of the independence of the form of theories from a chosen frame (covariance) etc. A research programme evolves through the application of the positive heuristic to existing hypotheses, some of which may initially have belonged to commonsense knowledge. In this way one obtains new theories which can be confronted with the 'facts', i.e., which could be empirically refuted and then corrected by reiteration of the same method. Note that empirical refutations, though very important, are not indispensable: since theories can satisfy certain metaprinciples to a greater or lesser degree, it is possible to improve such theories without reference to experimental results. For example, since hypotheses can be more or less unified, a step within a given programme leading from one hypothesis to the next can be motivated by the desire for a higher degree of unity. (Relativistic hypotheses are a case in point. See chapter eight.).

We have just remarked that mathematics often plays a central role in structuring the heuristics of research programmes. The role of mathematical reasoning, which is clearly illustrated by Newton's derivation of his inverse square law, will presently be examined in greater detail.

§1.2 The Role of Mathematics Within Physics

In the preface of their 'Methods of Mathematical Physics' Hilbert and Courant express their regret that, towards the beginning of the twentieth century, the fruitful interplay between mathematics and physics had almost stopped. They write:

> Since the Seventeenth Century physical intuition has served as a vital source for mathematical problems and methods. Recent trends and fashions have however weakened the connection between mathematics and physics; mathematicians, turning away from the roots of mathematics in intuition, have concentrated on refinement and emphasized the postulational side of mathematics, and at times have overlooked the unity of their science with physics and other fields. In many cases physicists have ceased to appreciate the attitudes of mathematicians. This rift is unquestionably a serious threat to science as a whole . . .

It is well known that empirical science has often exerted a beneficial influence on the development of mathematics. Physics sets problems for which mathematical solutions are urgently required; as a result, certain branches of pure mathematics receive a powerful impetus and develop rapidly. For a long time mathematics and physics were regarded as identical disciplines; or, rather, people were not clear about the differences between them. Euclidean geometry for example was taken to describe the properties of real, albeit idealised, space. Its postulates, as distinct from its axioms, were accepted because they supposedly expressed self-evident truths about physical space. In Einstein's words, geometry constituted one of the oldest physical theories (Einstein 1934, p. 119). In the preface to his *Principia,* Newton treats geometry as a branch of mechanics, i.e. as a branch of physics:

> Therefore geometry is founded in mechanical practice and is nothing but that part of universal mechanics which accurately proposes and demonstrates the art of measuring.

In many cases physics played the dominant role in that it was physics which gave rise to the mathematical problems and, to a large extent, dictated the type of mathematics to be used. In his theory of fluxions, Newton constructed the calculus specifically for the study of differentiable motions:

the fluent was the time variable and the fluxion the instantaneous velocity. Hence analysis, the discipline which dominated mathematical thinking for over two centuries, owes its origins to physics. Similarly, the study of differential equations and the development of what is now known as advanced calculus were closely connected with the articulation of the Newtonian programme in the eighteenth century. A similar process took place in the nineteenth century when Faraday, using 'line of force' as a new physical concept, enunciated laws of which Maxwell later gave a mathematical formulation. This powerfully contributed to the development of vector algebra and vector analysis. These examples illustrate the heuristic role of physics with regard to mathematics; in each of the cases so far considered, it was physics which led to the discovery of an appropriate mathematical instrument. It looks as if, whenever mathematics and physics have interacted, the latter set the pace. I would like to examine whether the reverse process has ever taken place, whether the solution of a mathematical problem has ever led to the discovery of a physical theory. Prima facie, this process ought to be possible. There is no reason why the mathematical rendering of some vague physical principle should not unexpectedly capture more of the truth than the scientist was initially aware of. This question thus concerns the heuristic role of mathematics vis-à-vis physics. There are two important ways in which mathematics further physical discovery.

§1.2.1 Increase of Content through Translation into a Mathematical Language

As mentioned above, the scientist may start from an intuitive physical principle. Through being translated into some mathematical theory available at the time, the principle is generally modified: it may in particular acquire some surplus structure and thus become a stronger physical assumption (see Redhead 1975, §4, for an excellent account of mathematical surplus structure). For example, Fresnel set out to give a mathematical formulation to his conjecture that light is a wave process through the ether. He instinctively resorted to the periodic function with which he was most familiar, namely $\sin x$. His original assumption that light is a wave phenomenon is obviously weaker than his final hypothesis that the wave is representable by the function $\sin(2\pi t/T)$.

It has often happened that certain physical hypotheses which require very specific mathematical theories for their formulation were proposed after, and sometimes immediately after, the mathematical theories were elaborated. This seemingly pre-established harmony between the mathematics and the subsequent physics has been a cause of puzzlement among historians and philosophers of science. One example of such a coincidence is the relationship between the mathematics of Hilbert space and the formulation of quantum mechanics. Another very good example is the development of Riemannian geometry prior to the emergence of GTR. It was as if Gauss, Riemann, Ricci and Levi-Civita had Einstein in mind when they constructed their non-Euclidean geometries. Einstein himself considered his discoveries as a continuation of Gauss's and Riemann's work. Yet Einstein was a physicist and Riemann a pure mathematician. In his 'Geometry and Experience' Einstein distinguished strictly between geometry qua mathematical discipline on the one hand and geometry qua physical theory on the other. It is prima facie strange that mathematicians should anticipate the work of later physicists. The solution of this puzzle is twofold. On the one hand, there is no pre-established harmony between mathematics and physics. The physicist forces, or impresses, his principles into an existing mathematical framework. The physical principles thus acquire the surplus structure of the mathematical system which is used. When Einstein decided that he could not account for gravity in special-relativistic terms, he turned to his friend Grossmann, a working mathematician steeped in the Riemannian tradition. It is small wonder that Grossmann used Riemannian geometry to cast the field equations for empty space in the form $R_{ij} = 0$. In this way Grossmann and Einstein geometrised gravitation (for more on this see §8.4). The harmony between mathematics and physics is thus forcibly established, not pre-established. On the other hand however, Riemannian geometry had to be of such a nature as to lend itself to the use to which it was put by Einstein and Grossmann. In other words, this geometry had to possess an empirical, or a quasi-empirical, character which allowed physics to be imbedded in it.

That this was in fact the case is clearly indicated by the following quotation from Riemann's famous dissertation *About the Hypotheses Which Lie at the Basis of Geometry:*

> It will emerge from this that a multiply extended magnitude [Grösse] is capable of containing within itself a special case of a three-dimensional magnitude [dreifach ausgedehnte Grösse]. From this it necessarily follows that geometrical propositions are not deducible from the concept of magnitude, but that those properties, through which space is distinguished from other conceivable three-dimensional magnitudes, are derivable only from experience. These facts, like all facts, are not necessary, they possess only empirical certainty, they are hypotheses; one can therefore examine their probability which is, within observational limits, indeed very high. (Riemann 1953, pp. 272–3, my translation)

Riemann's approach to geometry is quasi-empirical in that it is based on the notion of measured distance between neighbouring points. Riemann physicalised mathematical geometry, thus making it possible for Einstein and Grossmann to geometrise gravitation. We should however remark that empirical considerations alone, i.e. without a priori assumptions, do not determine the structure of Riemannian Geometry. Riemann himself realised that the differentiability condition is not derivable from experience:

> I will undertake this task only under certain restrictive assumptions and shall first confine myself to those lines on which the ratios between the magnitudes dx . . . vary continuously. The question of the validity of the geometrical presuppositions within infinitely small regions connects with the question of the foundations of all metrical relationships in space. To this question, which can be reckoned to the science of space, the above remark applies, namely that, for a discrete manifold, the principles of the metric relationships are already contained in the concept of the manifold itself, whereas, in the case of a continuous manifold, these principles have to be imported from outside. (Riemann 1953, p. 286)

There is another assumption of whose a priori character Riemann seems to be unaware, namely that space is determined by a metric, i.e. by one fundamental symmetric tensor g_{mn}. That this is an a priori assumption is shown by the fact that, after 1916, purely affine geometries were developed, which could accommodate the facts then known to Einstein (see §8.6).

Let me say that I am by no means committed to the Kantian notion of the a priori. I use 'a priori' in the sense of 'not imposed by experience'. This conception of the a priori is perfectly compatible with Mach's: Mach (1883, introduction and 1905, pp. 164–182) identified a priori knowledge with the instinctive or innate knowledge which may have arisen through the gradual adaptation of the mind to its surroundings, i.e. through natural selection. Thus, for Mach, a priori knowledge is inheritable, i.e. genetically imbedded in the organism.

Going back to Riemann, we see that metrizability and differentiability are a priori assumptions which belong to the surplus structure of geometry with respect to physics. Thus, forcing physics into the Riemannian mould imposed more constraints than are warranted by observational results. Does this imply that the later hypotheses of Weyl, Eddington and Schrödinger were more 'empirical', or more 'inductive', than GTR? By no means. Metrizability, or rather metrization, was not replaced by laws induced from experience, but by different *geometrical* assumptions. Weyl (1923, §40) linked the presence of the electromagnetic field to the change in *length* of a vector transported parallel to itself. Eddington (1923, §97) proposed to explain gravitation and electromagnetism by means of the symmetric and antisymmetric parts of the Ricci *tensor*. As is well known, the Ricci tensor is obtained by contracton from the curvature tensor B^i_{jmn}; and B^i_{jmn} is obtained from the affinity Γ^i_{jm} by considering the variation of a vector taken round an infinitesimal loop. Eddington derived the symmetry in j and m of Γ^i_{jm} from a *commutativity* condition. Schrödinger (1950) constructed his theory by applying a *variational principle* to a certain integral, where the integrand is what he and Eddington considered as the simplest *invariant density* built out of the curvature tensor, namely $\sqrt{(-\det[R_{ij}])}$.

I mention all these details in order to make the following point: length, parallel transport, contraction, commutativity, tensor densities and variations of integrals are *mathematical* notions not directly abstracted from experience. The physicist operates at the mathematical level, hoping that his operations mirror certain features of reality. However, he is not very clear as to how this mirroring takes place; so he lets himself be guided by the syntax, or by the symbolism, or some mathematical system.

Of course, these seemingly abstract unified field theories have novel empirical consequences. It was precisely because Weyl's hypothesis had some undesirable physical consequences that Einstein rejected it. This illustrates the point already made that physics is strengthened by mathematical surplus structure.

Mathematical speculation thus seems to have played a dominant role in the development of Einstein's programme. However, it can also be said that the problems which physics faced after 1916, e.g. the fact that GTR does not treat gravitation and electromagnetism on a par, gave rise to new types of non-Riemannian geometry. Can we not say, with equal right, that physics played a dominant heuristic role with regard to mathematics? In my opinion the relationship between mathematics and physics is best described, in dialectical terms, as a to and fro movement between two poles. One moves from physical principles to idealising mathematical assumptions; then back to some more physics; then forward to fresh mathematical innovations with ever increasing surplus structure. The so-called harmony between physics and mathematics is not a miracle but the result of an arduous process of mutual adjustment.

§1.2.2 Realistic Interpretation of Mathematical Entities

There is a second way in which mathematics can play a fundamental role in physical discovery. The usual method in theoretical physics is to give mathematical expression to some physical hypothesis, then use logico-mathematical techniques to draw consequences from the hypothesis. In doing this, the physicist may resort to any number of mathematical operations; these are sometimes in the nature of 'tricks' or 'gimmicks' which simplify the deductive process. As Duhem (1906, ch. 1) pointed out, it would be foolish to insist on giving physical interpretations to all mathematical quantities and operations used in a scientific theory. Duhem is evidently right: adding lengths corresponds to juxtaposing rods; multiplying lengths may correspond to the construction of rectangular areas, but multiplying t by $\sqrt{-1}$, though convenient, does not seem capable of any physical interpretation. However, through trying to find a realistic interpretation of certain mathematical entities which appear at first sight to be devoid of any physical meaning, the scientist may be led to a physical conjecture (on this see Popper 1963, pp. 117–119). This can occur in two different ways.

The first and straightforward way is through an increase in the empirical content of a given theory. By being realistically interpreted, part of the mathematical scaffolding becomes physically meaningful, hence in principle testable. The theory remains syntactically unchanged but its observational content, i.e., its contact with 'observable' reality, is extended. Dirac's equation for the electron, his explanation of spin and especially his discovery of the positron are cases in point. Dirac proposed a relativistic equation which was found to possess negative energy solutions. Prima facie, such solutions cannot be physically interpreted. By insisting on interpreting the negative solutions, Dirac predicted the existence of the positron; the absence of an electron of charge $(-e)$ and energy $(-E)$ was interpreted as the presence of a positron, that is of a particle of charge $(+e)$ and energy $(+E)$.

The second way in which the realistic interpretation of mathematical entities leads to new discoveries is both more complex and more interesting than the previous one. The theory itself is altered in this process; that is, the syntactical expression of the physical principles themselves does not escape unscathed. This usually occurs as follows. We start from some hypothesis H which we may, for one reason or another, want to modify. We express H in an equivalent form (on such equivalences see Feynman 1965, p. 168), H*(t), say, which brings out a certain mathematical entity t. t is then given a realistic interpretation which subsumes it under a philosophical category, e.g. the category of substance; this category may be governed by some very general laws, e.g. conservation or symmetry laws. It is found that H*(t) violates these laws. The breakthrough is achieved by H*(t) being modified into some theory H′(t) which obeys the laws in question. We shall see how, by using variational methods, Einstein re-expressed his original field equations F in the form F*(t) where t, despite its frame dependence, was interpreted as a gravitational energy matrix (see §8.4). Let T be the nongravitational energy tensor. Thus (T + t) must obey a conservation law and the field equations must be symmetric in T and t. Einstein noticed that F*(t) violates this condition; he modified F*(t) so as to obtain equations symmetric in T and t.

§1.3 Meyerson's Principle of Identity

In this section I shall give a brief exposition of Meyerson's philosophy of science. I shall argue that this philosophy accounts, *in terms of a single principle,* both for the applicability of mathematics to physics and for the role which conservation laws, together with certain philosophical notions like that of substance, play in the progress of science. This Meyersonian principle was briefly described above; it allows of diverse formulations which, though distinct, bear to one another a sort of family resemblance. According to Meyerson (1908 & 1921, ch. 5) all explanations, whether scientific or commonsensical, spring from one basic tendency of the human mind; namely the tendency to deny diversity and change; or to assert the existence of constants behind the fleeting appearances; or to explain the many in terms of the one; or to subsume the flux of becoming under the immutability of being. The best formulation is, in my opinion, that the human mind inevitably tends to deny diversity and assert sameness, or identity, in both space and time. For the human mind, only the undifferentiated one is real; everything else is appearance. This sounds like the worst kind of metaphysical doctrine which seems bound to be of no relevance to science. May I however ask the reader to be patient, suspend disbelief for the time being, and let me examine how much of the role played by mathematics and by conservation laws in physics can be derived from Meyerson's principle of identity? I have spoken of this principle as of a tendency inherent in the human mind. If this tendency went unchecked, it would assert the existence of a unique Parmenidean sphere which is undifferentiated in space and changeless in time; it would thus degrade all phenomena to the rank of mere sensory illusions. Becoming would be mere illusion and only being would be real. The whole of science can be conceived as a sequence of attempts to rescue as much of Paramenides's sphere as is compatible with the phenomena. It is well known that Greek atomism postulates the void, together with a plurality of particles each of which can be assimilated to a Parmenidean sphere. Both the atoms and the void are immutable in themselves. The flux of appearances is explained in terms of different configurations of the same particles within the same empty space. Let A and K stand for two consecutive states of affairs which, following Meyerson, will henceforth be referred to as the antecedent and the consequent respectively. According to atomism, the differences between A and K are to be attributed to different distributions of the same atoms within an unchanging void. Classical mechanics accentuates the parity between A and K by showing that the process leading from A to K is reversible. A reversal of all the velocities in K would lead back to A.

Every reduction in the number of different atoms, every theory according to which atoms differ only in mass and charge, are hailed by science as great steps forward. This is because each such step involves a reduction of diversity, an increase in homogeneity. In the same spirit, Einstein proposed to define simplicity in terms of the paucity of the primitive elements occurring in a given theory. In Meyersonian language, the number of primitives is a measure of diversity. Already at this stage, we recognise the close affinity between Meyerson's principle of identity and Einstein's unity principle which provided the heuristic of unified field theories.

Meyerson sees the progress of science as a struggle between the mind which tries to impose identity and a differentiated reality, which resists such an imposition. The mind cannot completely succeed in denying all diversity; but some aspects of reality allow of subsumption under the principle of identity; *which particular aspects* can be determined only a posteriori, e.g., by trial and error. The general identity principle however is a priori.

§1.3.1 Legal Explanations

The identity principle applies at two distinct levels: at the second-order level of laws and at the first-order level of things. This gives rise to what Meyerson calls legal explanations and causal explanations respectively. Legality means that the same law applies irrespective of place and time: the form of the law is both a-spatial and a-temporal. There is permanence, i.e. constancy of form: many diverse phenomena illustrate, or manifest, the same underlying law which constitutes their Platonic form. For example, the falling stone, the moon revolving round the earth and the passenger thrown forward in a decelerating train are all instantiations of one and the same gravitational law. This permanence of the rule of law enables us to predict, hence has high survival value. According to Meyerson, every animal species needs for its survival this ability to anticipate facts. Thus he gives an evolutionary reason for the a priori character of the principle of identity as applied to laws. This principle is a priori in the sense of being inborn in a species which needs it for survival.

§1.3.2 Causal Explanations and Conservation Laws

The human mind goes much further than asserting just the permanence of natural laws; it searches for causal explanations which Meyerson defines as those under which certain substances, or things, are conserved; i.e. the total quantity of these substances remains unchanged through time. Conservation laws thus flow from the principle of identity *qua* negation of the diversity of things in time. I have said that Meyerson calls this version of the principle the causal one: the effect is equal to, i.e., it contains as much substance as, its cause.

Meyerson doubts that the causal principle, as distinct from the merely legal one, has any survival value. He thinks that phenomenological laws are all that is needed for a species to gain control over nature. The causal tendency is a luxury which cannot be explained in biological terms. It can however be argued that the postulation of substances which perdure in time enables us to keep track of things, hence predict the sensations which these things supposedly cause in us. To assume that the tree exists even when I do not look at it enables me to explain, thus predict, what I shall perceive when I turn round and look at it again. Mach claimed that the concept of material object can be replaced by that of constant functional relationships between elements of sensation; but even *he* admitted that concepts like that of an electron having constant mass and charge effect great economy of thought and thus have great pragmatic value. Moreover, the notion of constant functional relationships, even if it could replace that of substance, is a cumbersome notion introduced merely for reasons of philosophical purism; it is far removed both from commonsense and from scientific realism, which remain our best guides for action. Thus, both the legal and the causal versions of the identity principle have survival value.

Meyerson describes the way in which we construct conservation laws as follows. The mind fixes on certain processes in nature which it hypostasizes; it turns these processes into substances or things, whose total quantity it then assumes to be constant. For example, commonsense anchors sense-data in material objects which are supposed to perdure in time. This commonsense hypothesis forms the basis of the classical law of the conservation of mass. Another example is provided by the principle of inertia (Meyerson 1908, ch. 3). For Aristotle motion was change and velocity a ratio describing this process of change. Multiplying this ratio by mass, Descartes turned it into a *thing*, which he called quantity of motion and whose sum total remains constant. This constancy was derived from God's 'immutability', which is another name for the principle of identity. Similarly, energy is hypostasized motion-cum-position within a field of force (kinetic and potential energies). This automatically leads to a new conservation law, the conservation of energy law. Note that in classical physics matter is the primary substance, while momentum and energy derive their existence from matter (and charge) in motion. Special relativity puts matter and energy on a par; so we can assume the existence of a unique substance which our senses apprehend, sometimes as radiant energy say and sometimes as hard impenetrable matter. Special relativity thus enhances the status of energy by making it a constituent part of physical reality. This establishes a link between the special theory and general relativity, where energy considerations played an important part in the discovery of the correct field equations (see §§8.2–4). In connection with relativity however, we should sound a note of warning: substances are generally described, not by single quantities, but by tensors; the fact that tensor equations are frame-independent reflects the absolute character of substance; conservation is expressed by the vanishing of the tensor's divergence.

Summing up, we can say that the method of applying Meyerson's identity principle to evolution in time is to single out some process common to the antecedent A and to the consequent K; then to hypostasize this process, calling it say $\varphi(A)$, viz. $\varphi(K)$; finally to assert that $\varphi(A) = \varphi(K)$, i.e., to assert a conservation law. $\varphi(A)$ may, for example, stand for the total mass in A, or the total momentum, or energy. We have seen that the human mind has a tendency to deny all forms of change; if it could have its own way, it would affirm the identify of A and K and thus abolish time. (Note that, for Meyerson, time is measured by change.) Since this proves impossible, the mind resorts to the next best thing, namely the affirmation that A and K are equivalent under certain aspects. The relation $\varphi(A) = \varphi(K)$ is obviously an equivalence relation.

This method of hypostasizing processes is also used in mathematics. For a long time, functions were taken to be processes in which the value of an independent variable x gave rise to the corresponding value of $y = f(x)$. This process is clearly referred to by the word 'transformation' which is often used synonymously with 'function'. By defining a function as a class of n-tuples, set theory reified this process into a thing. Hermann Weyl wrote: "Thus we are able to subordinate genetic construction to the static existence of relations" (Weyl 1949, p. 5). It is obvious that the atemporal character of mathematics makes it ideally suited to the expression of theories based on the identity principle.

§1.3.3 The Geometrisation of Nature

So far, we have considered the identity principle as applied to change in time, to temporal diversity. Let us now examine its application to diversity in space, more generally to diversity in some domain of simultaneous events. Consider for example the simultaneous perception of two qualities. Two so-called secondary qualities, like colour and sound, strike us as being irreducibly different in a way that two frequencies are not. There is continuity between two waves giving rise to two distinct sensations like those of colour and sound, whereas the sensations themselves seem to be separated by an unbridgeable gap. This is why the mind, which is intent on eliminating diversity, tries to reduce all secondary qualities to the primary ones. This, by itself, proves insufficient. Only spatio-temporal properties are deemed completely intelligible. Mass and charge are felt to be occult, or rather heterogeneous, relatively to space and time. Hence the Cartesian attempt to reduce matter to extension and explain all physical phenomena in terms of figure and movement, i.e., in kinematical terms. As is well-known, this Cartesian attempt failed. It was supplanted by the Newtonian system which postulates space, time, mass and force as the primitive concepts. With the emergence of electromagnetism, charge and ether were added to this list.

The rise and development of the relativity programme can be explained, in Meyersonian terms, as follows. General relativity and the sequence of unified field theories are the realisation of the old Cartesian ideal. According to this view, the Cartesian enterprise failed not because it was wrong-headed but because the underlying geometry was too simple, too poor, to account for the diversity of the observed phenomena. By complicating their geometries, Einstein, Weyl, Eddington and Schrödinger proposed to construct global explanations of all physical phenomena. Meyerson's philosophical framework provides an explanation of why mathematics played such an important role in the creation of the new systems and why the Relativists tried, time and again, to force physics into one geometry; they were trying to dissolve the qualitative heterogeneity, i.e. diversity, of the phenomena into the homogeneity of one unified geometry (Meyerson 1924, ch. 20). This homogeneity does not of course mean that the metric or that the affine connections of the geometry do not change from one point to the next; but this is the sort of continuous numerical change which offends least against the mind's dislike of diversity. Anyway, Hermann Weyl minimises the significance of this dependence of the metric on each point of space-time in the following terms:

> The nature of the metric is *one* and is absolutely given; only the mutual orientation in the various points is capable of continuous changes and dependent upon matter. Euclidean space may be compared to a crystal, built up of uniform unchangeable atoms in the regular and rigid unchangeable arrangement of a lattice; Riemannian space to a liquid, consisting of the same indiscernible unchangeable atoms, whose arrangement and orientation, however, are mobile and yielding to forces acting upon them. (Weyl 1949, pp. 87–8)

Mach's definition of mass (Mach 1909, notes) as negative inverse ratio of accelerations can be seen as a first step towards the geometrisation of physics. Einstein's explanation, in Riemannian terms, of gravity and inertia was of course the major success of the relativity programme. The field equations of GTR, namely $R_{mn} - \frac{1}{2}g_{mn}R = -kT_{mn}$, can be interpreted in two very different ways. Reading them from right to left we can say that energy determines, or at least modifies, geometry. However, we can also read the equations from left to right and say that, once the geometry is given, i.e., once the g_{mn}'s are given, then the energy content of space is defined. This is one way of eliminating the right hand side of the equations, i.e. the qualitative side, with which Einstein himself was dissatisfied. He wrote:

> By this formulation one reduces the whole mechanics of gravitation to the solution of a single system of covariant partial differential equations. The theory avoids all internal discrepancies which we have charged against the basis of classical mechanics. But, it is similar to a building one wing of which is made of fine marble (left part of the equation), but the other wing of which is built of low grade wood (right side of the equation). The phenomenological representation of matter is, in fact, only a crude substitute for a representation which would correspond to all known properties of matter. (Einstein 1950, p. 81)

Einstein was clearly giving the geometrical wing precedence over the material one. In the same spirit Eddington had written:

> According to the new point of view Einstein's law of gravitation does not impose any limitation on the basal structure of the world. $G_{\mu\nu}$ may vanish or it may not. If it vanishes we say that space is empty; if it does not we say that momentum or energy is present; and our practical test whether space is occupied or not—whether momentum and energy exist there—is the test whether $G_{\mu\nu}$ exists or not. (Eddington 1923, p. 120)

That this is not an idiosyncratic view of Eddington's is shown by the following quotation from Schrödinger's *Space-Time Structure:*

> I would rather you did not regard these equations [that is $-(R_{i\kappa} - \frac{1}{2}g_{i\kappa}R) = T_{i\kappa}$] as field equations, but as a definition of $T_{i\kappa}$, the matter tensor. Just in the same way as Laplace's equation $\text{div}\vec{E} = 0$ (or $\nabla^2 V = -4\pi\rho$) says nothing but: wherever the divergence of $\vec{E}$ is not zero we say there is a charge and call $\text{div}\vec{E}$ the density of charge. Charge does not cause the electric vector to have a non-vanishing divergence, it is this non-vanishing divergence. In the same way matter does not cause the geometrical quantity which forms the first member of the above equation to be different from zero, it is this non-vanishing tensor, it is described by it. (Schrödinger 1950, p. 99)

The relativists tried to subject electromagnetism to the treatment which they had meted out to gravitation: they tried to imbed the electromagnetic field into geometry. Maxwell himself had looked upon the charge density ρ as the divergence of the electric field $\vec{E}$; and we have just seen that this was also Schrödinger's view. By trying to derive the electromagnetic field from a geometry which would also explain gravitation, the relativists were, in Meyersonian terms, attempting to dissolve the specificity of electrical phenomena into the homogeneity of the same underlying kinematics. Matter and charge, which had appeared to us as possessing irreducibly different qualities, would now be fused into one global geometry. Putting it in theological language: God created one geometry whose various aspects are apprehended by our senses in different ways. This is nothing but the culmination of the Cartesian programme. The Meyersonian scheme can be summed up as follows:

Principle of Identity

(A) Identity of Form: Legal Explanations (Constancy of functional relationships, phenomenological laws).

(B) *Identity of Substance:* Causal Explanations.

- (B1) Identity of Substance in Time: Classical Conservation Laws (Matter, Momentum, Energy).
- (B2) Identity of Substance in Space: Unity of Matter (Cartesian Reduction of Matter to Space).
- (B3) Identity in Space-Time: Geometrisation of Nature (General Relativity, Unified Field Theories, Block Universe).

Whatever one might otherwise think of it, Meyerson's philosophy explains why scientists tried to force their physical principles into a geometrical mould and consequently injected into these principles some of the surplus structure of the mould.

We have now completed the circle: Meyerson's identity requirement figured among the most fundamental intuitive or pre-systematic, principles which constitute the heuristics of research programmes; such principles are made more powerful and more precise by the intervention of certain mathematical techniques; this effort towards increased mathematisation can in its turn be explained by means of the Meyersonian principle of identity.

Before leaving Meyerson one final question should be asked: is it in principle possible for us to explain the whole of reality in geometrical terms? Meyerson admits that it is not. First, the geometry itself, if it is to define matter, charge and energy, must contain arbitrary elements. If one adopts Eddington's or Schrödinger's approach, then one can give no reason why the metric g_{mn} or the affinity Γ^i_{mn} should assume any specific values at specific points; it is no longer possible to say it is because of the presence of matter or of charge, for this very presence is to be defined in terms of the g_{mn}'s or of the Γ^i_{mn}'s. One can, for example, give no reason why space should be curved and not flat. Thus the values of the metric, or of the affinity, are contingently given; they are simply posited. Secondly: although we have banished all qualities from physical theory, we have not thereby eliminated them but pushed them into a no man's land between physics, on the one hand, and physiology or psychology or both, on the other. We have simply shifted the burden onto the so-called mind-body problem. In Aristotle's philosophy we could, for example, make sense of the reason why certain bodies are *perceived* as hot; according to Aristotle, this is because they *are* hot. In relativity we have to accept that certain geometrical properties of the physical world surface as sensations at the level of consciousness. As pointed out by Meyerson however, the transition from kinematics to sensation constitutes an irrational jump which will always remain incomprehensible to us. Moreover, physics does not completely succeed in banning all qualities from its domain: in testing physical hypotheses, we have to interpret certain propositions in operationalist terms. Hence the necessary existence of bridging principles between certain physical statements on the one hand and sense-data on the other. Finally, in its attempt to establish reversibility, and thus a kind of parity between past and future, physics has utterly failed. The entropy law underlines the fundamental asymmetry between past and future. Thus, according to Meyerson, nature sets definite limits to the applicability of the identity principle, i.e. to the principle of the negation of diversity which remains *an a priori tendency* of the human mind. If one accepts that the identity principle has essential survival value, then nature has to comply with it only to the minimum extent of enabling the species to survive.

— 2 —

LORENTZ'S PROGRAMME

§2.1 Lorentz's Ether Programme

As indicated above, I shall appraise the progress of Lorentz's ether programme in the terms provided by MSRP. I therefore need to specify both the hard core and the heuristic of this programme. The hard core consists of Newton's laws of motion (with the possible exception of the third law), of the Galilean transformation and of Maxwell's equations augmented by the expression for the so-called Lorentz-force.

The heuristic of the programme arises from the overall *metaphysical* (see §1.5) principle that all physical phenomena are governed by actions transmitted by the ether. Applications of the heuristic to specific problems (which may or may not be set by refutations) generate a sequence of theories.

In this chapter we shall be mainly concerned with three consecutive theories belonging to Lorentz's ether programme. I shall refer to them as T_1, T_2, and T_3. T_1 consists of the hard core as defined above, together with the (tacit) assumptions: (i) that moving clocks are not retarded, and (ii) that material rods are not shortened by their motion through the ether. T_2 is obtained by substituting the LFC (that is, Lorentz-Fitzgerald contraction hypothesis) for assumption (ii). According to the LFC, a body moving through the ether with velocity $\vec{v}$ is shortened by the factor $\sqrt{1 - v^2/c^2}$ in the direction of $\vec{v}$. T_3 is the conjunction of the hard core, of the LFC and of the assumption that, contrary to (i), clocks moving with the velocity $\vec{v}$ are retarded by the same factor $\sqrt{1 - v^2/c^2}$.[1] I claim that both the shift from T_1 and T_2 and that from T_2 to T_3 were non ad hoc in all of the three senses defined above. This implies in particular that the introduction of the LFC, which took Lorentz from T_1 to T_2, was not an ad hoc manoeuvre.

[1]This is a slight simplification. We shall see that Lorentz deduced the LFC from the molecular forces hypothesis (MFH). However, though he used an instrumental notion of 'local time', he did not realise that the MFH entails the clock retardation hypothesis (i.e., proposition (i) in the text). This oversight enabled Lorentz to construct the hybrid hypothesis T_2 which implies rod-contraction but no clock-retardation. Only after Einstein published his results in 1905 did it occur to Lorentz that the MFH implies that moving clocks are retarded by the factor $\sqrt{1 - v^2/c^2}$. He thus arrived at T_3 which can, to all intents and purposes, be considered as observationally equivalent to special relativity.

Without resorting to any electromagnetic considerations, both T_1 and T_2 are easily seen to be non ad hoc. Arguing against Popper, Grünbaum showed that the LFC does not constitute an ad hoc modification of the ether theory, since its confirmation is possible through an experiment different from that of Michelson and Morley. In the Kennedy-Thorndike type of experiment described by Gruñbaum, the arms of the interferometer have different lengths, L and ℓ (see introduction, §3). By virtue of the contraction hypothesis, T_2 predicts that the difference $(t_2 - t_1)$ between the times it takes the two rays R_1 and R_2 to return to the half-silvered mirror is equal to:

$$\frac{2(L/\gamma)}{c}\gamma^2 - \frac{2\ell}{c}\gamma = \frac{2\gamma}{c}(L - \ell).$$

This value is clearly different from the one entailed by the strictly classical theory, namely from:

$$\frac{2L}{c}\gamma^2 - \frac{2\ell}{c}\gamma = \frac{2\gamma}{c}(L\gamma - \ell) \text{ (see introduction, §3).}$$

As for T_3 and for Einstein's STR, they both imply that the measured velocity of light is equal to c in all inertial frames; so that:

$$t_2 = \frac{2L}{c} \text{ and } t_1 = \frac{2\ell}{c}\text{; whence: } t_2 - t_1 = \frac{2}{c}(L - \ell).$$

In view of the inequalities:

$$\frac{2\gamma}{c}(L\gamma - \ell) \neq \frac{2\gamma}{c}(L - \ell) \neq \frac{2}{c}(L - \ell) \neq \frac{2\gamma}{c}(L\gamma - \ell),$$

the three predictions are all different from one another.

It is thus settled that neither Lorentz's T_2 nor his T_3 was ad hoc_1. It will be more difficult to decide whether T_2 was ad hoc_2 and/or ad hoc_3.

Was T_2 ad hoc_2 in the sense that until 1905 nobody bothered to test its novel predictions? Later, I shall show that it was not; but even had it been, this constitutes no damning criticism of the ether programme; it would only have meant that its empirical progressiveness had not yet been shown. Ad hocness in the second sense becomes a demerit of a research programme only if it is a lasting feature.

Is then T_2 ad hoc_3? In actual scientific practice a hypothesis is intuitively judged to be ad hoc if it looks arbitrary, i.e. if it fits poorly into its research programme. Thus the introduction of a theory which is ad hoc in this sense destroys the organic unity of the whole nexus since the various components of the resulting system are structured according to conflicting plans. If for example, within the ether programme, a theory postulating some new instantaneous action-at-a-distance were proposed, the new theory would be found intuitively ad hoc.

Ad hocness$_3$ is a good explication of this intuitive notion of ad hocness. To repeat, a theory is said to be ad hoc$_3$ if it conflicts with the heuristic of the research programme. Were it to be shown that T_2 is ad hoc$_3$, then two results would be achieved at one stroke: from a *methodological* point of view Lorentz's programme would be shown to have had serious defects in 1905; from an *historical* point of view, it would become plausible that Einstein, who attributed great importance to the criterion of 'internal perfection', was motivated to start his rival programme by the patched-up state of Lorentz's T_2.

Holton seems to be claiming that Lorentz's T_2 was both ad hoc$_2$ and ad hoc$_3$; and that Einstein's programme was triggered off by Einstein's recognition of these defects. Holton writes:

> This saving Hülfshypothese [the LFC] is introduced completely *ad hoc* . . . *No explicit comment is made which connects this assumed shrinkage with the Lorentz transformations in their still primitive form, as published earlier in this book* . . .
>
> The contraction hypothesis when it was made was clearly and quite blatantly *ad hoc*—or, if one prefers to use the *patois* of the laboratory, *ingeniously cooked up for the narrow purpose for which it was to serve* . . .
>
> The important point to note is that 'ad hoc' is not an absolute but a *relativistic* term. Postulates 1 and 2 [Einstein's two postulates in his [1905]] may be said to have been introduced *ad hoc* with respect to the Relativity Theory of 1905 as a whole . . . But these postulates were *not ad hoc with respect to the Michelson experiment,* for they were not specifically imagined in order to account for its results . . . (Holton 1969, pp. 177–181, my italics)[2]

Holton makes two distinct claims. His first claim is that the contraction hypothesis (LFC), which differentiates T_2 from T_1, was not connected with the rest of T_2, in particular with the Lorentz transformation; thus, on my terms, the LFC is allegedly ad hoc$_3$. His second claim is that the LFC was specifically engineered in order to account for Michelson's result; from this, together with the fact that for a long time the Michelson result was the only one which seemed to support the LFC, I conclude that, in Holton's view, the LFC was not independently tested and therefore was ad hoc$_2$. There are two main reasons why I cannot agree with Holton's claims.

---

[2]The term 'relativistic' was probably a slip of the pen and should read 'relative'. But I cannot make head or tail of Holton's sentence 'Postulates 1 and 2 may be said to have been introduced *ad hoc* with respect to the Relativity Theory of 1905 as a whole'. How can Einstein's two postulates, which constitute special relativity theory or are at any rate part of relativity theory, be ad hoc with respect to relativity theory?

First, Lorentz deduced the LFC from a deeper theory, namely from what I call the molecular forces hypothesis (MFH), which can be loosely formulated as follows: 'Molecular forces behave and transform like electromagnetic forces'. Moreover, in his deduction of the LFC, Lorentz made use of his famous transformation, as is clearly indicated by the following passage:

> For, if we now understand by S_1 and S_2 not, as formerly, two systems of charged particles but two systems of molecules—the second at rest and the first moving with velocity v in the direction of the axis x—between the dimensions of which the relation subsists as previously stated; and if we assume that in both systems the x-components of the forces are the same, while the y- and z-components differ from one another by the factor $\sqrt{1 - v^2/c^2}$, then it is clear that the forces in S_1 are in equilibrium whenever they are so in S_2 . . . The displacement would naturally bring about this disposition of the molecules of its own accord and thus effect a shortening in the direction of motion in the proportion of 1 to $\sqrt{1 - v^2/c^2}$, *in accordance with the formulae given in the above-mentioned paragraph*. (Einstein et al. 1923, p. 7, my italics)

I therefore maintain that for anybody prepared to accept the assumption of an ether at rest, the MFH is a plausible auxiliary hypothesis which introduces no alien elements into Lorentz's programme. Putting it more objectively, the theory T_2 proposed by Lorentz is not ad hoc_3; for the MFH is structured in accordance with the heuristic of an ether programme requiring that physical phenomena be explained in terms of actions propogated through the ether.

Moreover, *the MFH arose out of considerations which had nothing to do with Michelson's experiment*. The MFH arose out of mathematical considerations pertaining to the transformation properties of Maxwell's equations. *Hence Michelson's null result is a novel fact relative to the MFH*. The MFH is consequently not ad hoc_2; it constituted both theoretical and empirical progress.

Let me now briefly turn to the implications of my methodological theses for the various historical accounts of the Einsteinian revolution. My claim is that the LFC is not ad hoc, and that Michelson's result, far from providing an obstacle to Lorentz's programme, in fact supported it. This clearly rules out all explanations of the genesis of STR which depend on the assumption that Einstein *correctly* assessed the ad hoc character of the LFC relative to Michelson's result. One such explanation is Holton's, even though he attributes only an indirect role to Michelson's result. He alleges that Einstein was dissatisfied with the LFC because it was blatantly ad hoc; it was 'cooked up for the narrow purpose which it was to serve'. But ad hoc relative to what?[3] Obviously Holton's claim is that the LFC was ad hoc relative to

[3]Holton correctly points out that 'a statement may be *ad hoc* relative to one context but not *ad hoc* relative to another'. (Holton 1969, p. 181.)

Michelson's result, since the 'narrow purpose which it was to serve' was precisely an explanation of the null result of the 'crucial' experiment. Holton adds that:

> the problem Einstein saw was not the logical status of the Contraction Hypothesis, not Michelson's experiment result (for it could be accommodated, even if not 'ohne Weiteres') but the inability of Lorentz's theory to fulfil the criterion of 'inner perfection' of a theory. (Holton 1969, pp. 184–5)

So Lorentz's theory lost its 'inner perfection' on the introduction of the LFC, which was contrived for the sole purpose of explaining Michelson's result. On this account, the 'crucial' experiment did play an important, if indirect, role in the genesis of STR: the search for an explanation of Michelson's result compelled Lorentz to resort to an hypothesis whose *ad hoc* character provided Einstein with a good reason for starting his revolutionary new programme.

One might defend this Holtonian account against my arguments by assuming that Einstein appraised the LFC incorrectly.[4] Perhaps Einstein mistakenly regarded Lorentz's LFC as *ad hoc* in the sense of being engineered simply for the purpose of neutralising Michelson's result. But this assumption is highly implausible. Einstein read the *Versuch*, in which Lorentz proposed the MFH and derived the LFC from it. Further Einstein could hardly have regarded the LFC as more ad hoc than his own Principle (P2). For first, on Einstein's own criteria, STR, as presented in 1905, was far from being 'internally perfect'. It consisted of two heterogeneous parts which were on totally different levels: on the one hand a high-level, universal covariance principle and, on the other, a so-called light postulate which was both low-level and prima facie incompatible with covariance. Secondly, whereas Lorentz *explained* why a moving rod contracts, Einstein bluntly asserted that the speed of light is an invariant, an assumption from which Michelson's result trivially follows. Reichenbach was at least partially correct when he wrote that:

> it would be mistaken to argue that Einstein's theory gives an explanation of Michelson's experiment since it does not do so. Michelson's experiment is simply taken over as an axiom. (Reichenbach 1958, p. 201)

Reichenbach was right in the following sense: while (P2) can be *intuitively* regarded as a low-level generalisation of Michelson's result, the latter is prima facie unconnected with the MFH. (The fact that (P2) is both low-level and counter-intuitive was recognised by many of Einstein's contem-

---

[4]No doubt Holton would argue that the fact that my methodology admits of this possibility shows the folly of trying to appraise the actions of great scientists with the help of explicit general criteria. Fortunately, on my account, Einstein did not make such a mistake.

poraries; it understandably gave rise to the myth that Einstein was a *positivist*, who unquestioningly obeyed the dictates of experience.)

My view that Einstein could hardly have judged Lorentz's MFH as more ad hoc than his own principle (P2) is supported by the following fact. The proposition that, in all inertial frames, the measured speed of light must be equal to the same constant c is deducible from Lorentz's pre-1905 system; and this system includes the MFH.[5] In other words, *Lorentz's theory explains not only Michelson's null result but also the invariance of* c. Whatever meaning is attached to 'ad hoc relative to a context', it cannot allow that the MFH should *both* imply the light postulate *and*, unlike the light postulate, be ad hoc relative to Michelson's experiment. Lorentz was justified in asserting that:

> . . . the chief difference [is] that Einstein *simply postulates* what we have *deduced*, with some difficulty and not altogether satisfactorily, from the fundamental equations of the electromagnetic field. (Lorentz 1909, p. 230, my italics)

Einstein's programme eventually proved superior to Lorentz's in a strictly objective sense, but this superiority does not rest on the ad hoc character of Lorentz's system.

§2.2 The Genesis of the Lorentz Transformation

It has by now become clear that, in order to vindicate my claim that T_2 is not ad hoc—in the second and third senses—I have to go into actual history and examine the methods by which Lorentz's theories were constructed. Hence the necessity of filling in some technical details, starting with Lorentz's derivation of his famous transformation.

In the TEM (published in 1892) Lorentz posits an immobile ether which permeates all ponderable bodies without being affected by their motion. The ions are charged particles which move while keeping their charge and mass unaltered; the ions can be conceived as small regions within the ether in which the charge and mass densities do not vanish. Lorentz writes Maxwell's equations in the absolute frame as follows:

$$\text{(I)} \quad \nabla \cdot \vec{D} = \rho$$

$$\text{(II)} \quad \nabla \cdot \vec{H} = o$$

$$\text{(III)} \quad \nabla \times \vec{D} = -\frac{1}{4\pi c^2}\frac{\partial \vec{H}}{\partial t}$$

[5]Admittedly Lorentz performed the deduction only in 1909 (§2.6 below). However, I am enough of a Polanyiite to find it implausible that Einstein failed to realise, prior to 1905, that, from an *intuitive* point of view, the MFH cannot be more ad hoc than (P2).

$$\text{(IV)}\ \nabla \times \vec{H} = 4\pi\left(\frac{\partial \vec{D}}{\partial t} + \rho\vec{u}\right),$$

where ∇ is the vector operator $\left(\frac{\partial}{\partial x}, \frac{\partial}{\partial y}, \frac{\partial}{\partial z}\right)$; $\vec{D}$ is the electric displacement; $\vec{H}$ the magnetic field, ρ the charge density, $\vec{u}$ the charge velocity and c the speed of light.

The so-called Lorentz-force $\vec{E}$ which acts on a charge e moving with velocity $\vec{w}$ is given by

$$\text{(V)}\ \vec{E} = e(4\pi c^2\vec{D} + \vec{W} \times \vec{H}).$$

In the TEM Lorentz deals with each scalar component separately. For purposes of convenience I am using vector notation, which simplifies the presentation without affecting the structure of Lorentz's mathematical arguments.

An important problem to which Lorentz addressed himself was how to determine the field from the charge and velocity distributions in the ether.

Let us assume that ρ vanishes for large values of $(x^2 + y^2 + z^2)$; in other words there exists a sphere outside which no ions are present. Taking curls of both sides of (3) we have:

$$\nabla \times (\nabla \times \vec{D}) = -\frac{1}{4\pi c^2} \cdot \nabla \times \frac{\partial \vec{H}}{\partial t}$$

In other words:

$$\nabla(\nabla \cdot \vec{D}) - \nabla^2\vec{D} = -\frac{1}{4\pi c^2}\frac{\partial}{\partial t}(\nabla \times \vec{H}).$$

By (I) and (IV):

$$\textbf{(1)}\ \left[c^2\nabla^2 - \frac{\partial^2}{\partial t^2}\right]\vec{D} = c^2\nabla\rho + \frac{\partial}{\partial t}(\rho\vec{u})$$

Since the right hand side is a known function of t,x,y,z, we now face the purely mathematical problem of solving a partial differential equation of the form:

$$\textbf{(2)}\ \left[\nabla^2 - \frac{1}{c^2}\frac{\partial^2}{\partial t^2}\right]f = G(t,x,y,z)$$

where f is the unknown and G is a given function of four variables.

Under the heading 'Théorèmes Mathématiques' Lorentz proves the following proposition. Let τ' be a given region of space to whose interior the point $\vec{r} = (x,y,z)$ belongs. The function f of t,x,y,z defined by:

$$f(t,x,y,z) = -\frac{1}{4\pi}\int \frac{1}{|\vec{r} - \vec{r}'|} \cdot G\left(t - \frac{|\vec{r} - \vec{r}'|}{c}, x', y', z'\right) \cdot d\tau'$$

where $\vec{r}' = (x',y',z')$ and $d\tau' = dx' \cdot dy' \cdot dz'$, satisfies equation (2).

As they stand, these solutions are not very useful for we ordinarily measure our coordinates in a frame of reference fixed relatively to a moving ponderable body, like the earth. For any short time interval any slowly accelerating body can be looked upon as moving uniformly along a straight line. We therefore face the problem of expressing Maxwell's equations in a system moving with constant velocity through the ether. Without loss of generality, suppose that the motion takes place along the x-axis. Let (t_1,x_1,y_1,z_1) be the 'absolute' coordinates in a frame S_1 fixed in the ether. If (t,x,y,z) are the coordinates in a system S moving with constant velocity $\vec{v} = (v,o,o)$ along the x-axis, then: $x = x_1 - vt_1$; $y = y_1$; $z = z_1$; $t = t_1$. Note that we are using the Galilean transformation, which Lorentz took to be the only 'real' one until after 1905.

$$\frac{\partial}{\partial x_1} = \frac{\partial}{\partial x}, \frac{\partial}{\partial y_1} = \frac{\partial}{\partial y}, \frac{\partial}{\partial z_1} = \frac{\partial}{\partial z}, \text{ but } \frac{\partial}{\partial t_1} = \frac{\partial}{\partial t} - v \cdot \frac{\partial}{\partial x}.$$

If we let $\vec{u}$ stand for the charge velocity in the moving S frame, the absolute velocity will equal $\vec{v} + \vec{u}$ and Maxwell's equations become:

(Ia) $\nabla \cdot \vec{D} = \rho$

(IIa) $\nabla \cdot \vec{H} = 0$

(IIIa) $\nabla \times \vec{D} = -\frac{1}{4\pi c^2} Q\vec{H}$

(IVa) $\nabla \times \vec{H} = 4\pi Q\vec{D} + 4\pi\rho(\vec{v} + \vec{u})$

(Va) $\vec{E} = 4\pi c^2\vec{D} + \vec{v} \times \vec{H} + \vec{u} \times \vec{H}$

where Q is the scalar operator $\left[\frac{\partial}{\partial t} - v\frac{\partial}{\partial x}\right]$ and $\vec{E}$ the Lorentz force per unit charge.

Once again, given the charge and velocity distributions, we have to find $\vec{D}$ and $\vec{H}$. Taking curls of both sides of (IIIa) and making use of (Ia) and (IVa), we obtain:

(3) $[c^2\nabla^2 - Q^2]\vec{D} = c^2\nabla\rho + Q[\rho(\vec{v} + \vec{u})]$

and similarly

(4) $[c^2\nabla^2 - Q^2]\vec{H} = -4\pi c^2\nabla \times [\rho(\vec{v} + \vec{u})]$

We therefore have to solve a partial differential equation of the form:

(5) $\Box f = G(t,x,y,z)$

where:

$$\Box = c^2\nabla^2 - Q^2 = c^2\left(\frac{\partial^2}{\partial x^2} + \frac{\partial^2}{\partial y^2} + \frac{\partial^2}{\partial z^2}\right) - \left(\frac{\partial}{\partial t} - v\frac{\partial}{\partial x}\right)^2$$
$$= c^2\left[\left(1 - \frac{v^2}{c^2}\right)\frac{\partial^2}{\partial x^2} + \frac{\partial^2}{\partial y^2} + \frac{\partial^2}{\partial z^2}\right] - \frac{\partial^2}{\partial t^2} + 2v\frac{\partial^2}{\partial x\partial t}$$

Lorentz starts by considering a special case of this last equation, namely $\Box f = 0$. In view of the availability of a general solution for $\left[c^2\nabla^2 - \frac{\partial^2}{\partial t^2}\right] f = 0$, it is desirable, from a purely mathematical point of view, to find a transformation which reduces the equation $\Box f = 0$ to the form:

$$\left[A^2\nabla^2 - B^2\frac{\partial^2}{\partial t^2}\right]f = 0$$

(Dividing this last equation by B^2 yields:

(6) $$\left[k^2\nabla^2 - \frac{\partial^2}{\partial t^2}\right]f = 0, \text{ where } k = \frac{A}{B}\Bigg]$$

The solution can be systematically reached in two steps consisting of the following changes of variables:

First step We want $\left(1 - \frac{v^2}{c^2}\right)\frac{\partial^2}{\partial x^2}$ to be of the form $\frac{\partial^2}{\partial \xi^2}$.

Hence:

$$\frac{\partial}{\partial x} = \gamma\frac{\partial}{\partial \xi}, \text{ where } \gamma = \left(1 - \frac{v^2}{c^2}\right)^{-1/2}.$$

We therefore take $\xi = \gamma x$, from which it follows that:

(7) $$\Box = c^2\left(\frac{\partial^2}{\partial \xi^2} + \frac{\partial^2}{\partial y^2} + \frac{\partial^2}{\partial z^2}\right) - \frac{\partial^2}{\partial t^2} + 2v\gamma\frac{\partial^2}{\partial \xi\partial t}.$$

Second step We express $\Box$ in the form:

$$\Box = c^2\left[\left(\frac{\partial}{\partial\xi} + v\frac{\gamma}{c^2}\frac{\partial}{\partial t}\right)^2 + \frac{\partial^2}{\partial y^2} + \frac{\partial^2}{\partial z^2}\right] - \left(1 + \frac{v^2\gamma^2}{c^2}\right)\frac{\partial^2}{\partial t^2}$$

(Note that, in view of $\gamma = \left(1 - \frac{v^2}{c^2}\right)^{-1/2}$, we have: $1 + \frac{v^2\gamma^2}{c^2} = \gamma^2$).

We propose to determine new variables x' and t' such that:

$$\frac{\partial}{\partial x'} = \frac{\partial}{\partial\xi} + \frac{v\gamma}{c^2}\frac{\partial}{\partial t}, \quad \frac{\partial}{\partial t'} = \frac{\partial}{\partial t}.$$

Therefore:

$$\frac{\partial\xi}{\partial x'} = 1, \quad \frac{\partial t}{\partial x'} = \frac{v\gamma}{c^2}, \quad \frac{\partial t}{\partial t'} = 1 \text{ and } \frac{\partial\xi}{\partial t'} = 0,$$

from which it follows that:

$$\xi = x' \text{ and } t = t' + \frac{v\gamma}{c^2}x'.$$

Solving for x' and t', we obtain:

$$x' = \xi \text{ and } t' = t - \frac{v\gamma}{c^2}\xi$$

Hence:

$$\textbf{(8)}\quad \Box = c^2\left(\frac{\partial^2}{\partial x'^2} + \frac{\partial^2}{\partial y'^2} + \frac{\partial^2}{\partial z'^2}\right) - \gamma^2\frac{\partial^2}{\partial t'^2} = c^2\nabla'^2 - \gamma^2\frac{\partial^2}{\partial t'^2}$$

where $y' = y$, $z' = z$ and $\nabla' = \left(\frac{\partial}{\partial x'}, \frac{\partial}{\partial y'}, \frac{\partial}{\partial z'}\right)$.

The product of these two transformations is:

$$x' = \xi = \gamma x,\ y' = y,\ z' = z,\ t' = t - \frac{v\gamma}{c^2}\xi = t - \frac{v\gamma^2}{c^2}x.$$

In terms of the absolute coordinates:

$$\textbf{(9)}\quad x' = \gamma x = \gamma(x_1 - vt_1),\ y' = y = y_1,\ z' = z = z_1 \text{ and}$$
$$t' = t - \frac{v\gamma^2}{c^2}x = t_1 - \frac{v\gamma^2}{c^2}(x_1 - vt_1) = \gamma^2\left(t_1 - \frac{v}{c^2}x_1\right).$$

These equations, which transform $\Box$ into $\left[c^2\nabla'^2 - \gamma^2 \frac{\partial^2}{\partial t'^2}\right]$ constitute the so-called Lorentz Transformation, to within an extra factor γ in the expression of t'. We may wonder why Lorentz did not put $t' = \gamma t''$ and thus obtain the full invariance of the operator $\left[c^2\nabla^2 - \frac{\partial^2}{\partial t^2}\right]$.[6] The net effect of all these transformations would have been to carry $\left[c^2\nabla^2 - \frac{\partial^2}{\partial t^2}\right]$ into $\left[c^2\nabla'^2 - \frac{\partial^2}{\partial t''^2}\right]$. The answer is simply that Lorentz was not, at this stage, interested in invariance for its own sake, but only as a means of solving a mathematical problem, and a very particular one at that. All the evidence points to the accidental character of a device introduced in order to solve a specific equation, namely $\Box f = 0$. It is also puzzling to find that Lorentz did not use his transformation when he turned to the more general equation:

$$\Box f = G(t,x,y,z)$$

The transformation would have given:

$$\left[c^2\nabla'^2 - \gamma^2 \frac{\partial^2}{\partial t'^2}\right] f = G\left(t' + \frac{v\gamma}{c^2} x' , \frac{1}{\gamma} x', y', z'\right) \text{ ; i.e.}$$

$$\left[\nabla'^2 - \left(\frac{\gamma}{c}\right)^2 \frac{\partial^2}{\partial t'^2}\right] f = \frac{1}{c^2} G\left(t' + \frac{v\gamma}{c^2} x' , \frac{1}{\gamma} x', y', z'\right)$$

Let $\sigma = [(x' - x'')^2 + (y' - y'')^2 + (z' - z'')^2]^{1/2}$.

By Lorentz's previous theorem, an immediate solution of the last differential equation is given by:

$$f = -\frac{1}{4\pi c^2} \int_{\tau''} \frac{1}{\sigma} G\left(t' - \frac{\gamma}{c}\sigma + \frac{v\gamma}{c^2} x'', \frac{1}{\gamma} x'', y'', z''\right) \cdot d\tau'',$$

where $d\tau'' = dx'' \cdot dy'' \cdot dz''$.

Putting $x'' = \gamma\xi$, $y'' = \eta$, $z'' = \zeta$, we obtain:

$$d\tau'' = \gamma \cdot d\xi \cdot d\eta \cdot d\zeta = \gamma\, d\kappa \text{ (say).}$$

[6]For an approach to relativity theory using the invariance of $\left[c^2\nabla^2 - \frac{\partial^2}{\partial t^2}\right]$ see Stephenson & Kilmister's 'Special Relativity for Physicists' (1958).

Hence:

$$f = -\frac{\gamma}{4\Pi c^2}\int_\kappa \frac{1}{\sigma} G\left(t' - \frac{\gamma\sigma}{c} + \frac{v\gamma^2}{c^2}\xi, \xi, \eta, \zeta\right) d\kappa.$$

Since $x' = \gamma x$, $t' = t - \frac{v\gamma^2}{c^2}x$, $y' = y$, $z' = z$, we have:

$$f = -\frac{\gamma}{4\pi c^2}\int_\kappa \frac{1}{\sigma} G\left(t + \frac{v\gamma^2}{c^2}(\xi - x) - \frac{\gamma\sigma}{c}, \xi, \eta, \zeta\right) d\kappa,$$

where $\sigma = [(x'' - x')^2 + (y'' - y')^2 + (z'' - z')^2]^{1/2}$
$= [\gamma^2(\xi - x)^2 + (\eta - y)^2 + (\zeta - z)^2]^{1/2}$.

The point (x', y', z'), i.e. $(\gamma x, y, z)$, must belong to τ''. Since $\kappa = \{(\xi, \eta, \zeta):(\gamma\xi, \eta, \zeta) \epsilon \tau''\}$, it follows that $(x, y, z) \epsilon \kappa$. Lorentz obtains the same result but, instead of using his own transformation as I have just shown, he repeats the lengthy arguments through which he established his previous theorem. (LCP, vol. 2 p. 321) This lends further support to the view that the discovery of the Lorentz transformation was accidental and that Lorentz initially put equation (9) to a very specific and narrow use.

§2.3 Physical Interpretation of the Lorentz Transformation: First Version of the Theory of Corresponding States

In the TEM there is no mention of the Michelson-Morley experiment. To repeat: the Lorentz transformation was originally used as an uninterpreted mathematical device in much the same way as we might use the expression ict without attaching any physical meaning to the multiplication of the distance ct by $i = \sqrt{-1}$. However, later in that same year of 1892, Lorentz realised that equations (9) lent themselves to a physical interpretation; this he outlined in his paper "The Relative Motion of the Earth and the Ether" (LCP, vol. 4 p. 219) (henceforth referred to as RMEE) and later expounded in greater detail in the *Versuch* (LCP, vol. 5 p. 1).

Let us take a look at the equations: $x' = \gamma x$, $y' = y$, $z' = z$, and $t' = t - \frac{v\gamma^2}{c^2}x$, where: $\gamma = \left(1 - \frac{v^2}{c^2}\right)^{-1/2}$ and x, y, z, t are the Galilean coordinates in the moving frame S. x' is simply obtained by expanding x by the factor γ, which is greater than 1. The variable t' is more difficult to interpret physically because it involves both time and position. Fortunately, by considering a system of particles at rest in S, i.e. a system of particles all moving with the same velocity $\vec{v}$ through the ether, Lorentz was able to simplify his problem. Now the field depends only on x, y, z, or, alternatively, on x′, y′, z′; so the 'local' time t′ can be safely ignored. To the moving system S, Lorentz made correspond a system S′ at rest in the medium; S′ is obtained by expanding S by the factor γ along the x-axis, while keeping both the charge and the other two dimensions unaltered. Conversely, S is a contracted image of S′; the connection between this physical interpretation and the contraction hypothesis, which Lorentz later inferred, thus becomes obvious.

The next step was to calculate the forces acting at corresponding points of S and S′. Lorentz found that each component of one force is proportional to the corresponding component of the other. Consequently, the vanishing of one of the forces entails that of the other.

We shall now give Lorentz's derivation in some detail. Let us go back to equations (3) and (4) and consider an electrostatic system in the *moving* frame S. In this case, $\vec{u} = o$, and all quantities are independent of the time. Hence:

$$\Box = \left[c^2\nabla^2 - v^2\frac{\partial^2}{\partial x^2}\right] = c^2\left[\left(1 - \frac{v^2}{c^2}\right)\frac{\partial^2}{\partial x^2} + \frac{\partial^2}{\partial y^2} + \frac{\partial^2}{\partial z^2}\right]$$

The equations for $\vec{D}$ and $\vec{H}$ become:

$$(10)\begin{cases}\left[\nabla^2 - \dfrac{v^2}{c^2}\dfrac{\partial^2}{\partial x^2}\right]\vec{D} = \nabla\rho - \dfrac{v}{c^2}\,\vec{v}\,\dfrac{\partial\rho}{\partial x}\\[2ex] \left[\nabla^2 - \dfrac{v^2}{c^2}\dfrac{\partial^2}{\partial x^2}\right]\vec{H} = -4\pi\nabla\times(\rho\vec{v}) = 4\pi\vec{v}\times\nabla\rho\end{cases}$$

Let w be any solution of the equation:

$$(11)\quad \left[\nabla^2 - \frac{v^2}{c^2}\frac{\partial^2}{\partial x^2}\right]w = \rho.$$

We see that a solution of (10) is given by:

$$(12)\quad \vec{D} = \nabla w - \frac{v}{c^2}\frac{\partial w}{\partial x}\vec{v},\ \vec{H} = 4\pi\vec{v}\times\nabla w.$$

From (Va) it follows that:

$$(13)\quad \vec{E} = 4\pi c^2\vec{D} + \vec{v}\times\vec{H} = \left[4\pi c^2\nabla w - 4\pi v\frac{\partial w}{\partial x}\vec{v} + 4\pi(\vec{v}\cdot\nabla w)\vec{v} - 4\pi v^2\nabla w\right]$$

$$= 4\pi(c^2 - v^2)\nabla w = \frac{4\pi c^2}{\gamma^2}\nabla w,$$

$$\text{since } \vec{v} \cdot \nabla w = (v,o,o) \cdot \left(\frac{\partial w}{\partial x}, \frac{\partial w}{\partial y}, \frac{\partial w}{\partial z}\right) = v \cdot \frac{\partial w}{\partial x}.$$

The problem therefore reduces to finding a solution of (11), i.e. of the equation:

$$\left(1 - \frac{v^2}{c^2}\right)\frac{\partial^2 w}{\partial x^2} + \frac{\partial^2 w}{\partial y^2} + \frac{\partial^2 w}{\partial z^2} = \rho.$$

If we now resort to the Lorentz transformation restricted to the spatial variables x, y, z, and put: $x' = \gamma x$, $y' = y$, $z' = z$, we obtain:

(14) $\nabla'^2 w = \rho$, where: $\nabla' = \left(\frac{\partial}{\partial x'}, \frac{\partial}{\partial y'}, \frac{\partial}{\partial z'}\right)$.

Lorentz proceeds to interpret the transformation as follows. Let S be a moving electrostatic system, S′ a second system, which is fixed in the ether and obtained by expanding the dimension of S along the direction of the x-axis by the factor γ. In other words, to the point (x,y,z) in S corresponds the point (x′,y′,z′) of S′ such that:

(15) $x' = \gamma x,\ y' = y,\ z' = z.$

Corresponding volume elements are to have equal charges so that, if ρ' denotes the density in S′, we have:

(16) $\rho' = \frac{1}{\gamma}\rho$, since $dx' \cdot dy' \cdot dz' = \gamma \cdot dx \cdot dy \cdot dz$.

In the electrostatic system S′, the displacement $\vec{D}'$ is the gradient of a scalar function w′ satisfying:

(17) $\nabla'^2 w' = \rho' = \frac{1}{\gamma}\rho.$

The force $\vec{E}'$ per unit charge is:

(18) $\vec{E}' = 4\pi c^2 \vec{D}' = 4\pi c^2 \cdot \nabla' w'.$

Comparing (17) and (14), we obtain:

(19) $w = \gamma w'.$

Therefore:

$$\vec{E}' = 4\pi c^2 \nabla' w' = \frac{4\pi c^2}{\gamma}\left(\frac{\partial w}{\partial x'}, \frac{\partial w}{\partial y'}, \frac{\partial w}{\partial z'}\right)$$

$$= \frac{4\pi c^2}{\gamma}\left(\frac{1}{\gamma}\frac{\partial w}{\partial x}, \frac{\partial w}{\partial y}, \frac{\partial w}{\partial z}\right).$$

By (13):

$$\vec{E} = \frac{4\pi c^2}{\gamma^2}\left(\frac{\partial w}{\partial x}, \frac{\partial w}{\partial y}, \frac{\partial w}{\partial z}\right).$$

Hence:

if $\vec{E} = (E_1, E_2, E_3)$ and $\vec{E}' = (E_1', E_2', E_3')$,

then $E_1 = E_1'$, $E_2 = E_2'/\gamma$, $E_3 = E_3'/\gamma$, i.e.

(20) $\vec{E} = \left(1, \frac{1}{\gamma}, \frac{1}{\gamma}\right) \vec{E}'$

Equation (20) plays a fundamental role in the theory of corresponding states. Since corresponding ions have equal charges, the forces acting on them are related by the last equation. Note that if $\vec{E}' = 0$ at some point of S′, $\vec{E}$ also vanishes at the corresponding point of S.

It is not surprising that the first step towards interpreting the Lorentz formation should have been taken in electrostatics, where the time variable can be ignored. As is well known, the time coordinate and, more generally, the kinematical aspect of the transformation caused Lorentz difficulties which were finally settled by Einstein.

I have already mentioned that this interpretation of the transformation equations was first explained at length in the *Versuch* of 1895; but Lorentz had already used it in 1892 in order to derive the contraction hypothesis and thus account for the null result of Michelson's experiment. In his RMEE Lorentz refers to the TEM in the following context:

> Let A be a system of material points carrying electrical charges and at rest with respect to the ether; B the system of the same points while moving in the direction of the x-axis with the common velocity p through the ether. *From the equations developed by me one can deduce which forces the particles in the system B exert on one another.* The simplest way to do this is to introduce still a third system C which, just as A, is at rest, but differs from the latter as regards the location of the points. System C namely can be obtained from system A by a simple extension, by which all dimensions in the direction of the x-axis are multiplied by the factor $(1 + p^2/2V^2)$ and all dimensions perpendicular to it remain un-

altered. Now, the connection between the forces in B and in C amounts to this, that the x-components in C are equal to those in B, whereas the components at right angles to the x-axis are $(1 + p^2/2V^2)$ times larger than in B. (Note that: V = speed of light = c and p = speed of moving frame)

Both the transformation equations and the contraction hypothesis were proposed in 1892. Thus a mathematical device led to the discovery of a physical hypothesis in a way which will be explained in greater detail in the next paragraph. In 1899, after developing a large part of his theory of corresponding states, Lorentz described the starting point of his investigations as follows.

> In previous researches I have assumed that all electrical and optical phenomena present in ponderable bodies are caused by small charged particles (electrons) . . . *certain mathematical artifices have permitted me to arrive, by a concise argument, at conclusions to which, without these artifices, I should not have arrived except by considerably lengthier developments.* (Lorentz, LCP vol. 5 p. 189, my translation and italics)

§2.4 The Molecular Forces Hypothesis (MFH), the Lorentz-Fitzgerald Contraction Hypothesis (LFC), and Michelson's Result

Let us now examine in greater detail how the above considerations bear on Michelson's result.

In 1892 Lorentz put forward the molecular forces hypothesis (MFH) to which he was led by the coordinate transformation proposed in the TEM. The MFH asserts that, in passing from the stationary system S′ to the moving system S, the molecular forces transform like the electrostatic ones; in other words: the moving and stationary molecular forces are also connected by the equation:

$$\vec{F} = (1, 1/\gamma, 1/\gamma)\vec{F}'$$

I shall now reconstruct Lorentz's deduction of the contraction hypothesis from the MFH. He implicitly used an additional assumption $\mathcal{U}$ which states that the equilibrium configuration of a system of particles is unique. In 1892 he does not bother to mention $\mathcal{U}$ explicitly, but in his famous 1904 paper he refers to it as an independent hypothesis.

Let O′A′ = L′ be the length of the rod in the stationery system S′. Let OA = L be the length of the same rod in the moving system S obtained

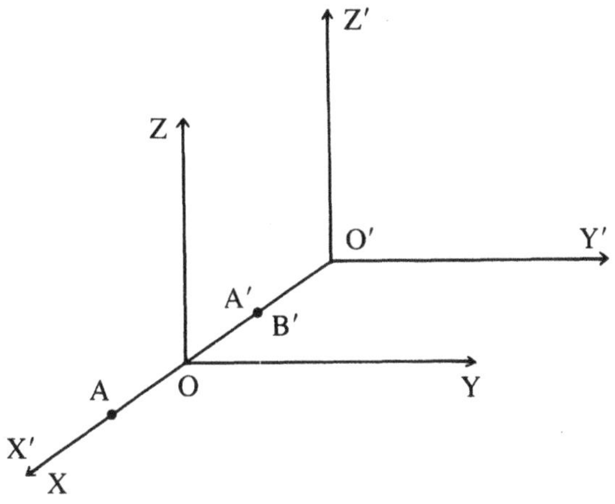

by imparting to S′ a uniform rectilinear motion with speed v along O′X′. We leave open the question of whether L = L′ or L ≠ L′. Let O′B′ = γ L; i.e. O′B′ is obtained by expanding the rod OA by the factor γ, whereby the charge and mass of corresponding elements are to remain the same. Let $\vec{G}$ be the sum of all the forces, molecular as well as electromagnetic, acting at some point of OA and let $\vec{G}'$ be the force exerted at the corresponding point P′ of O′B′. Since P is in equilibrium, $\vec{G} = 0$. By the MFH: $\vec{G} = (1, 1/\gamma, 1/\gamma)\ \vec{G}'$. Hence $\vec{G}'$ also vanishes and P′ is in equilibrium. Since P′ is an arbitrary point of O′B′, the whole of the rod O′B′ must be in equilibrium. By assumption, the same holds for O′A′; so, by the uniqueness hypothesis $\mathcal{U}$, O′B′ = O′A′; i.e. γL = L′ or L = (1/γ)L′; OA is therefore contracted by the factor 1/γ.

Applying this result to Michelson's experiment, we find that, if BD = L, then BE equals, not L, but L/γ. Hence:

$$t_2 - t_1 = [(2L/\gamma c)\gamma^2 - (2L/c)\gamma] = O,$$

so no shift of the fringes will occur.

Thus, we see that Michelson's experiment did not refute the conjunction of Newton's laws and Maxwell's equations but indirectly led Lorentz to put forward a plausible auxiliary hypothesis, namely the MFH. An interesting feature of this development is that *the MFH did not arise from a consideration of the experimental result*. After all, Lorentz knew of the Michelson-Morley result since 1887 and he admitted that it had been worrying him for some time (LCP, vol. 4 p. 221). Not until 1892, after the discovery of his mathematical transformation equations, did he think of putting forward the MFH. In 1887 he could, quite bluntly, have postulated that one dimension of a body contracts in the direction of its motion through the ether. That he waited for five years in order to find a suitable solution shows that he would have considered such a simplistic contraction hypothesis unacceptable. This is why I cannot agree with Holton that ". . . the contraction hypothesis, when it was made, was clearly and quite blatantly ad hoc—or, if one prefers to use the patois of the laboratory, 'ingeniously cooked up' for the narrow purpose which it was to serve". As against this assertion, I claim that, since the discovery of the MFH was independent of Michelson's experiment whose null result follows from a consequence of the MFH, namely from the LFC, the experiment strongly supports Lorentz's theory. The Michelson result is, according to the amended definition of novelty, a 'novel' or 'unexpected' prediction of the MFH. This hypothesis is consequently not ad hoc_2.

Was the MFH 'implausible' in the sense of fitting badly into Lorentz's system taken as a whole? In other words: was the MFH ad hoc_3? Lorentz explained why it was not. Commenting on the plausibility of the MFH, he wrote in 1892:

> What determines the size and shape of a solid body? Evidently the intensity of the molecular forces; any cause which would alter the latter would also influence the shape and dimensions. Nowadays we may safely assume that electric and magnetic forces act by means of the intervention of ether. It is not far-fetched to suppose the same to be true of the molecular forces.

In the *Versuch* of 1895 he stated:

> Surprising as this hypothesis [i.e. the LFC] may appear at first sight, yet we shall have to admit that it is by no means far-fetched, as soon as we assume that molecular forces are also transmitted through the ether, like the electric and magnetic forces of which we are able at the present time to make this assumption definitely. If they are so transmitted, the translation will very probably affect the action between two molecules or atoms in a manner resembling the attraction or repulsion between charged particles.

The MFH is therefore not ad hoc (in the third sense) within the ether programme, whose positive heuristic requires that physical phenomena be explained in terms of contiguous actions through the medium. The molecular forces which determine the shape of a given body are transmitted by the same carrier as that of the electromagnetic field; since both types of force are states of the same substratum, they can be expected to behave and transform in the same way.

Thus, concerning the effect which Michelson had on Lorentz's scientific thinking, the following points emerge:

(1) Lorentz did nothing about the 'crucial' experiment until he discovered his transformation equations; the latter were not discovered under the impact of Michelson's result, of which they are independent. We have here the example of a research programme which could use 'facts' only after its theoretical apparatus had been developed to the point where it could absorb anomalies. Significantly, it was not the refutation as such but the discovery of certain mathematical methods which brought about progress. Neither inductivism nor falsificationism, but MSRP, enables us to understand this process.

(2) Once the transformation equations and their physical interpretation had been found, Michelson's experiment had a psychological—i.e., external—effect on Lorentz's thinking. In the light of the transformation equations, the experiment made him aware of the following implausible assumption which he had *tacitly* accepted all the way along: 'the dimensions of a body moving through the ether are not altered by the motion; hence molecular forces behave very differently from, in fact exactly compensate for, the effects of electromagnetic forces; for the latter tend to shrink the body along its direction of motion.' In other words, Lorentz had taken for granted the postulate P that motion through the ether does not affect the dimensions of a ponderable body; he did not realise what strong assumptions *P* actually involves. His new theory is not obtained by adding an ad hoc hypothesis to his previous system,[7] but by substituting a plausible theory, the MFH, for the implausible postulate P. It can be argued that Lorentz could have anticipated Michelson's result on the basis of the invariance properties of Maxwell's equations; in particular: after the discovery of the transformation equations, he could quite easily have expected a moving body to be flattened along the direction of its motion in the ether.

We can now deal with the controversy as to whether or not Einstein knew of Michelson's experiment when he proposed the special theory of relativity. The fact that he admits having read the *Versuch* shows that he knew of Lorentz's account of Michelson's result; it shows at the same time that Michelson's experiment could hardly have played any significant role in the genesis of relativity theory, since Lorentz had satisfactorily explained its null result with the help of the MFH. In order to start his relativity programme, Einstein had to have, independently of Michelson's result, of the LFC and of the MFH, reasons for questioning the whole classical approach. We shall see that, through criticising the classical programme as a whole, Einstein was led to his two postulates from which Michelson's result

[7] No *mere* addition of an auxiliary hypothesis can save a theory from refutation; for the mere addition of a hypothesis either increases or else does not alter the set of logical consequences of a given theory.

trivially follows. In this sense the 'crucial' experiment was superfluous from Einstein's viewpoint as well.

§2.5 From the *Versuch* to the "Théorie Simplifiée"

Lorentz's fully fledged theory of corresponding states will be subsequently presented and explained in some depth. At this stage however, it suffices to describe in general terms the overall structure and aim of this theory: it sets out to establish the covariance of Maxwell's equations; i.e., it proposes to show, by a judicious choice of the so-called 'effective' variables, that these equations assume the same form in all frames moving with uniform rectilinear velocity in the ether. By virtue of this form-invariance, the theory is then seen to entail that a whole host of experiments designed to detect absolute motion, eg. Michelson's experiment, must yield null results. As is well known, covariance plays a fundamental role in relativity physics. It is therefore of great historical importance to examine how much of Lorentz's theory of corresponding states is already contained in his 1895 *Versuch*. This will give us an estimate of how much, or of how little, Einstein, who is known to have read the *Versuch*, owed Lorentz. We have seen that in one part of the *Versuch* Lorentz applied his transformation

$$x' = \gamma(x_1 - vt_1),\ y' = y_1,\ z' = z_1$$

to an electrostatic system in a moving frame[8]; he then physically interpreted this mathematical operation and was thereby led to the contraction hypothesis (LFC). For this he had only needed to take account of the spatial variables x', y', z'. In the same paper he also pursued a different line of thought which issued in a new version of the theorem of corresponding states. Once again he examined the operator $[c^2\nabla^2 - Q^2]$ and let himself be guided by considerations of mathematical convenience. This time he decided to neglect all terms of the second order in (v/c).

$$\begin{aligned} c^2\nabla^2 - Q^2 &= c^2\left(\frac{\partial^2}{\partial x^2} + \frac{\partial^2}{\partial y^2} + \frac{\partial^2}{\partial z^2}\right) - \left(\frac{\partial}{\partial t} - v\frac{\partial}{\partial x}\right)^2 \\ &\doteqdot c^2\left[\left(\frac{\partial}{\partial x} + \frac{v}{c^2}\frac{\partial}{\partial t}\right)^2 + \frac{\partial^2}{\partial y^2} + \frac{\partial^2}{\partial z^2}\right] - \frac{\partial^2}{\partial t^2} \end{aligned}$$

He wrote:

The form of this expression suggests that, instead of t, we introduce a

[8] See p. 88. In this notation: x_1, y_1, z_1, t are the absolute coordinates and x, y, z, t are the Galilean coordinates in the moving frame.

new independent variable $t' = t - \frac{v}{c^2} x$ . . . (LCP vol. 5 p. 49)[9]

He therefore put:

$$x' = x,\ y' = y,\ z' = z,\ t' = t - \frac{v}{c^2} \cdot x.$$

This is nothing but a limiting case of the following transformation, which he had already used in the TEM:

$$x' = \gamma x,\ y' = y,\ z' = z,\ t' = t - \frac{v\gamma^2}{c^2} x, \text{ where}$$
$$\gamma = (1 - v^2/c^2)^{-1/2} \doteqdot 1 + v^2/2c^2 \doteqdot 1.$$

t′ is called the 'local time' *(Ortszeit)* because, unlike t, it depends on the location of the point (x, y, z).

Let us incidentally note the purely mathematical origin of this notion of local time, which eventually led to Poincaré's and Einstein's concepts of frame-dependent simultaneity. Thus the introduction of a mathematical device is, once more, seen to be the starting point of new revolutionary developments in physics.

Let us now go back to the *Versuch*. Whereas in the TEM only the independent variables are changed, the *Versuch* introduces the new idea of a simultaneous change in the field quantities: this yields the covariance of Maxwell's equations for the case where the density vanishes and where we neglect all second order terms.

In the following, I shall slightly alter Lorentz's arguments by reconstructing them in what seems to me a simpler and more unified form.[10] Going back to equations (Ia)–(IVa) and putting $\rho = 0$, we obtain:

$$(\text{I}a^\circ)\ \nabla \cdot \vec{D} = 0$$
$$(\text{II}a^\circ)\ \nabla \cdot \vec{H} = 0$$
$$(\text{III}a^\circ)\ \nabla \times \vec{D} = -(1/4\pi c^2)Q\vec{H}$$
$$(\text{IV}a^\circ)\ \nabla \times \vec{H} = 4\pi Q\vec{D}$$

where Q is the operator $[(\partial/\partial t) - v(\partial/\partial x)]$. Let us now carry out the following transformations simultaneously:

---

[9] I have taken the liberty here of writing vx for $(P_x \cdot x/V^2 + P_y \cdot y/V^2 + P_z \cdot z/V^2)$, which is nothing but $\vec{v} \cdot \vec{r}/c^2$, where $\vec{v} = \vec{p} = (p_x, p_y, p_z)$, $\vec{r} = (x,y,z)$ and $c = V$.

[10] For more detail the reader is referred to Sections 30, 52 and 56 of the *Versuch* (LCP vol. 5 pp. 48, 75, 81, 84).

$$x' = x;\ y' = y;\ z' = z,\ t' = \left(t - \frac{v}{c^2}x\right);\ \vec{D}' = \vec{D} + (1/4\pi c^2)\vec{v} \times \vec{H} \text{ i.e. } \vec{D}' = (D_1, D_2 - vH_3/4\pi c^2, D_3 + vH_2/4\pi c^2);\ \vec{H}' = \vec{H} - 4\pi\vec{v} \times \vec{D}, \text{ i.e. } \vec{H}' = (H_1, H_2 + 4\pi vD_3, H_3 - 4\pi vD_2), \text{ where } \vec{v} = (v,o,o).$$ [11]

Hence:

$$\frac{\partial}{\partial x} = \left[\frac{\partial}{\partial x'} - \frac{v}{c^2}\frac{\partial}{\partial t'}\right];\ \frac{\partial}{\partial y} = \frac{\partial}{\partial y'};\ \frac{\partial}{\partial z} = \frac{\partial}{\partial z'};\ \frac{\partial}{\partial t} = \frac{\partial}{\partial t'};$$

$$\frac{\partial}{\partial x'} = \left[\frac{\partial}{\partial x} + \frac{v}{c^2}\frac{\partial}{\partial t}\right].$$

It can now be verified that:

(I′*a*) $\nabla' . \vec{D}' = 0$

(II′*a*) $\nabla' . \vec{H}' = 0$

(III′*a*) $\nabla' \times \vec{D}' = -(1/4\pi c^2)\partial\vec{H}'/\partial t'$

(IV′*a*) $\nabla' \times \vec{H}' = 4\pi\partial\vec{D}'/\partial t'$,

where $\nabla' = (\partial/\partial x', \partial/\partial y', \partial/\partial z')$.

As an example, let us derive $\nabla' . \vec{H}' = o$ from (I^{o}_{a}) − (IV^{o}_{a}):

$$\nabla' . \vec{H}' = \frac{\partial H_1'}{\partial x'} + \frac{\partial H_2'}{\partial y'} + \frac{\partial H_3'}{\partial z'}$$

$$= \left[\left(\frac{\partial}{\partial x} + \frac{v}{c^2}\frac{\partial}{\partial t}\right)H_1 + \frac{\partial}{\partial y}\left(H_2 + 4\pi vD_3\right) + \frac{\partial}{\partial z}\left(H_3 - 4\pi vD_2\right)\right]$$

$$= \left[\nabla . \vec{H} + \frac{v}{c^2}\frac{\partial H_1}{\partial t} + 4\pi v\left(\frac{\partial D_3}{\partial y} - \frac{\partial D_2}{\partial z}\right)\right]$$

By (II°a):

$$\nabla . \vec{H} = 0$$

---

[11] These are, to within 2nd order terms, Einstein's transformation equations for the electromagnetic field. (We simply have to replace H by 4πcH.) With hindsight, it is hardly surprising that we should obtain the covariance, to within 2nd order terms, of Maxwell's equations. Cf. A Einstein, H A Lorentz, et al., 1923 p. 51.

By (III'a):

$$\frac{\partial D_3}{\partial y} - \frac{\partial D_2}{\partial z} = -\frac{1}{4\pi c^2} QH_1 = -\frac{1}{4\pi c^2}\left(\frac{\partial}{\partial t} - v\frac{\partial}{\partial x}\right)\mathrm{H}_1$$

Hence:

$$\nabla'\vec{H}' = \frac{v}{c^2}\frac{\partial H_1}{\partial t} + 4\pi v \cdot \left(-\frac{1}{4\pi c^2}\right)\left(\frac{\partial}{\partial t} - v\frac{\partial}{\partial x}\right) H_1 = \frac{v^2}{c^2}\frac{\partial H_1}{\partial x} \doteqdot 0$$

since we neglect terms in $\frac{v^2}{c^2}$.

(I'a) – (IV'a) have exactly the same form as Maxwell's equations. Lorentz concludes that, if in a system at rest the vectors $\vec{D}$ and $\vec{H}$ are certain functions of t,x,y,z, then in the same system moving with velocity v through the ether, the vectors $\vec{D}'$ and $\vec{H}'$ are the same functions of the new variables t', x', y', z'. He calls these two states of the system corresponding states, and thus obtains the proposition that the earth's motion causes no first-order effects on experiments carried out with terrestrial light sources (see §2.6).

In the 'Théorie Simplifiée' published in 1899, Lorentz fused the following two ideas, which had been developed separately in the *Versuch:*

(1) Transformation of the spatial variables x,y,z

(2) Transformation of the time t together with the introduction of new field quantities $\vec{F}'$ and $\vec{H}'$.

He carries out (1) and (2) by going back to, or rather by rediscovering, his transformation equations of 1892:

$$x' = \gamma x = \gamma(x_1 - vt_1);\ y' = y = y_1;\ z' = z = z_1;\ \text{and}$$
$$t' = t - \frac{v\gamma^2}{c^2}x = t_1 - \frac{v\gamma^2}{c^2}(x_1 - vt_1) = \gamma^2\left(t_1 - \frac{v}{c^2}x_1\right),$$

where t_1, x_1, y_1, z_1 are the absolute coordinates.

The vectors $\vec{F}' = (F_1', F_2', F_3')$ and $\vec{H}' = (H_1', H_2', H_3')$ are defined as follows:

$$F_1' = 4\pi c^2 D_1;\ F_2' = 4\pi c^2\gamma D_2 - \gamma v H_3;\ F_3' = 4\pi c^2\gamma D_3 + \gamma v H_2;$$

$$H_1' = \gamma H_1;\ \mathrm{H}_2' = \gamma^2 H_2 + 4\pi\gamma^2 v D_3;\ H_3' = \gamma^2 H_3 - 4\pi\gamma^2 v D_2.$$

It is easily verified that equations (Ia) – (IVa) go over into:

$$\nabla_{\cdot}'\vec{F}' = (4\pi/\gamma)c^2\rho - 4\pi\gamma v\rho u_1;\ \nabla_{\cdot}'\vec{H}' = 0;$$

$$\nabla' \times \vec{H}' = + (\gamma^2/c^2)\partial \vec{F}'/\partial t' + 4\pi\gamma\rho \underset{(u_1,\,0,\,0)}{\vec{u}} + 4\pi\rho\gamma(\gamma - 1).$$

$$\nabla' \times \vec{F}' = - \partial\vec{H}'/\partial t'.$$

Mainly because of the extra factor γ in the expression of t', Lorentz obtains neither the full covariance of Maxwell's equations nor the invariance of the operator $\left[\nabla^2 - \frac{1}{c^2}\frac{\partial^2}{\partial t^2}\right]$. However, both in the *Versuch* and in the 'Théorie Simplifiée', he says enough to give us a fairly good idea of the theory he eventually put forward in 1904.

§2.6 The Final Version of Lorentz's Theory of Corresponding States

The last version of the theory of corresponding states appeared in 1904 under the revealing title: "Electromagnetic Phenomena in a System Moving with any Velocity Less than that of Light" (henceforth referred to simply as "Electromagnetic Phenomena"). This theory will interest us both from the historical and from the strictly philosophical points of view.

Historically, although Einstein did not read the "Electromagnetic Phenomena" until after 1905, the basic ideas underlying this work were, as explained above, already to be found in the *Versuch* with which he was familiar. The *philosophical* significance of the theory of corresponding states is that it was, or could easily have been made into, a system observationally equivalent to special relativity. Thus, in order to compare the merits of the two rival theories, non-empirical criteria have to be invoked. We shall see that the criteria take into account the heuristic power of competing research programmes.

Before turning to history proper, let me briefly outline Lorentz's project for a theory of corresponding states. This will answer the philosophical question posed above by showing that Lorentz's last hypothesis can be taken to be equivalent to STR. Note that I use the word 'project' on purpose, because Lorentz did not carry out his entire plan as laid down in 1904.

Maxwell's equations, we recall, hold in a frame of reference fixed in ether. Consider a moving system $\overline{S}$ and an observer O carried along by the motion of $\overline{S}$ through the ether. If the instruments of O were unaffected by this motion, then the measured lengths, time-intervals and field intensities would bear to one another relations more complicated than Maxwell's equations; in particular, the measured velocity of light in $\overline{S}$ would not be the same in all directions and the observer would thus realise that he was moving relatively to the medium. The theory of corresponding states asserts that the instruments are distorted in such a way that the measured quantities (i.e. t', x', y', z', $\vec{E}'$, $\vec{H}'$, ρ') do satisfy Maxwell's equations; hence the measured velocity of light is constant in $\bar{S}$. The observer will thus imagine himself to be within a system S' at rest in the ether, a system in which the *real* coordinates would be the quantities t', x', y', z' measured by the distorted instruments. (The fictitious system S' is called the state corresponding to $\bar{S}$. This informal description of Lorentz's theory is not to be taken as literally presupposing the presence of a conscious observer. The 'observer' is introduced for purposes of exposition and could well be replaced by a set of measuring instruments.)

Put in more technical language, Lorentz's programme[12] for a theory of corresponding states consisted of the following moves:

1. Consider a frame at rest and a frame $\bar{S}$ moving at constant velocity $\vec{v} = (v,o,o)$ through the ether. In $\bar{S}$: $\bar{x} = x - vt, \bar{y} = y, \bar{z} = z, \bar{t} = t$. (Galilean transformation). The components (D_1, D_2, D_3) and (H_1, H_2, H_3) of the electric and magnetic fields are the same in S and $\bar{S}$. However, in $\bar{S}$, the relations connecting $\vec{D}$, $\vec{H}$, ρ, $\bar{x}$, $\bar{y}$, $\bar{z}$, $\bar{t}$ are more complicated than Maxwell's equations, which only hold in S.
2. The first problem is to determine in the frame $\bar{S}$ 'effective' variables t', x', y', z', and $\vec{D}'$, $\vec{H}'$, ρ' in such a way that, with respect to t', x', y', z', the field quantities $\vec{D}'$, $\vec{H}'$, and ρ' satisfy Maxwell's equations.
3. Construct a system S' fixed in the ether and in which the *real* variables are the 'effective' variables of $\bar{S}$. In other words, using Einstein's terminology, we consider the following correspondence between events:

$$\begin{array}{ccccc} (t, x, y, z) & \mapsto & (\bar{t}, \bar{x}, \bar{y}, \bar{z}) & \mapsto & (t', x', y', z') \\ \text{in } S & & \text{in } \bar{S} & & \text{in } S'. \end{array}$$

   To the moving system $\bar{S}$ corresponds the immobile system S'. Since $\vec{D}'$ and $\vec{H}'$ satisfy Maxwell's equations relatively to t', x', y', z', it follows that $\vec{D}'$ and $\vec{H}'$ are, respectively, the real electric and the real magnetic field in S'.
4. Examine hypotheses (such as the MFH) which imply that, if the system S' is set in motion with constant velocity $\vec{v}$, it will rearrange itself so as to produce system $\bar{S}$. One object of the exercise is to show that the results of certain types of experiment, or possibly of any experiment whatever, are not affected by the motion of the earth through the ether. This is generally achieved by showing that, if certain quantities P',

[12] 'Programme' here is to be distinguished from 'research programme'. Lorentz's theory of corresponding states forms part of the larger classical ether programmes.

F′, . . . vanish in S′, then the corresponding quantities will also vanish in $\overline{S}$.

Let me immediately point out that neither in his 1904 paper nor in the "Theory of Electrons" did Lorentz completely achieve all these aims, except in the case of electrostatics. In all other cases he neglected second-order quantities in $\frac{v}{c}$. A great difficulty was posed by the problem of simultaneity: two events simultaneous in $\overline{S}$ may occur at different 'effective', i.e. 'local', times.

Let us now consider Lorentz's theory as actually presented in his "Electromagnetic Phenomena". He writes Maxwell's equations as follows:

$$\text{(i)}\ \nabla \cdot \vec{D} = \rho$$

$$\text{(ii)}\ \nabla \cdot \vec{H} = 0$$

$$\text{(iii)}\ \nabla \times \vec{H} = \frac{1}{c}\left(\frac{\partial \vec{D}}{\partial t} + \rho \vec{v}\right)$$

$$\text{(iv)}\ \nabla \times \vec{D} = -\frac{1}{c}\frac{\partial \vec{H}}{\partial t}$$

where ρ is the density and $\vec{\nu} = (\nu_1, \nu_2, \nu_3)$ is the velocity vector at time t at the point (x, y, z).

In a frame of reference moving with velocity $\vec{v} = (v, o, o)$ through the ether, the 'effective' coordinates are given by the Lorentz transformation equations as we know them today:[13]

$$\textbf{(1)}\ x' = \gamma\,(x - vt),\ y' = y,\ z' = z,\ t' = \gamma\,(t - vx/c^2).$$

The vectors: $\vec{H}' = (H_1', H_2', H_3')$ and $\vec{D}' = (D_1', D_2', D_3')$ are defined as follows:

$$\textbf{(2)}\quad \begin{aligned} D_1' &= D_1;\ D_2' = \gamma\,(D_2 - \frac{v}{c}H_3);\ D_3' = \gamma\,(D_3 + \frac{v}{c}H_2); \\ H' &= H_1;\ H_2' = \gamma\,(H_2 + \frac{v}{c}D_3);\ H_3' = \gamma\,(H_3 - \frac{v}{c}D_2). \end{aligned}$$

Let us note, in passing, that equations (2) are identical with Einstein's transformation equations for the electric and magnetic fields.

[13]However, Lorentz still interposes the Galilean coordinates: $\overline{x} = x - vt$, $\overline{y} = y$, $\overline{z} = z$, $\overline{t} = t$, between (t, x, y, z) and (t′, x′, y′, z′).

Under the transformations (1) and (2), Maxwell's equations become:

(i′) $\nabla' \cdot \vec{D}' = \left(1 - v\frac{u_1'}{c^2}\right) \rho'.$

(ii′) $\nabla' \cdot \vec{H}' = 0.$

(iii′) $\nabla' \times \vec{H}' = \frac{1}{c}\left(\frac{\partial \vec{D}'}{\partial t'} + \rho'\vec{u}'\right).$

(iv′) $\nabla' \times \vec{D}' = -\frac{1}{c}\frac{\partial \vec{H}'}{\partial t'},$

(3) where $\vec{u}' = (\gamma^2(\upsilon_1 - v), \gamma\upsilon_2, \gamma\upsilon_3)$ and $\rho' = \rho/\gamma$.

Except for the *intended interpretation* of ρ' and $\vec{u}'$, equations (i′) – (iv′) are exactly those obtained by Einstein in the EMB. Although the *equation* $\nabla' \cdot \vec{D}' = (1 - vu_1'/c^2)\,\rho'$ is correct, the *correct transform* of the charge density is not ρ' but $\sigma' = (1 - vu_1'/c^2)\,\rho'$. In fact:

$$\sigma' = \frac{1}{\gamma}\left(1 - \frac{vu_1'}{c^2}\right)\rho = \frac{1}{\gamma}\left[1 - \frac{v}{c^2}\gamma^2(\upsilon_1 - v)\right]\rho =$$

$$\gamma\left(1 - \frac{v\upsilon_1}{c^2}\right)\rho,$$

which is the value given by Einstein.

Here, Lorentz seems to have made an easily corrigible mistake. Of course, in terms of the Galilean coordinates $\bar{x}, \bar{y}, \bar{z}$, we have: $x' = \gamma\bar{x}$, $y' = \bar{y}$, $z' = \bar{z}$, from which it appears to follow that: $dx' \cdot dy' \cdot dz' = \gamma \cdot d\bar{x} \cdot d\bar{y} \cdot d\bar{z}$, hence ρ' = transformed density = $\frac{1}{\gamma}\rho$. But so to interpret ρ' is to forget that, in general, *ρ depends on the time, so that the last equation holds only for an electrostatic system*. In order to determine the density ρ' in S′ at the space-time point (t′, x′, y′, z′) we consider an infinitesimal volume dV′ of point-charges *all at the same time t′*. We therefore have to take into account a number of events which, while being simultaneous in S′, may not be simultaneous in S. *Lorentz's mistake in transforming ρ is therefore deeply significant. It stems from the difficulties presented by the physical interpretation of local time.* In 1904 Lorentz had not yet realised that his local or 'effective' time was in fact the time measured by a moving clock synchronised in accordance with Einstein's convention. Yet, the transformed equations obtained by Lorentz in 1904 are so similar to Maxwell's that one wonders why he did not simply postulate: $\left(1 - \frac{vu_1'}{c^2}\right)\rho' = \sigma'$ = transformed density, and thus obtain the full covariance of Maxwell's equations. However, if he had taken this step, he would still have had to face the problem posed by equation (iii′), which now becomes:

$$\nabla' \times \vec{H}' = \frac{1}{c}\left[\frac{\partial \vec{D}'}{\partial t'} + \sigma' \vec{u}' \Big/ \left(1 - \frac{vu_1'}{c^2}\right)\right].$$

He would have had to interpret $\vec{u}'/(1 - vu_1'/c^2)$ as an *effective velocity,* and for this he needed to have developed a general kinematical framework. This is where Einstein's approach, which starts with general kinematical considerations, proved far superior to Lorentz's. *Whereas Einstein sorts out his kinematics before imposing the condition of Lorentz-covariance on all physical laws and in particular on electromagnetic theory, Lorentz painfully tries to arrive at a new kinematics via electromagnetism.*

The connection between STR and the theory of corresponding states can now be described as follows. Einstein looks upon the 'effective' variables t′, x′, y′, z′, as the only real coordinates in $\bar{S}$. The equations discovered by Lorentz, namely:

$$x' = \gamma\,(x - vt),\; y' = y,\; z' = z,\; t' = \gamma\left(t - \frac{v}{c^2}x\right)$$

are used by Einstein in his EMB in order to connect the coordinates in S directly with those in $\bar{S}$. The variable t′ is the time given by clocks synchronised according to a convention, later called the Einstein synchronisation convention, which is as follows: let C and C′ be two identical clocks at rest at two points *O* and *P* of some inertial frame I. Let a light signal leave *O* at C-time t_1 and reach *P* at C′-time t′; the light beam is instantaneously reflected at *P* and goes back to *O* where it arrives at C-time t_2. The two clocks are said to be synchronised if, for all values of t_1, the ensuing values of t′ and t_2 satisfy: $t' = \frac{1}{2}(t_1 + t_2)$.

After 1905 Lorentz realised that his 'effective' variables were in fact the *measured* lengths, time intervals and field intensities in the moving system $\bar{S}$. For example, an experimenter in $\bar{S}$ obtains as measure of his x-coordinate, not $\bar{x} = x - vt$, but $x' = \gamma\bar{x} = (x - vt)/\sqrt{1 - v^2/c^2}$. This is because, by virtue of the contraction hypothesis, his rods are shortened by the factor $1/\gamma$.

Lorentz also realised—after 1905—that his contraction hypothesis entails the retardation of moving clocks by the same factor $1/\gamma$. Like measuring rods, clocks also consist of ponderable bodies which are contracted in the direction of their absolute motion. Hence, it can no longer be taken for granted, as Lorentz had done in both his theories T_1 and T_2, that the measured time is independent of the state of motion of the measuring clock.

In order to examine how time is affected by motion, let us consider a very simple clock which consists of a rigid rod and of two mirrors fixed at the two extremities of the rod.

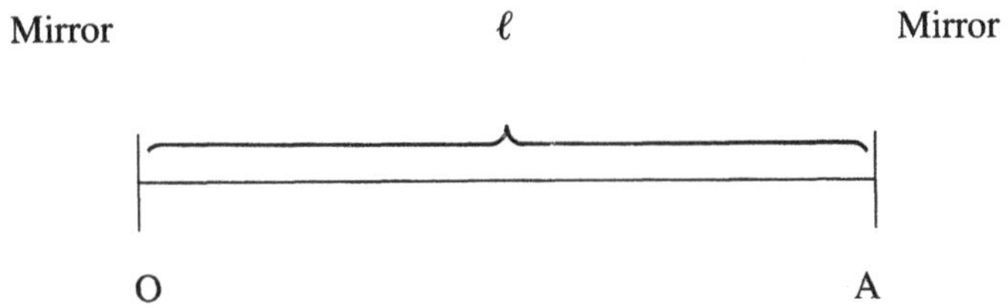

Let us choose as our unit the time it takes a light signal to complete the round-trip OAO; i.e. the time it takes a light signal to leave O, reach A, be instantaneously reflected at A and finally arrive back at O. In the absolute frame our unit of time is thus equal to $2\ell/c$, where ℓ is the length of the rod OA.[14]

Now suppose our clock-rod O′A′ to be at rest in a frame X′O′Y′ which is moving with velocity $\vec{v}$ along the x-axis of the ether frame. Let $\vec{v}$ make the angle φ with O′A′.

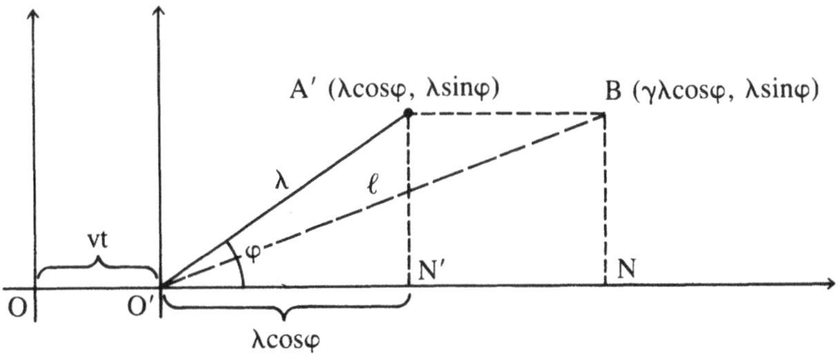

O′A′ thus represents the rod as already shortened by the factor $\frac{1}{\gamma}$ along $\vec{v}$. Putting O′A′ = λ, we have: O′B = ℓ, i.e.

(1) $\ell^2 = (\gamma\lambda cos\varphi)^2 + \lambda^2 sin^2\varphi = \lambda^2 (sin^2\varphi + \gamma^2 cos^2\varphi)$ where

$\gamma = (1 - v^2/c^2)^{-1/2}$. Hence,

---

[14]For an approach to Relativity based on this type of time-measurement see S. Prokhovnik's *The Logic of Special Relativity* (1967), which contains a cogent defense of the absolutist interpretation of STR.

$$(2)\begin{cases} \ell^2 = \dfrac{\lambda^2}{c^2}\,(c^2 - v^2 sin^2\varphi)/(1 - v^2/c^2);\ \text{i.e.} \\ \ell = \dfrac{\lambda}{c}\sqrt{c^2 - v^2 sin^2\varphi}/\sqrt{1 - v^2/c^2}. \end{cases}$$

Let us now calculate the time T, which a light beam takes to complete the round trip O′A′O in the moving frame. Remember that T is, by definition, the unit of time in X′O′Y′.

In the moving frame light travels between O′ and A′ at a velocity $\vec{u}$ determined by $|\vec{u} + \vec{v}| = c$

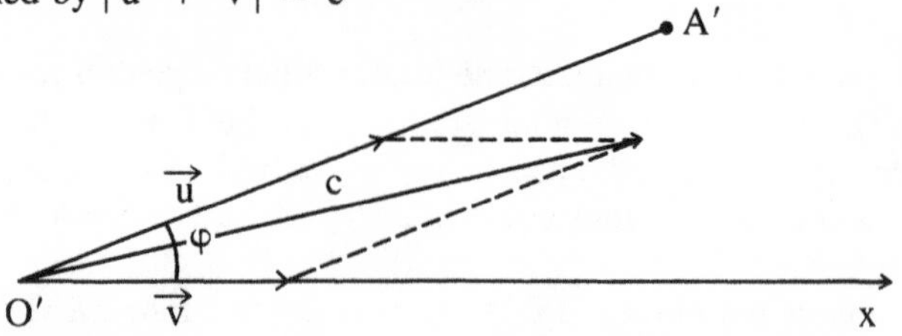

Thus, $c^2 = u^2 + v^2 + 2uv \cos \varphi$, i.e. $u^2 + (2v \cos \varphi)u - (c^2 - v^2) = 0$. Solving this quadratic in u, we obtain:

$$(3)\quad \begin{aligned} u &= -v \cdot cos\varphi + \sqrt{v^2cos^2\varphi + (c^2 - v^2)};\ \text{i.e.} \\ u &= -v \cdot cos\varphi + \sqrt{c^2 - v^2 \cdot sin^2\varphi}. \end{aligned}$$

In order to calculate the speed u′ of the (reflected) ray between A′ and O′, we substitute $(\Pi - \varphi)$ for φ. Thus:

$$(4)\ u' = v \cdot cos\varphi + \sqrt{c^2 - v^2 sin^2\varphi}$$

We have:

$$(5)\ T = \frac{\lambda}{u} + \frac{\lambda}{u'} = \frac{2\lambda\sqrt{c^2 - v^2 \cdot sin^2\varphi}}{(c^2 - v^2)}\ \text{(by (3) and (4)).}$$

It follows from (2) and (5) that

$$(6)\ T = \frac{2\ell}{c}\frac{1}{\sqrt{1 - v^2/c^2}} = \frac{2\ell}{c}\gamma.$$

Remembering that $2\ell/c$ is our unit of time in the stationary frame, we conclude from (6) that, if we use the same clock in the moving frame, our measured time will be dilated by the factor γ; i.e., the moving clock will be retarded by the factor $1/\gamma = (1 - v^2/c^2)^{-1/2}$.

Thus, we have derived both the time dilatation and the contraction effects from one assumption, namely the LFC, which in turn follows from Lorentz's unique auxiliary assumption, namely the MFH.

It remains for us to show that the contraction and retardation effects imply the Lorentz transformation, in which the effective variables can now be identified as the *measured* coordinates in the moving frame $\bar{S}$.

It proves advantageous to establish a more general result which will be needed in later chapters. Our assumptions and methods generalise those used by Lorentz in the ad hoc derivation of his transformation.[15] Let us suppose that a body moving with velocity $\vec{u}$ through the ether is contracted both along the vector $\vec{u}$ and transversally to it by the factors $\mu(u)$ and $\psi(u)$ respectively. Also assume that moving clocks are retarded by some factor $\Omega(u)$. Note that μ, ψ and Ω are taken to be any positive, well-behaved, functions of the speed u (they need not even be smaller than 1). We propose to solve the following problem: What is the measured round trip speed of light along a rod which remains at rest in a frame moving with velocity $\vec{v}$ through the ether?

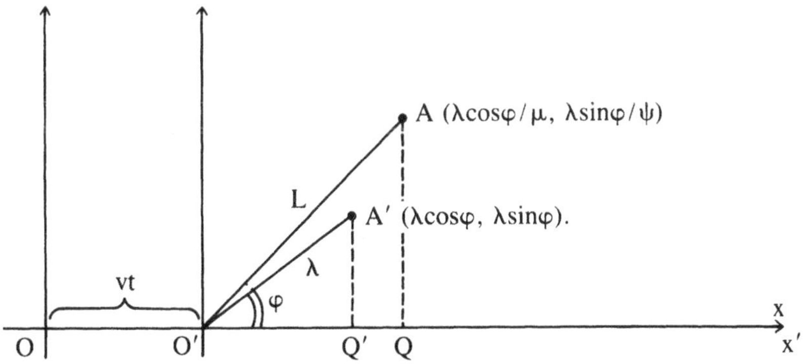

As above, let $\overrightarrow{O'A'}$ represent the rod as already contracted by its absolute motion. We choose our X-axis in the direction of the velocity $\vec{v}$ of the moving frame. Let $O'A' = \lambda$ and $X\hat{O}'A' = \varphi$. Thus the 'real' components of $\overrightarrow{O'A'}$ are $\lambda\cos\varphi$ and $\lambda\sin\varphi$. However, due to the longitudinal and transverse shortenings of his measuring rods, an observer at rest in X′O′Y′

[15]The results we are about to prove were first established by Sjödin, who used somewhat different methods from the ones applied here. The reader is referred to the brilliant work of Podlaha and Sjödin, especially to the latter's 'Synchronization in Special Relativity and Related Theories' (1979).

ascribes the values $\frac{\lambda\cos\varphi}{\mu}$ and $\frac{\lambda\sin\varphi}{\psi}$ to the components of $\overrightarrow{O'A'}$. Thus the *measured* length L of the rod and the angle θ which it *seems* to make with the X-axis are given by:

$$\text{(7)}\ L\cos\theta = \frac{\lambda\cos\varphi}{\mu},\ L\sin\theta = \frac{\lambda\sin\varphi}{\psi}$$

(Obviously, were the rod to be stationary in the ether, then its length would equal L. That is: L is the real length of the absolutely stationary rod.)

We can immediately write down the equations connecting the absolute spatial coordinates (x, y) of A′ with its 'effective' coordinates (x′, y′) as measured in the moving frame:

$$x' = L\cos\theta = \frac{\lambda\cos\varphi}{\mu} = \frac{OQ' - OO'}{\mu} = \frac{1}{\mu}(x - vt)$$

$$y' = L\sin\theta = \frac{\lambda\sin\varphi}{\psi} = \frac{Q'A'}{\psi} = \frac{1}{\psi}y.$$

Since z is also a transverse coordinate, we can add $z' = \frac{1}{\psi}z$ to the above equations. The transformation of the spatial coordinates is thus as follows:

$$\text{(8)}\ x' = \frac{1}{\mu}(x - vt);\ y' = \frac{1}{\psi}y;\ z' = \frac{1}{\psi}z.$$

In the rest of this section 'time' will, unless otherwise stated, denote real or absolute time.

Consider a light ray which, relatively to the moving frame, goes from O′ to A′, is reflected at A′ and then returns to its source O′. Let u and u′ denote the forward and return velocities of the ray respectively. By eqs. (3) and (4) above:

$$\text{(9)}\ \begin{cases} u = -v\cdot\cos\varphi + \sqrt{v^2\cdot\cos^2\varphi + (c^2 - v^2)} \\ \quad = -v\cdot\cos\varphi + \sqrt{c^2 - v^2\sin^2\varphi} \\ u' = v\cdot\cos\varphi + \sqrt{v^2\cdot\cos^2\varphi + (c^2 - v^2)} \\ \quad = v\cdot\cos\varphi + \sqrt{c^2 - v^2\sin^2\varphi} \end{cases}$$

The time it takes the light ray to cover the distance O′A′ in the moving frame is therefore:

$$\text{(10)}\quad \begin{cases} T_1 = \lambda/u = \lambda/(-v\cos\varphi + \sqrt{c^2 - v^2\sin^2\varphi}) \\ \quad = \dfrac{\lambda}{(c^2 - v^2)} (\sqrt{c^2 - v^2 \cdot \sin^2\varphi} + v \cdot \cos\varphi) \\ \quad = \\ \quad = \dfrac{\lambda\gamma^2}{c^2} (\sqrt{c^2 - v^2\sin^2\varphi} + v \cdot \cos\varphi) \end{cases}$$

Similarly:

$$\text{(11)}\quad T_2 = \frac{\lambda\gamma^2}{c^2} (\sqrt{c^2 - v^2\sin^2\varphi} - v \cdot \cos\varphi)$$

where T_2 is the time the light ray takes to return from A′ to O′. Thus the total duration of the round trip is:

$$\text{(12)}\quad T_1 + T_2 = \frac{2\lambda\gamma^2}{c^2} \sqrt{c^2 - v^2 \cdot \sin^2\varphi}$$

Let us remember that the length and angle as measured in the moving frame are L and θ respectively, *not* λ and φ. It is therefore desirable to express T_1, T_2 and $(T_1 + T_2)$ in terms of L and θ. It follows from (7) that:

$$\text{(13)}\quad \lambda^2 = \mu^2L^2 \cdot \cos^2\theta + \psi^2L^2\sin^2\theta;\ \text{i.e. } \lambda = L\sqrt{\mu^2 \cdot \cos^2\theta + \psi^2 \cdot \sin^2\theta}$$

Substituting back in (7):

$$\text{(14)}\quad \begin{cases} \cos\varphi = \dfrac{\mu L\cos\theta}{\lambda} = \dfrac{\mu\cos\theta}{\sqrt{\mu^2\cos^2\theta + \psi^2 \cdot \sin^2\theta}} \\ \text{and:} \\ \sin\varphi = \dfrac{\psi\sin\theta}{\sqrt{\mu^2 \cdot \cos^2\theta + \psi^2 \cdot \sin^2\theta}} \end{cases}$$

Substituting in (10)–(12):

$$\text{(15)}\quad \begin{cases} T_1 = \dfrac{L\gamma^2\mu}{c} \left[\left(1 - (1 - \psi^2 \cdot \mu^{-2} \cdot \gamma^{-2})\sin^2\theta\right)^{1/2} + \dfrac{v}{c}\cos\theta\right] \\ T_2 = \dfrac{L\gamma^2\mu}{c} \left[\left(1 - (1 - \psi^2 \cdot \mu^{-2} \cdot \gamma^{-2})\sin^2\theta\right)^{1/2} - \dfrac{v}{c}\cos\theta\right] \\ T_1 + T_2 = \dfrac{2L\gamma^2\mu}{c} \cdot \left(1 - (1 - \psi^2 \cdot \mu^{-2} \cdot \gamma^{-2})\sin^2\theta\right)^{1/2} \end{cases}$$

Note that $T_1 + T_2$ is the 'real' time it takes the light signal to come back to O′. We now recall that moving clocks are retarded by the factor $\Omega = \Omega(v)$. Thus the clock at O′ will have registered the time interval $\Omega \cdot (T_1 + T_2)$. Remembering that the measured length of $\overrightarrow{O'A'}$ is L, we conclude that the round trip speed of light as measured in X′O′Y′ is:

$$\textbf{(16)}\ \bar{c}'(\theta) = \frac{2L}{\Omega \cdot (T_1 + T_2)}$$
$$= c \cdot \Omega^{-1} \cdot \gamma^{-2} \cdot \mu^{-1} [1 - (1 - \psi^2 \cdot \mu^{-2} \cdot \gamma^{-2}) \sin^2\theta]^{-\frac{1}{2}}$$

This expression of $\bar{c}'(\theta)$ generally depends on the angle θ, i.e., on the direction of $\overrightarrow{O'A'}$. Through varying this direction and thus obtaining different values for the speed of light, the observer will realise that he is moving with respect to the ether; that is, he will detect his own absolute motion; unless, of course, the coefficient $(1 - \psi^2 \cdot \mu^{-2} \cdot \gamma^{-2})$ of $\sin^2\theta$ in the expression of $\bar{c}'(\theta)$ vanishes, i.e. unless:

$$\textbf{(17)}\ \psi = \mu \cdot \gamma \text{ or } \mu = \psi \cdot \gamma^{-1}.$$

Should (17) be satisfied—as in the case of the Lorentz-Fitzgerald contraction effect—then the round trip velocity of light will be the same along all directions in X′O′Y′. That is: within each uniformly moving frame this velocity will be a constant, which generally differs from c. It is interesting to note that the constancy of the measured speed of light does not depend on Ω; it is independent of the time dilatation effect. Of course, the actual value of $\bar{c}'$ varies with Ω, as is clear from (16).

Let us now return to the more general case, to the case where we do *not* assume $\psi = \mu\gamma$. We have so far succeeded in transforming only the spatial coordinates x, y, z (Cf. eg (8) above). This is because, in order to transform the time variable t, we have to adopt a convention for synchronising clocks in the moving frame. True, we managed to determine $\bar{c}'(\theta)$ independently of any such convention; but this is because round trip velocities involve a single clock, namely a clock placed at the source of light. The transformation of the coordinate time t however involves a consideration, not only of real physical effects like contractions and retardations, but also of a stipulative convention. In this section we shall adopt Einstein's synchronisation procedure, leaving the study of the more general case to later chapters (see chapter four).

Consider three identical clocks at O, O′ and A′. The clocks at O and O′ are supposed to have been synchronised at time t = 0. (We recall that, unless otherwise stated, 'time' means 'absolute time'.)

At t_1 a light beam is sent from O′ towards A′ where it arrives at t; the beam is reflected at A′ and sent back to O′, which it reaches at t_2. Thus:

(18)
$$\begin{cases} t = t_1 + T_1 \text{ and } t_2 = t + T_2 \\ \text{whence: } \frac{1}{2}(t_1 + t_2) = t + \frac{1}{2}(T_2 - T_1) \end{cases}$$

Because the clock at O′ is retarded by the factor, it gives, instead of t_1 and t_2, the values

(19) $t'_1 = \Omega t_1$ and $t'_2 = \Omega t_2$ respectively.

The clock at A′ is said to be synchronous with the one at O′ in accordance with Einstein's convention iff:

(20) $t' = \frac{1}{2}(t'_1 + t'_2)$,

where t′ is the clock-time at A′ corresponding to the absolute time t.

Einstein's synchronisation procedure ensures that the forward velocity of light, i.e., its speed from O′ to A′, is equal to its return velocity, i.e., to its speed from A′ back to O′. It tells us however nothing about the measured round trip velocity $\bar{c}'(\theta)$; in classical ether theories, $\bar{c}'(\theta)$ is determined exclusively by real physical causes, e.g. by dimension contraction and clock retardation (see chapter four).

It follows from (15) and from (18) – (20) that:

(21)
$$\begin{cases} t' = \frac{\Omega}{2}(t_1 + t_2) = \Omega\left[t + \frac{1}{2}(T_2 - T_1)\right] \\ \quad = \Omega\left[t - \frac{L\gamma^2\mu v}{c^2}\cos\theta\right] \end{cases}$$

By (7) and by the figure above:

$$L\mu\cos\theta + vt = \lambda\cos\varphi + vt = x.$$

Hence:

$$L\mu\cos\theta = x - vt.$$

Substituting in (21) and remembering that $\gamma = (1 - v^2/c^2)^{-1/2}$, we obtain:

$$t' = \Omega\gamma^2\left(t - \frac{v}{c^2}x\right).$$

In conjunction with (8) we have the following transformation for all the coordinates:

(22) $x' = \frac{1}{\mu}(x - vt);\ y' = \frac{1}{\psi}y;\ z' = \frac{1}{\psi}z;\ t' = \Omega\gamma^2\left(t - \frac{v}{c^2}x\right)$

These equations generally hold for any theory which entails the retardation coefficient $\Omega(v)$ together with longitudinal and transverse contraction coefficients equal to $\mu(v)$ and $\psi(v)$ respectively; all this, *provided* Einstein's synchronisation method is used.

Let us now turn to the specific case of Lorentz's theory T_3 in which $\psi(v) = 1$ and $\mu(v) = \Omega(v) = 1/\gamma = (1 - v^2/c^2)^{-1/2}$.

Substituting in (22) we obtain:

$$\textbf{(23)}\ x' = \gamma(x - vt),\ y' = y,\ z' = z,\ t' = \gamma\left(t - \frac{v}{c^2}x\right),$$

which is the Lorentz transformation. By (16) it follows that $\bar{c}'(\theta) = c$; the measured round trip speed of light is the same in all inertial frames. Since Einstein's convention identifies the round trip with the one-way velocity of light, it follows that c is a (measured) universal invariant.

Lorentz was therefore in a position to *explain* why the *measured* velocity of light must be the same in all inertial frames. However, this *explanation* is carried out against a fixed kinematical background which is taken for granted, i.e. regarded as self-evident; namely: the background provided by absolute space and time.

As already mentioned, Einstein differs from Lorentz in that he treats the 'effective' variables as the 'real' ones and totally abolishes the Galilean transformation. That is, for Einstein there is no contrast between 'effective' or 'measured' on the one hand, and 'real' on the other. Lorentz's theory of corresponding states is observationally equivalent to STR because experimental results involve only measured, i.e., effective, magnitudes. Since the latter are connected by Maxwell's equations—as Lorentz himself had shown in the case of first order quantities—we are unable, whether we adopt Lorentz's or Einstein's theory, to decide, *on empirical grounds*, whether our reference frame is in motion or at rest in the ether.

I do not wish to suggest that, by identifying the 'effective' with the 'real', Einstein was adopting a positivist stance. In connection with STR, he had after all accepted the 'metaphysical' existence of unobservable inertial frames together with a realist event-ontology. He must however have realised that the acceptance both of Lorentz's standpoint and of a universal relativity principle entails the further acceptance of a conspiracy theory: although length contractions and clock retardations are real physical effects, we are systematically prevented from ever detecting them experimentally. Worse still: we can choose any inertial frame as our stationary system and judge, with impunity, that it is the rods of the absolute frame which are 'really' shortened. (This is *logically* possible, because length-contraction is a three-place relation $A(K_1,B_1,B_2)$ between an inertial frame K_1, a body B_1 at rest in K_1, and an identical body B_2 moving with respect to K_1. Thus, it is perfectly possible for both $A(K_1,B_1,B_2)$ and $A(K_2,B_2,B_1)$ to hold.)

In chapter six we shall see that Lorentz balked at going over to relativity mainly because of his belief that Maxwell's equations are *not* fully covariant. This means that for a long time, Lorentz did not adhere to a strict Lorentz-covariance principle.

§2.7 The Rationality of Lorentz's Pursuing his Own Programme after 1905

Looking at Lorentz in the light of later developments can be very misleading. The subsequent success of relativity theory may give the impression that he was wrong-headed, not to say crankish, in not immediately accepting Einstein's ideas; that he was too slow to see the light. But Lorentz was the champion of the classical electromagnetic programme started by Faraday and articulated by Maxwell. Its hard core, as already explained, consists of Newton's laws of motion and of Maxwell's equations. Newton had a classical principle of relativity, according to which all mechanical laws have the same form in any two frames moving with uniform rectilinear velocity with respect to each other. However, for reasons of philosophical intelligibility, Newton assumed the existence of absolute space, i.e., of a preferred frame, although the latter could not be determined on the basis of mechanics alone. The absolute space hypothesis was an idle (metaphysical) component of Newtonian theory, in the sense that any inertial frame could be considered as immobile in absolute space. With the wave theory of light, which seemed to presuppose a medium of transmission, arose the possibility of turning the absolute space hypothesis into an empirically testable theory.[16] Lorentz's ontology consisted of an infinite immobile ether in which charge was continuously distributed. The electrons are spherical regions of the ether where the charge and mass densities may differ from zero. The total amount of charge remains constant, but the movement of the electrons creates a field which travels in free space (i.e., in the ether) at a constant finite speed c. Lorentz tentatively assumed that the electron possesses no material mass; the electron engenders a field which, by acting back on its source, decelerates its motion; this capacity for resisting change of motion is the electromagnetic mass, which varies with the speed and accounts for the total inertia of the particle. Thus we see how fundamental is the role played by charge, absolute time and absolute space (i.e., the ether) in Lorentz's approach. These are

[16] I do not mean that the absolute space hypothesis was to become testable in isolation, but that its addition to existing theories would increase the number of testable consequences.

the ultimate constituents of a physical world closely resembling Newton's: the classical world.

Thus, for Lorentz to have switched to the relativity programme would have involved a major change in his metaphysical outlook. But why should he have made the change? If Einstein's theory had immediately thrown up new facts which Lorentz's system either could not account for or could only account for in an ad hoc way, then Lorentz's adherence to the classical ontology could not be characterised as rational. But, on the contrary, Lorentz's own approach, based on his classical ontology, enabled him to make theoretical and empirical progress—often in advance of Einstein. Another point, which will be taken up at greater length in chapter six, ought now to be mentioned. We have seen that, because of his faulty interpretation of ρ', Lorentz did not establish the full covariance of Maxwell's equations. He nevertheless managed to explain all the null results known to him. In other words, Lorentz accounted for the phenomena without going to full covariance. Consequently, as long as there remained the prospect of retaining some formal asymmetry between the ether and other frames, there was no good reason for Lorentz to accept a relativity principle based on the complete reciprocity of all inertial systems. Thus his continued adherence to his own programme after 1905 was completely rational.

The ontology presupposed by Einstein's theory is radically different from the classical one. Some positivists, among them Bridgman (1936, ch. 2), have claimed that Einstein had no metaphysical commitments at all, relativity being a mere description of actual physical operations. But this is an illusion. In his EMB, Einstein implicitly posits a domain of events, each of which can be described by 4 coordinates (t,x,y,z) in any one of infinitely many equivalent inertial frames. *Events* are therefore the basic constituents of the Einsteinian universe. At the beginning it was very difficult for Lorentz to acquiesce in this radical change of world view, especially since, as was said above, his theory and Einstein's explain the same facts. Because of the observational equivalence of the two theories, one might be tempted to think that 'metaphysics' is irrelevant to this whole issue. We shall, however, see that ontologies, which might be unimportant in the case of individual theories, can play an essential heuristic role in the development of research programmes by providing different regulative principles.[17] This can be properly appreciated only after examining Einstein's programme in greater detail. What has so far been established is that one cannot explain the success of STR in terms of the *demerits* of Lorentz's rival theory. Thus, if the eventual

[17]For the relationship between metaphysics and heuristics see §1.14.

acceptance by the scientific community of Einstein's programme in preference to Lorentz's was rational, if there are acceptable general criteria according to which Einstein's approach was objectively better than Lorentz's, that rationality must lie in the *extra merits* of Einstein's theory. These will be examined in the next chapter; but let us end the present one by summing up the points which have so far been made.

(1) Michelson's result, like any experimental result, gives no inductive proof, whether of Einstein's light principle, of the LFC, or of the MFH.

(2) The Lorentz transformation had a purely mathematical origin. It was introduced as a device for solving a partial differential equation, namely $\Box\varphi = 0$. Through this device Lorentz was led to the MFH. This illustrates the creative role which mathematics can play in the development of a research programme.

(3) The MFH implies both the LFC and the clock retardation hypothesis, but Lorentz was initially unaware of the second implication.

(4) Michelson's result, which refuted Lorentz's original system T_1, is explained by his amended theories T_2 and T_3, both of which include the LFC. Neither T_2 nor T_3 is ad hoc_1; both theories make novel predictions.

(5) That the LFC is ad hoc_2 in the sense that nobody bothered to test it independently of Michelson's result constitutes *no relevant criticism* of the hypothesis as such.

(6) The MFH (which entails the LFC) is not ad hoc_3; it fits in well with Lorentz's programme taken as a whole. The MFH is based on the general argument that, since all forces are states of the *same* medium, they must all be affected in the *same* way by the motion of bodies through the medium.

(7) If we interpret 'ad hoc' to mean 'cooked up for the narrow purpose for which it was to serve' (see Holton 1969), in this case for explaining away Michelson's result, then Einstein's principle (P2) fares worse than the MFH. The MFH is a high-order theory, which yields (P2) while being seemingly unrelated to Michelson's result.

(8) In view of (4), (5) and (6), the ether programme was progressing when Einstein proposed STR.

(9) By establishing the first-order covariance of Maxwell's equations, Lorentz systematically accounted for all first-order null results. He also went a long way towards demonstrating the full covariance of electromagnetic theory. He was prevented from reaching his goal by his mistaken interpretation of the transformed electric density ρ'. This mistake ultimately stemmed from his inability (or rather from his reluctance) to interpret his own concept of 'local time' in realist terms. To the very end, he equated 'real' time with 'absolute' time.

— 3 —

EINSTEIN'S PROGRAMME

§3.1 Einstein's Heuristics

I have shown that Lorentz's ether programme was progressive until after 1905—the year in which Einstein published his EMB. In this chapter we shall try to deal with the following two questions. First: what were Einstein's reasons for objecting to the classical programme and hence for starting his own? (We have already seen that these reasons could not have been of an empirical kind.) Secondly: once special relativity was launched, why did scientists like Planck, Lewis and Tolman choose to work on Einstein's programme rather than on Lorentz's?

§3.1.1 Einstein's Appraisal of Classical Physics

Why did Einstein object to classical physics? It should immediately be said that this question will *not* be answered in a psychologistic way. We shall not, for example, indulge in speculations about Einstein's childhood. It will be shown that certain metaphysical beliefs held by Einstein correspond to heuristic prescriptions which, when skillfully applied to particular cases, become powerful tools for the construction of scientific theories. Metaphysics can thus play an important role in starting a new programme, especially when the existing one is empirically successful. Of course, the triumph of a programme over its rivals can be achieved only by empirical means. However interesting its metaphysics, the programme will ultimately be judged by its ability to anticipate facts.

I should now like to formulate, as clearly as I can, three important methodological rules which formed part of Einstein's heuristics. To these methodological prescriptions correspond strong metaphysical beliefs which Einstein articulated in his later years. My task will then consist in showing that he effectively followed these prescriptions both in criticising classical physics and in constructing STR. In doing this, I shall closely follow Einstein's own advice:

> If you want to learn something about the methods used by theoretical physicists, I propose that you keep to the following fundamental principle: do not listen to the physicists' words but consider their actions. (Einstein 1934, p. 113; my translation)

Let us now turn to the three rules:

(α) Theories which conflict with 'observational' results should be rejected. *"The first point of view is obvious: the theory must not contradict empirical facts."* (Einstein 1949, p. 21)

It is difficult to assess how seriously Einstein took this falsificationist view. On the one hand, being a thoroughgoing realist, he admitted that if a theory clashed with certain empirical results, then the theory—or rather a whole conjunction of hypotheses comprising the theory under test as well as certain observational theories—must be false. At times, he seemed effectively to hold a very strong falsificationist position:

> The great attraction of the theory [general relativity] is its logical consistency. If any deduction from it should prove untenable, it must be given up. A modification of it seems impossible without destruction of the whole. (Einstein 1950, p. 110)

On the other hand, empirical refutations as such were not starting points of Einstein's scientific investigations. We have seen that he attributed little importance to Michelson's experiment. Similarly, Miller's results in the 1920's seemed to have made very little impact on him. It is true that his friend Max Born inspected the equipment used by Miller and came to the conclusion that Miller was a sloppy experimenter whose results could not be trusted (Einstein & Born 1969, pp. 107, 128). So, typically, Einstein left it to other physicists to sort out 'trivial' empirical details while he was engaged in work of a more theoretical character. (At about the same time as Miller carried out his experiments, Einstein was trying to go beyond *general* relativity to a field theory which would unify gravitation and electromagnetism.) (Einstein & Born 1969, p. 110)

At times, Einstein gave the impression that he was prepared to ignore empirical refutations altogether.

> The testing of the theory (General Relativity) is unfortunately too difficult for me. Man is after all a poor devil. However, Freundlich's results do not affect me in the least. Even if no bending of light rays, no motion of the perihelion (Mercury) and no red shift were known to exist, the gravitational field equations would still be very convincing because they rid us of the inertial systems (this ghost which acts on everything but on which nothing else acts back). It is actually remarkable that men are deaf as regards the strongest arguments while they always tend to overrate the value of precise measurement. (Einstein & Born 1969, p. 258; my translation)

Einstein's decision not to take empirical refutations too seriously was motivated by his (correct) belief that *"it is often, perhaps even always, possible to adhere to a general theoretical foundation by securing the adaptation of the theory to the facts by means of artificial additional assumptions.*" (Einstein 1950, p. 62)

(β) Apart from any external requirement of consistency with the 'facts', theories have to fulfill a so-called internal requirement. Science should present us with a coherent, unified, harmonious, simple and organically compact picture of the world. The mathematics used in the theory should reflect the degree of internal perfection of the latter.

> The aim of science is, on the one hand, a comprehension as complete as possible of the connections between sense experiences in their totality and, on the other hand, the attainment of this aim by the use of a minimum of primary concepts and relations. (Seeking as far as possible logical unity in the world picture, i.e. paucity in logical elements). (Einstein 1934, p. 16; my translation)

This requirement of internal perfection fits in with Einstein's pantheism, according to which Nature reveals the presence of a higher *immanent* intelligence.

> The individual feels the nothingness of human wishes and aims at the sublimity and wonderful order which reveal themselves in Nature as well as in the world of thought. Such a person considers his own individual existence as a kind of prison and wants to experience the totality of all being as a unified and intelligible whole. (Einstein 1934, p. 16; my translation)

Einstein went as far as asserting that reality, though independent of the mind, was nonetheless knowable a priori. His so-called aestheticism was not meant in any subjective sense but was linked to a definite metaphysical position. Because nature *is* simple, scientific hypotheses ought to be organically compact. Simplicity, or coherence, are not aimed at because they please our minds or because they effect economy of thought, but because they provide an index for *verisimilitude*.

> If it is true that the axiomatic foundations of theoretical physics cannot be derived from experience but have to be freely invented, can we at all hope to find the right way? Or worse still: does this 'right way' exist only as an illusion . . . To this I answer with complete confidence that this right way exists and that we are capable of finding it. In view of our experience so far we are justified in feeling that Nature is the realisation of what is mathematically simplest . . . It is my conviction that we are able, through pure mathematical construction, to find those concepts and the law-like connections between them which yield the key to the understanding of natural phenomena . . . The really creative principle is in mathematics. In a certain sense I consider it therefore to be true—as was the dream of the Ancients—that pure thought is capable of grasping reality. (Einstein 1934, p. 116; my translation)

(γ) Before formulating this third rule, which is more in the nature of a heuristic device than a methodological prescription, it is worth pointing out that Einstein was not an inductivist; he believed neither in the validity of a rule of induction nor in the existence of pure observational statements free of theoretical admixtures.

In *Out of My Later Years* he wrote: "*There is no inductive method which could lead to the fundamental concepts of physics. Failure to understand that constituted the basic philosophical error of the 19th century.*" (Einstein 1950, p. 76) In his Autobiography he said that "*all concepts, even those which are closest to experience, are from the point of view of logic freely chosen conventions, just as is the case with the concept of causality.*" (Einstein 1949, p. 12) However, in the light of metaphysical or empirical assumptions, the scientist can interrogate facts and isolate in them certain features, which sometimes consist in 'observed' symmetries between different experimental situations. These features might arouse suspicion because they are subsumed by theories which ought, prima facie, to predict very different results. The scientist may thus be led to the discovery of a general principle of which the feature in question is a direct manifestation. Principles obtained in this way are aptly called *theories of principle* and special relativity is one of them. Einstein wrote:

> . . . there is another group consisting of what I call theories of principle. These employ the analytic, not the synthetic method. Their starting point and foundation are not hypothetical constituents but empirically observed general properties of phenomena, principles from which mathematical formulae are deduced of such a kind that they apply to every case that presents itself. (Einstein 1950, p. 54)

The metaphysical doctrine underpinning the heuristic device we have just described is that there exist no accidents in nature. Observable symmetries are the signet of a more fundamental symmetry obtaining at the ontological level. If symmetries are observed throughout some given domain and if the extant explanatory theory postulates asymmetries at a deeper level, then the old theory ought to be replaced by a new one in which the symmetries appear at the most fundamental level. In this way, observed symmetries become direct manifestations of a deeper symmetry, or even of an identity.

Put in a slightly different way, this heuristic device consists in discovering, within a given system, an incoherence between asymmetric high-level propositions and symmetric observation statements; the latter are both empirically confirmed and predicted by the hypothesis under scrutiny; so this is no straight case of *empirical* refutation. The incoherence is then removed by substituting a 'symmetric' theory for the hypothesis under consideration.

Rule (γ) is in effect tantamount to the rejection of a conspiracy theory. Such a rejection is based on the following *intuitive* probability argument: it is highly unlikely that a deep asymmetry postulated by a theory should be systematically concealed by effects which exactly compensate for it throughout the observational domain (see §1.6).

Let us now examine how Einstein applied the above rule in analysing the phenomenon of electromagnetic induction.

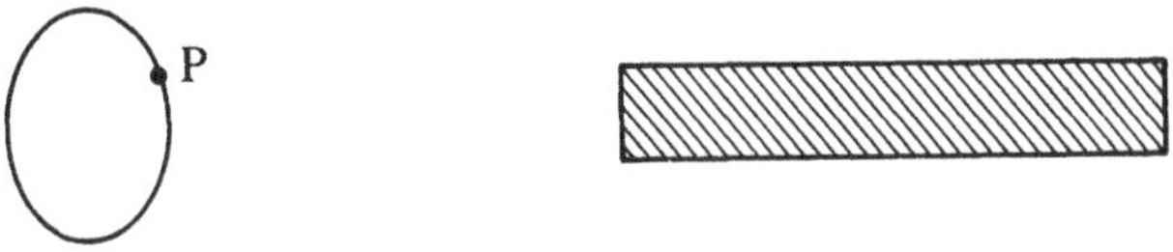

If we move a magnet with respect to the ether while keeping some conductor fixed, then, due to the variation of the magnetic field, an electric field arises at each point of space. Let P be any point of the conductor at which an electron may be situated. In view of the Lorentz formula:

$$\vec{F} = e\left(\vec{D} + \frac{\vec{v}}{c} \times \vec{H}\right), \text{ where } \vec{D} \neq 0 \text{ but } \vec{v} = 0,$$

the electron will experience a force which consequently generates a current in the conductor.

We now keep the magnet fixed and move the conductor with velocity $\vec{v}$ relatively to the medium. No electric field is created since $\vec{H}$ is static, i.e. independent of the time. The situation is thus very different from the previous one; so we expect the current either not to arise at all or, at any rate, to be different from what it was in the first case. However, in view of

$$\vec{F} = e\left(\vec{D} + \frac{\vec{v}}{c} \times \vec{H}\right), \text{ where } \vec{D} = 0 \text{ but } \vec{v} \neq 0$$

a current does arise and, if the relative motion between the conductor and the magnet is the same as in the previous case, the current also turns out to be the same. The result is therefore subsumed by Maxwell's theory. In other words: if we assume the existence of a preferred frame and accept Maxwell's equations, we can infer that the outcome of the experiment depends solely on the relative motion between magnet and conductor, not on their absolute motion with respect to the ether. Hence, this time without the aid of any auxiliary hypothesis, an ether theory yields the undetectability of the ether.

Thus, in classical electromagnetism, there is a basic *ontological* difference between a situation in which a magent moves in the ether (presence both of a magnetic and of an electric field) and one in which the same magnet is stationary (presence of a magnetic field alone). However, when we *apply* Maxwell's equations in order to compute the current arising from the motion of a conductor in the magnet's field, the result depends only on the relative motion between the magnet and the conductor. Hence, at the 'observational' level, there is symmetry between the following two situations: (a) magnet moving towards the conductor, and (b) conductor moving towards the magnet. This conflicts with the asymmetry obtaining at a higher level. Special relativity eliminates this asymmetry; equations of exactly the same form apply, whether we choose the magnet or the conductor as our frame of reference. There are no two separate electric and magnetic fields but a single anti-symmetric tensor transforming globally. The considerations applying to the classical account of the induction experiment apply also to Lorentz's explanation of Michelson's result. Once we accept the existence of an ether as carrier of the electromagnetic field, we are led to regard the latter as a state of the substratum. Molecular forces are transmitted by the same medium, so they also form part of its state; thus we have a good reason for supposing that molecular and electromagnetic forces are similar, i.e. for accepting the MFH. From this assumption follow the Lorentz-Fitzgerald contraction hypothesis and Michelson's null result. But there is something paradoxical in that, through postulating a universal medium, we are driven to conclude that it must be undetectable. Was it not dissatisfaction with this paradox so closely connected with the crucial experiment which caused Einstein to look for another explanation? The answer is that Einstein had become aware of the paradox independently of Michelson, as is indicated in the first paragraph of his EMB where the induction experiment is mentioned. What, from Einstein's viewpoint, was an unsatisfactory feature of classical physics had already been evinced by the Maxwellian account of the induction experiment and was, in this sense, completely independent of Michelson.

I leave open the question whether Einstein's three rules, which he certainly used as regulative principles, lend themselves to a more precise formulation. Einstein himself thought that:

> a sharper formulation would be possible. In any case, it turns out that among the augurs there usually is agreement in judging the inner perfection of the theories and even more so the degree of external confirmation. (Einstein 1949, p. 23)

Whatever the case may be, the lack of a more accurate rendering in no way entails that these rules, together with their metaphysical counterparts, should be given a subjective meaning. Einstein's metaphysical statements are admittedly vague; yet, they may still correspond to real properties of an external world independent of the scientist's mind, of his private feelings about perfection, harmony and the like. My main object will now consist in examining the role of the above prescriptions in the genesis of STR. In this specific context, it turns out that these otherwise vague rules and propositions assume a very precise form, leaving no doubt as to their intended objective meaning.

§3.2 The Discovery of Special Relativity (STR): Removal of the Asymmetry between Classical Mechanics and Electrodynamics

Let us look at the more general features of Einstein's objections to classical physics.

According to Einstein, one of Maxwell's and Faraday's greatest contributions to science was the introduction of the field as a new component of physical reality, which ought to be treated on a par with other constituents like corpuscles and electric charge (Einstein 1934, p. 160). However, Lorentz's electromagnetic theory confronts us with a dualism, to whose removal Einstein was to devote much of his life: on the one hand, there are discrete charged particles whose motions are governed by Newton's Laws and, on the other hand, a continuous field obeying Maxwell's equations. It is true that the charged corpuscles and their motions generate the field; but, once started, an electromagnetic disturbance propagates itself, independently of its source and with constant velocity c; the field may act back on the particles and thereby modify their motion. Fields and particles are therefore ontologically on a par. One way of resolving this dualism was to explain the behaviour of the field in terms of mechanical properties possessed by some all-pervading medium. Lorentz clearly recognised that, by the end of the nineteenth century, all such attempts had failed; he was about to try a solution in the opposite direction, in particular to explain inertia in electromagnetic terms.

Einstein was clearly dissatisfied with this dualism, as is apparent from the following passage:

> If one views this phase of the development of the theory critically, one is struck by the dualism which lies in the fact that the material point in Newton's sense, and the field as continuum are used as elementary concepts side by side. Kinetic energy and field-energy appear as essentially different things. This appears all the more unsatisfactory in as much as, in accordance with Maxwell's theory, the magnetic field of a moving electric charge represents inertia. Why not then *total* inertia? Then only field-energy would be left and the particle would be merely an area of special density of field-energy. In that case one could hope to deduce the concept of the mass-point together with the equations of the motions of the particles from the field equations,—the disturbing dualism would have been removed. (Einstein 1949, p. 36)

This lack of unity in the physical foundations, which violates prescription (β), was reflected in the mathematical formulation of the theory. Einstein explains:

> The weakness of the theory lies in the fact that it tried to determine the phenomena

> by a combination of partial differential equations (Maxwell's field equations for empty space) and total differential equations (equations of motion of point), which procedure was obviously unnatural! (Einstein 1950, p. 75)

The dualism was made far worse by Newton's classical principle of relativity (Newton 1686, p. 20 Corollary 5), which applies to mechanics but apparently not to electrodynamics. In view of the Galilean transformation which physicists took for granted, Maxwell's equations seem to presuppose an ether, or at any rate a unique frame in which they would hold good. Assessing Lorentz's work, Einstein wrote:

> For him [i.e., for Lorentz], Maxwell's equations concerning empty space applied only to a given system of coordinates, which, on account of its state of rest, appeared excellent in comparison to all other existing systems of coordinates. This was a truly paradoxical situation since the theory appeared to restrict the inertial system more than classical mechanics. (De Haas-Lorentz 1957, p. 7)

The absolute space hypothesis, the assumption that among all inertial frames there exists a privileged one, was an idle metaphysical component of classical mechanics. That its elimination does not reduce the empirical content of classical dynamics was clearly recognised by Newton himself. He wrote:

> The motion of bodies included in a given space are the same among themselves, whether that space is at rest or moves uniformly forward in a right line without any circular motion. (Newton 1686, p. 20 Corollary 5)

One could further maintain that the absolute space hypothesis was utterly useless in that it was, even in principle, impossible to define the absolute frame as that in which Newton's laws hold good; for, if these laws are true in any one of the inertial frames, they automatically hold in all. (In the *Science of Mechanics*, which Einstein carefully read, Mach attacked the concept of absolute space and went as far as proposing that even the distinction between inertial and non-inertial frames ought to be abolished.)

With the advent of the wave theory of light, more particularly with Fresnel's and Lorentz's postulation of a stationary ether, the situation changed dramatically. One could now define absolute space as that in which Maxwell's equations are true. Given the old kinematics and, in particular, the Galilean transformation, this definition singles out a unique frame in which Maxwell's equations hold and, consequently, light is propagated in all directions with the same speed c. The ether frame was taken to be inertial so that in all other frames, whether inertial or accelerated, light would have no constant velocity; thus, there arose the possibility of devising experiments in order to detect the 'absolute' motion of ponderable bodies. The experiments would be such that their outcomes depend on the absolute motion of some material body and thus tell us whether the body was moving or at rest in the ether. From this point of view, Michelson's experiment is typical: a null outcome would tell us that the earth is at rest in the ether; and from a shift in the fringes it could be concluded that the earth moves. In this particular case however, we 'know' that the earth changes its velocity with respect to the frame determined by the stars; so the earth must, at one point of its trajectory, be moving in the ether; we can thus predict that the experimental result must be positive.

Why did Einstein find such developments in the evolution of physics 'paradoxical'? We have seen that he disliked the dualism of particles and fields. The fact that the laws of mechanics governing the motions of particles obey the relativity principle while Maxwell's equations, which govern the behaviour of the field, do not, makes the dualism much worse. Given this problem-situation, there were to my mind two courses of action open to a unificationist like Einstein: he could maintain either that the relativity principle applies *neither* to mechanics *nor* to electrodynamics; or else that it applies to *both* at the same time. In the first case he could have modified mechanics so as to make it hold only in the ether frame; in the second case, he would have to extend the relativity principle to include electrodynamics. In its Galilean form however, relativity does not apply to electromagnetic theory. At this point Einstein's critique of the induction experiment proved crucial, in that it tipped the balance in favour of extending relativity to electrodynamics and *thereby* of modifying classical kinematics.[1]

To repeat: in the induction experiment there is, between two experimental results, a complete symmetry which is at odds with the asymmetry introduced by the theoretical explanation. To put it more pedantically: the observation statements describing the behaviour of the electric currents are identical in the two cases, but the explanations in terms of the accepted theory differ widely. There would be nothing intrinsically wrong in this state of affairs, had the asymmetry not been introduced by considerations of absolute motion which the relativity principle forbids. Seen in this light however, the experiment suggests that an extension of the relativity principle to include electrodynamic phenomena might abolish the 'theoretical' asymmetry. Such an extension promises to make the symmetry between the two experimental outcomes appear, not as a fortuitous result, but as the direct manifestation of a general principle, the principle of Lorentz-covariance. In this, Einstein was following prescription (γ). In the EMB he concluded that:

> examples [like the induction experiment] together with the unsuccessful attempts to discover any motion of the earth relatively to the light medium, suggest that the phenomena of electrodynamics as well as of mechanics possess no properties

---

[1]I do not of course mean that the experiment was 'crucial' in the traditional sense of refuting one theory while confirming another.

> corresponding to the idea of absolute rest. They suggest rather, as has already been shown to the first order of quantities, that the same laws of electrodynamics and optics will be valid for all frames of reference for which the equations of mechanics hold good. We will raise this conjecture (the purport of which will hereafter be called the Principle of Relativity) to the status of a postulate and also introduce another postulate, which is only apparently irreconcilable with the former, namely that light is always propagated in empty space with a definite velocity *c* which is independent of the state of motion of the emitting body. (EMB, p. 38)

Note the two references made to mechanics, underlining the important part which classical relativity played in Einstein's thinking prior to 1905.[2]

In this passage Einstein alludes to the absence of any first-order effects of absolute motion, a result explained by Lorentz through an early version of his theory of corresponding states. This first-order equivalence between observers, which runs counter to the preference given to a unique frame, must have increased Einstein's suspicion that the relativity principle applies to electrodynamics as well as to mechanics; under the new theory, the absence of first-order effects, instead of being a stray fact, would directly reveal the presence of a universal principle.

What commended the relativity principle was therefore its universality, its unifying role in subsuming mechanics and electrodynamics under the same law, finally in providing a unified explanation for various features of phenomena, such as the observed symmetry in the induction experiment and the absence of all first-order effects resulting from the earth's motion.

The phrase: "*. . . together with the unsuccessful attempts to discover any motion of the earth relatively to the light medium*" has given historians and philosophers of science some problems.[3] It also seems inconsistent with the thesis, put forward in §3.1, that the Michelson experiment played a negligible role in the genesis of STR.

In the quoted phrase, Einstein might of course be referring to Michelson's experiment which must have been at the back of his mind, if only through Lorentz's *Versuch*. This is perfectly compatible with his assertion that the experiment came to his *attention* only after 1905. To my mind, the above phrase is no more than a casual allusion to a number of a results, which he

---

[2]This was later confirmed in his more philosophical writings. See De Haas-Lorentz (1957), quoted above, and Einstein 1950, p. 55.

[3]Grünbaum, for example, says: "Unless they provide some other constant explanation for the presence of the latter statement in Einstein's text of 1905, it is surely incumbent upon all those historians of the Relativity Theory who deny the inspirational role of the Michelson-Morley experiment to tell us specifically what other "*unsuccessful attempts to discover any motion of the earth relatively to the light medium*" Einstein had in mind here." (Pearce Williams 1968, p. 114)

had registered without any surprise since they followed from his own conjectures.[4] On this point I agree with Holton (1969, pp. 164-5); otherwise, Einstein would certainly have cited Michelson's result in support of his second postulate, or rather of principle (P2). This principle presents us with a new difficulty.

Unlike his first postulate (P1), principle (P2) is thrown out with no justification whatsoever. Moreover, on the face of it, (P2) runs counter to Einstein's prescription (β): there seems to be no connection at all between the fundamental properties of space-time and those of light. Why should purely kinematical considerations involve c? The light principle is quite a low-level statement which is not as yet integrated into a more general system. It is precisely for this reason that philosophers and scientists supposed that Einstein was obeying the dictates of experience by basing his second postulate on Michelson's result. Later on, Einstein does say that (P2) is in agreement with experience (Einstein 1905, p. 40): he had after all heard of various experiments trying to detect the earth's absolute motion. Nowhere however does he assert that experience had either suggested principle (P2) or even made it look plausible to him.

I think the problem can be solved by examining more carefully Einstein's later writings (in particular his *Autobiography*) and then comparing them with his EMB. In his *Weltbild* Einstein wrote:

> Then came the Special Theory of Relativity with its recognition of the physical equivalence of all inertial systems. In conjunction with electrodynamics or the law of propagation of light, it implied the inseparability of space and time. (Einstein 1984, p. 143; my translation)[5]

Perhaps the most illuminating passage occurs in *Out of My Later Years*:

> The second principle on which the Special Relativity theory rests is that of the constancy of the velocity of light in the vacuum. Light in a vacuum has a definite and constant velocity, independent of the velocity of its source. *Scientists owe their confidence in this proposition to the Maxwell-Lorentz theory of electrodynamics*. (Einstein 1950, p. 56, my italics)

Also, in his *Autobiography,* Einstein tells us about a thought-experiment in which, at the age of sixteen, he imagined himself to be following a ray of light at speed c:

---

[4]He admitted to Shankland that "he had also been conscious of Michelson's result before 1905, partly through his readings of the papers of Lorentz and more because he had simply assumed this result of Michelson to be true" (Holton 1969, p. 154.)

[5]In 1949 Einstein again says "The Special Theory of Relativity owes its origin to Maxwell's equations of the electromagnetic field. Inversely, the latter can be grasped formally in satisfactory fashion only by way of the Special Theory of Relativity. Maxwell's equations are the simplest Lorentz invariant field equations which can be postulated for an anti-symmetric tensor derived from a vector field." (Einstein 1949, p. 62)

> If I pursue a beam of light with a velocity c (velocity of light in a vacuum), I should observe such a beam of light as a spatially oscillatory electromagnetic field at rest. However, there seems to be no such thing, whether on the basis of experience or according to Maxwell's equations. From the very beginning *it appeared to me intuitively clear that, judged from the standpoint of such an observer, everything would have to happen according to the same laws as for an observer who, relatively to the earth, was at rest*. For how, otherwise, should the first observer know, i.e. be able to determine, that he is in a state of fast uniform motion. (Einstein 1949, p. 53, my italics)

What is most striking about this passage is the conclusion which Einstein draws from his thought-experiment. He does not restrict himself to what seems warranted by the experiment, namely that c is an unattainable speed or that the addition law of velocities must break down. He immediately jumps to a general conclusion, or rather he puts forward a sweeping conjecture: the laws of physics and more specifically those of electromagnetism would have to be the same for the moving as for the stationary observer. Both historically and epistemologically, Einstein's second starting point (the first one being the relativity postulate) is not principle (P2), but the proposition:

(P3) *Maxwell's equations express a law of nature*

In virtue of (P1), these equations must assume the same form in all inertial frames. Maxwell's equations imply that, within each coordinate system in which they hold, electrodynamic disturbances are propagated with velocity c; which velocity must therefore be an invariant (see, however, §4.5). Thus (P1) and (P3) imply (P2).

In the electromagnetic section of the EMB, Einstein does in fact assume that Maxwell's equations are Lorentz-covariant and then derives the transformation laws for $\vec{E}$ and $\vec{H}$. He does not try to infer (P3) from (P1) and (P2). By exhibiting a transformation which makes Maxwell's equations covariant, Einstein established that the latter are compatible with the Relativity Postulate. This is a consistency proof. Although (P3) is a stronger statement than (P2), it is more plausible and, incidentally, less counter-intuitive. In accordance with (β), (P3) derives its plausibility from being a unified, well-knit theory in which the primitive concepts (electric field, magnetic field, charge density) are all closely connected; it had also been tested for a whole generation before 1905. Thus, the logical order is reversed by *a priori* heuristic considerations: (P3) is more plausible, though stronger, than (P2).

Another piece of evidence confirming the view that Einstein approached the problem of relativity through the covariance of Maxwell's equations is to be found in Lorentz's *Versuch*. In a part of this work which is completely independent of Michelson's experiment and of the contraction hypothesis,

Lorentz had proved that no first-order effects of the earth's motion can be detected. Neglecting all terms in $(v/c)^2$, he used a limiting case of his coordinate transformation; he than found transformation laws for the fields $\vec{E}$ and $\vec{H}$ under which the equations take on a form very similar to and in some cases identical with their form in the ether frame. It is not far-fetched to suppose that Lorentz's techniques made a strong impression on Einstein; they might well have led him to wonder whether a more general transformation would yield both complete covariance and the result that no effects whatever arise from uniform rectilinear motion. Lorentz himself was to attempt this solution in his 1904 paper. Einstein, however, did not read this paper before publishing his own EMB.

Nonetheless, Lorentz's programme for a theory of corresponding states was outlined in the *Versuch* and one is struck by the similarity between Lorentz's methods and those used later by Einstein. In the EMB the latter determined a transformation law for the coordinates x, y, z, t; assuming the covariance of Maxwell's equations, he then derived the transformation laws for $\vec{E}$, $\vec{H}$ and ρ. Thus, Lorentz's influence on Einstein cannot be overrated; it was not Michelson the experimentalist, but Lorentz the theoretician, who played a considerable inspirational role in the genesis of special relativity. This is indicated, in the all-too-brief second paragraph of Einstein's EMB, by the clause: *"as has already been shown to the first order of small quantities."*

We have seen that Einstein rejected Lorentz's *classical* approach; he nevertheless made use of Lorentz's tremendous technical achievement, albeit under very different kinematical assumptions. I have mentioned that the TEM of 1892 already contains the Lorentz transformation to within a constant factor in the expression of t'. Einstein's greatest contribution was to extend Lorentz's methods and give the transformed quantities a *realistic* interpretation.

It is by now abundantly clear why Einstein could rightly claim that Michelson's experiment had been irrelevant to his work and that he could easily have anticipated its null outcome. That almost nothing is cited in support of principle (P2) may be due to the fact that it follows from a well-corroborated hypothesis, namely the Maxwell-Lorentz equations, taken together with the relativity postulate (P1) for whose acceptance Einstein had already argued.

I should finally like to examine a powerful argument advanced by John Stachel (1982) in support of the view that Michelson's experiment did after all play an important role in the genesis of relativity. Let us denote by (CL) the law that, in *some* inertial frame K, the speed of all light beams in every direction is a constant c independent of the motion of the light source. Note: the proposition called principle (P2) above is much stronger than (CL). (P2) is the consequence of applying the relativity postulate (P1) to (CL) and thus concluding that c is independent both of the light source *and of the chosen inertial frame*. John Stachel pointed out that (P1), but not (CL), yields Michelson's result: on the one hand, the Maxwell-Lorentz theory, though inconsistent with Michelson's result, entails (CL) for the ether frame; on the other hand, the relativity postulate (P1) on its own, i.e., independently of (CL), implies the undetectability of the earth's motion, hence the null outcome of Michelson's experiment. In the experiment, the light source is at rest with respect to the apparatus; its null outcome could therefore be explained by any *relativistic* emission theory which would make the velocity of light depend on the motion of its light source. Thus, a priori, Einstein's assumption (CL) is not implicated in Michelson's result.

We have seen that, before 1905, Einstein must have known about Michelson's work if only through reading Lorentz's *Versuch*. Michelson's experiment was mentioned in several papers written by Einstein not long after the publication, in 1905, of the EMB. This leads John Stachel to the conjecture that, although Michelson's results is not explicitly referred to in the EMB, it might well have played *some* role in the discovery of principle (P1) but, of course, not in that of proposition (CL). As for Faraday's law of electromagnetic induction, it is explicitly mentioned in the EMB and everybody agrees that *it* played a crucial role in the discovery of (P1). It is therefore plausible, according to John Stachel, that both Michelson's and Faraday's results played a role in the emergence of STR.

At first sight, John Stachel's position seems very different from mine. It is my claim, however, that the difference is slight.

To begin with, since it incorporates the relativity postulate, principle (P2) is, on John Stachel's own account, strongly supported by Michelson's result. Moreover, both the relativity principle and the constancy of c are implicitly contained in Faraday's law of electromagnetic induction taken together with Lorentz's theorem of corresponding states, i.e., with the first-order covariance of Maxwell's equations, especially when the latter is extended to cover effects of *all* orders; and we have already seen how prone Einstein was to such sweeping generalisations. Although Einstein knew about Michelson's experiment and may even have regarded some account of its null outcome as an adequacy requirement for every new electromagnetic theory, *he* did not need Michelson in order to construct his own hypothesis; for he had already analysed Faraday's law in the light of Lorentz's theorem of corresponding states and thus obtained the relativity principle (P1). Thus, for Einstein, Michelson's result was a 'novel' fact which lent strong independent support to STR. Putting it a little differently: STR and, more particularly, (P2) are strongly supported by the convergence, towards the same conclusion, of Faradays's law on the one hand and of Michelson's result on the other.

One might still wonder why Einstein did not start by assuming (P1) and (P3) instead of (P1) and (P2). I think that he had at least two good reasons for presenting the new theory in the way he did. It is on the one hand preferable, from a logical point of view, to use (P2) which, though weaker than (P3), suffices both for developing a new kinematics and for deriving the Lorentz transformation. On the other hand, Einstein had come to the conclusion that Maxwell's equations, though true of macroscopic phenomena, could not provide ultimate foundations for the whole of physics. He might thus have wanted to make his spatio-temporal framework independent of electrodynamics. In 1955 he maintained:

> I knew only of Lorentz's works in 1895—*La Théorie Electromagnetique de Maxwell* [this is the 1892 work] and *Versuch einer Theorie der electrischen and optischen Ersheinungen in bewegten Koerpern*—but not Lorentz's later works, nor the consecutive investigations by Poincaré. In this sense my work of 1905 was independent. The new feature of it was the realisation of the fact that the bearing of the Lorentz transformation transcended their connection with Maxwell's equations and was concerned with the nature of space and time in general. A further new result was that the Lorentz-invariance is a general condition for any physical theory. This was for me of particular importance because I had already previously found that Maxwell's theory did not account for the microstructure of radiation and could therefore have no general validity. (Cited in Born 1956, p. 248)

In this passage, Einstein clearly indicates that there was a connection between Maxwell's equations and the Lorentz transformation, but that this connection was then transcended. The logical picture seems to be as follows:

[(*P*1) Λ (*P*3)] ⇒[(*P*1) Λ (*CL*)]; [(*P*1) Λ (*CL*)]⇒[*P*1) Λ (*P*2)];

[(*P*1) Λ (*P*2)]⇒(New kinematics and Lorentz transformation equations);

[(*P*1) Λ (Lorentz transformation equations)]⇒(Requirement of Lorentz-covariance for the whole of physics).

The connection between Maxwell's theory and the Lorentz transformation is given by:

[(*P*1) Λ (*P*3)]⇒(Lorentz transformation equations)

This connection is transcended by the result that the relativity principle, taken together with the Lorentz transformation, entails a new structure of space-time and a condition of Lorentz-covariance applying, not only to Maxwell's equation, but to the whole of physics.

One last question remains unanswered. Einstein faced a problem forced on him by the incompatibility, if taken together, of the following three hypotheses:

(P1) The relativity postulate
(N) Newton's second law of motion
(P3) The Maxwell-Lorentz equation

His solution consisted in modifying N, i.e. in replacing N by a new theory N′, such that [(P1) Λ (N′) Λ (P3)] is consistent. Considering that the relativity principle was first shown by Newton to apply to mechanics, it is puzzling that Einstein seems never to have envisaged keeping N and replacing (P3) by a new set of equations (P3′) covariant under the Galilean transformation. (P3′) would of course have to yield (P3) as a limiting case. This is all the more intriguing since he realised that Maxwell's equations were not as fundamental as Lorentz had taken them to be. In other words, the question is: why did Einstein throw in his lot with Maxwell rather than with Newton?

First of all, we have seen that Einstein was dissatisfied with the dualism between fields and particles which beset Lorentz's theory. In virtue of prescription (β), it seemed obvious that one component of this dual structure ought to be reduced to the other. But which one? As explained in his *Autobiography,* Einstein found that all attempts at a mechanical explanation of the behaviour of the field had failed.

> If mechanics was to be maintained as the foundation of physics, Maxwell's equations had to be interpreted mechanically. This was zealously but fruitlessly attempted, while the equations were proving themselves fruitful in mounting degree. (Einstein 1949, p. 25)

Secondly: Planck's quantum hypothesis, with which Einstein was fully acquainted, led to the following conclusion:

> This form of reasoning does not make obvious the fact that it contradicts the mechanical and electrodynamic basis, upon which the derivation otherwise depends. Actually, however, the derivation presupposes implicitly that energy can be absorbed and emitted by the individual resonator only in quanta of magnitude h, i.e. that the energy of a mechanical structure capable of oscillations as well as the energy of radiation can be transferred only in such quanta—in contradiction of the laws of mechanics and electrodynamics. *The contradiction with dynamics was here fundamental; whereas the contradiction with electrodynamics could be less fundamental. For the expression for the density of radiation-energy, although it is compatible with Maxwell's equations, is not a necessary consequence of these equations.* (Einstein 1949, p. 45)

These passages indicate why Einstein gave precedence to Maxwell over Newton and so took his starting point with electro-dynamics rather than with mechanics. They also show that, having accepted Planck's quantum hypothesis, Einstein could not regard Maxwell's equations as fundamental. He had therefore enough reservations about electromagnetism to avoid making it into a cornerstone of his kinematics. Thus: although (P3) both entails and lends plausibility to principle (P2), the latter is more fundamental in the sense of applying both to micro- and to macroscopic phenomena. Einstein's inspired and lucky guess was that the invariance of c was a universal principle which transcends its initial dependence on Maxwell's theory.

§3.3 THE HEURISTIC SUPERIORITY, IN 1905, OF THE RELATIVITY PROGRAMME: EINSTEIN'S COVARIANCE VERSUS LORENTZ'S ETHER

So far, we have spoken only of the emergence of relativity, not of any collapse of classical physics. Since, as shown above, Lorentz's theory T_3 proves observationally equivalent to STR, there arises the question whether the ether programme did really break down. After all, Einstein's transformed coordinates can be reinterpreted as the coordinates effectively measured in a moving Lorentzian frame. True, in the latter the 'real' coordinates are still the quantities: x_r, y_r, z_r, t_r, given by the Galilean transformation:

$$x_r = x - vt;\ y_r = y;\ z_r = z;\ t_r = t;$$

but, due to time-dilatation, length-contraction and the synchronisation of clocks by means of light-signals, the *measured* coordinates are:

$$x' = \gamma(x - vt),\ y' = y,\ z' = z,\ t' = \gamma(t - (v/c^2)x), \text{ where}$$

$$\gamma \underset{\text{Def}}{=} (1 - v^2/c^2)^{-1/2}$$

Moreover, Lorentz had tentatively assumed that matter was nothing more than a distribution of charge in the ether; so-called neutral matter was a mixture of equal amounts of positive and negative electricity. He had explained inertia in purely electromagnetic terms and proposed a law governing the variation of the mass of moving electrons with their speed. In 1905 he was therefore in a position to account for the increase of inertia with velocity. Distinguished physicists like Ehrenfest held that Lorentz's and Einstein's theories were *distinct, but parallel,* ways of looking at the same phenomena. Ehrenfest wrote:

> These researches of Lorentz's have become classic for their methods as well as for their results. As one sees they furnish as a conclusion, as a theorem, that which, in generalised form, Einstein placed as a postulate at the beginning of his relativity theory in 1905. (Ehrenfest 1923, pp. 471-8)

Thus, as he indicated at the end of his *Theory of Electrons,* Lorentz was in a position so to reformulate his theory that no crucial experiment between his system and Einstein's could have been devised in 1905. For example, Lorentz could have corrected his transformation law for the charge density and thus obtained the full covariance of Maxwell's equations. (This is precisely what Poincaré did in his paper: "On the Dynamics of the Electron" discussed in chapter five.)

In view of this situation, why did brilliant mathematicians and physicists, like Minkowski and Planck, abandon the classical programme in order to work on special relativity? Was Lorentz about to face insuperable difficulties which were known to his contemporaries? We have seen that he used the MFH in order to obtain the appropriate laws about rod-contractions and clock-retardations. He had thereby assumed a transformational similarity between electromagnetic and molecular forces. In view of his programme, his next most natural step would have been to give a precise law of force for atomic interactions. We have also mentioned that, in order to explain the variation of inertia with speed, Lorentz accounted for the mass of the electron in purely electromagnetic terms. That is: in producing the revolutionary results which either matched or even anticipated Einstein's, Lorentz had to give a classical account of elementary particles. By initially setting up a new kinematics, Einstein was able to remain agnostic about the ultimate structure of matter; this is why his approach in 1905 is strikingly—but deceptively (see chapter four)—more positivistic than Lorentz's. For example, Einstein managed to derive the equivalence of mass and energy without making extra assumptions about the atomic constitution of matter (see chapter seven). Lorentz was therefore unlucky in his choice of problems: he was straight away involved in difficulties which were to defeat Einstein himself, but at a much later stage. With hindsight, we can see that Lorentz would probably have failed anyway; he was overtaken by the quantum theorists who realised that classical laws, more particularly Maxwell's equations, could not explain atomic stability. However, not only would Einstein have been confronted with similar difficulties but, in 1905, there was hardly any indication that Lorentz could go no further in developing his programme; i.e. that no satisfactory classical account of elementary particles could be arrived at. Yet, already in 1906, a physicist of Planck's stature *and conservatism* abandoned the classical approach altogether, with the full knowledge that STR and, of course, also Lorentz's theory T_3 might well have been refuted by Kaumann's experiment (see chapter six).

Given the parity between Einstein's and Lorentz's hypotheses, a Kuhnian explanation of Planck's 'conversion' in terms of bandwagon effect may seem tenable. But the idea of a new bandwagon is highly implausible. First, Einstein was a relatively unknown figure while Lorentz was a recognised authority. Secondly, Lorentz's theory was eminently intelligible whereas Einstein's involved a major revision of our most basic notions of space and time. Thirdly, there was no build-up of anomalies which Einstein's theory resolved better than Lorentz's. Moreover, at the time when Planck was converted, that is in 1906, no bandwagon effect had started. Nor did the leading protagonists of the old paradigm die out unconverted. Lorentz himself eventually adopted the new outlook. In the *Theory of Electrons*, first published in 1909, he gives essentially the same account of the theory of corresponding states as in his 1904 paper. However, the footnotes added to later editions of his book clearly indicate that, by 1915, he had already accepted the relativity principle.

Let us now turn to the most commonly held view, according to which Einstein's theory represented the success of positivism in ridding science of redundant metaphysics.

I shall both develop this positivist claim and present my answer to it by comparing the Einsteinian revolution with the Copernican (or rather with the Keplerian) one. This comparison will point to a feature which has often been a symptom, if not the cause, of decline in the heuristics of some research programmes. This feature consists in a divorce between the empirical content and the mathematical formulation of certain hypotheses: these hypotheses contain a large number of physically uninterpreted mathematical entities.[6] According to positivists like Mach, the mere elimination of such entities increases simplicity and thereby constitutes progress. My claim is that such eliminations are *by-products* of new research programmes whose heuristic rids them of certain 'unnecessary' entities. This may be accompanied by a *contingent* increase in simplicity. Let us take the example of the Copernican Revolution.

The Platonic programme of saving the phenomena by the use of circular motions was initially successful: to each mathematical entity corresponded a physical one. Each planet was fixed on a physically real crystalline sphere which performed a number of axial rotations. It was however discovered that the distance between the earth and a given planet varied, so the astronomers resorted to eccentrics, epicycles and equants in order to account for the new phenomena. The physical problem was to determine the motion of the heavenly bodies with respect to the earth. Since the paths of the planets are non-circular and since their motion is non-uniform, a widening gap opened between the physical problem and the permitted mathematical methods which allowed only for circular motions. Although the earth supposedly occupied the centre of the universe, the paths of the planets about the earth were not directly dealt with; epicycles, deferents and equants, *all of which had no physical reality,* were introduced in order to predict astronomical data; both the centre of an epicycle and the *punctum equans* are empty points in space.

[6]This is so to speak the obverse of the point made earlier about the second heuristic role of mathematics in physics. There we saw how new physical theories can be constructed by interpreting hitherto uninterpreted mathematical entities (see §3.1).

Copernicus did not heal this rift between the physical picture and the mathematical description. True, he got rid of the equant; but, although his problem was to determine the motion of the planets with respect to a fixed sun, he interposed between the sun and the planets roughly as many epicycles, with as many empty centres, as were involved in the rival system. It was left to Kepler to investigate the direct relation between the sun and the planets, to abolish epicycles and, finally, to find that the planets describe ellipses with one focus at the centre of the sun.

Let us now return to Lorentz. We have seen (in §3.1) that the Lorentz-transformation is always carried out in two steps. The first step yields the Galilean coordinates:

$$x_r = x - vt,\ y_r = y,\ z_r = z,\ t_r = t.$$

The second one gives us the effective coordinates:

$$x' = \gamma(x - vt),\ y' = y,\ z' = z,\ t' = \gamma\,(t - vx/c^2).$$

The Galilean coordinates are interposed between the absolute coordinates and the effective ones in the same way that various epicycles were placed between the earth, or the sun, on the one hand and the planets on the other.[7]

In a moving frame only the Galilean co-ordinates are taken by Lorentz to be ontologically 'real' in the same way that, before Kepler, only circular motions were considered metaphysically permissible. These metaphysical assumptions were naturally reflected in the mathematics: in the Galilean transformation used by Lorentz and in the epicycles used by Ptolemy and Copernicus. The Galilean transformation is a vestige of the original aim of the classical programme, namely the aim of giving to the ether frame a privileged status. (Because of the Galilean transformation, Maxwell's equations hold good only in the ether frame.) The assumption of an ether frame no longer has any observational cash-value. Similarly, the Ptolemaic epicycles were reminders of a hope which had long vanished, the hope of finding that the motions of the planets are both uniform and circular.

Copernicus was aware that the motions of the planets are neither circular nor uniform and Lorentz later realised that the effective co-ordinates, not the Galilean ones, are the *measured* quantities in the moving frame.

---

[7]By drawing a parallel between the Galilean transformation on the one hand and a system of epicycles on the other, I do not want to suggest that, in Lorentz's theory T_3, the Galilean transformation is uninterpreted. In fact, even the epicycles can be interpreted in the following trivial way: God, in contemplating his creation, sees it as a huge system of interlocking circles. Similarly, in Lorentz's case, God would perceive an infinite extended substance, the ether, in which any two events are separated by an absolute time interval. Such interpretations, which do not increase the empirical content of existing theories, could conceivably be made useful by indicating how they are to be heuristically exploited in order to construct new physical theories. Lorentz did not give such an indication in connection with the Galilean transformation.

I have drawn a parallel between Copernicus and Lorentz. Kepler and Einstein can be similarly compared. Kepler's greatest contribution to astronomy allegedly consisted in eliminating epicycles and in showing that the 'real' paths of the planets are ellipses with one focus at the sun. Similarly, according e.g. to Bridgman and to von Laue, Einstein's chief merit lay in abolishing the Galilean transformation and in identifying the effective, i.e. *measured,* coordinates as the only real ones. In equating 'to be' with 'to be perceived or measured' Einstein is supposed to have carried out a positivistic revolution in physics. However, if the merit *of* both Kepler and Einstein consisted *only* in ridding physics of unnecessary 'epicycles', then the importance of these two physicists is grossly overrated: Copernicus and Lorentz did all the creative work, while Kepler and Einstein merely applied Occam's razor in order to demolish the expendable metaphysical scaffolding used by their predecessors. Moreover, Copernicus knew that the paths of the planets were not circular, that his epicycles were part of the scaffolding; similarly Lorentz realised that he did not need the Galilean coordinates in order to deduce the null results which he set out to explain. *If so, Kepler and Einstein contributed to the economy of thought and not to the growth of knowledge.*

This is an unacceptable conclusion. Let us start with Kepler. Copernicus's account of the motion of heavenly bodies had been largely Aristotelian in character: because the planets are perfect spheres, their natural motion is both uniform and circular. Through trying to give a dynamical explanation of the motion of heavenly bodies, Kepler provided classical astronomy with its heuristic. He proposed to determine the forces which, emanating from the sun, act directly on the planets. He abolished all epicycles *because* he wanted nothing but forces to mediate between the sun and the planets. Circles centred on empty points concealed the 'true' relation linking one heavenly body to another. Kepler proposed a *dynamic* theory, which is now largely forgotten because it was contradicted and supplanted by Newtonian astronomy. But, in forgetting Kepler's dynamics, we should not forget that Kepler created the programme which culminated in the Newtonian synthesis; Kepler's method consisted in trying to discover the law of force responsible for the periodic motion of the planets round the sun. *Getting rid of Copernican epicycles was not an end in itself: it was subordinate to the needs of the new heuristic.*

Einstein, like Kepler, created a programme, not only an isolated theory. We shall see that Einstein's heuristic is based on a general requirement of Lorentz-covariance for all physical laws. We recall that the Lorentz-transformation sends (x,y,z,t) directly into (x′,y′,z′,t′) without passing by the Galilean coordinates x_r, y_r, z_r, t_r. *The new heuristic therefore requires the abolition of the Galilean transformation, which plays the role of a cumbersome epicycle. The parallel with Kepler is complete.*

After these criticisms of the Kuhnian and positivist 'explanations' of the

Einsteinian revolution, let me venture my own. In my view, the main difference between Lorentz and Einstein lies in the difference between the heuristics of their respective programmes. *The ether programme did not collapse but was superseded by a programme of greater heuristic power.* This greater heuristic power explains why Planck and others joined Einstein's programme *before* it became empirically progressive. The difference between the two theories cannot be appreciated by taking an instantaneous look at Lorentz's and Einstein's systems. One has first to imbed them in their respective programmes. In this way one realises that the two theories are similar because they stand at the intersection of two research programmes which later diverged. It will further be shown that the difference between the two approaches did not emerge with hindsight, but guided the deliberate choice of scientists at the beginning of this century.

I have reached the seemingly paradoxical conclusion that both Einstein (and Planck) on the one hand and Lorentz on the other were perfectly rational in doing what they did, i.e. in doing opposite things. Let me immediately add that they were rational, *given* their metaphysical positions. The conflict between Lorentz and Einstein is, among other things, the age-old conflict between two metaphysical doctrines which, Polanyi notwithstanding, do not belong to the tacit component but can be articulated. Lorentz held that the universe obeys intelligible laws (e.g. wave processes presuppose a medium, there exists an absolute 'now' etc.) and Einstein held that the universe is governed by principles which can be given a mathematically coherent form. (e.g. *all* laws are covariant.) All major scientific revolutions were accompanied by an increase of mathematical coherence together with a (temporary) loss of intelligibility. (This applies to the Copernican, to the Newtonian, to the Einsteinian and to the quantum-mechanical revolutions.) It can moreover be argued that intelligibility is a time-dependent property, while mathematical coherence is not. We still consider Newtonian astronomy more coherent than Ptolemaic astronomy; but action-at-a-distance was unintelligible before Newton, became perfectly intelligible at the end of the eighteenth century and again unacceptable after Maxwell.

Let us go back to Lorentz who, unlike Einstein, did not create the heuristic of his own programme. The heuristic of Lorentz's programme consisted in endowing the ether with such properties as would explain the behaviour both of the electromagnetic field and of as many physical phenomena as possible. In view of the overwhelming success of Newtonian dynamics, it is hardly surprising that the ether was supposed to possess primarily mechanical properties. *The ether programme developed rapidly in certain respects; yet towards the end of the nineteenth century its positive heuristic was running out of steam.* A succession of mechanical models for the ether were proposed and abandoned. One serious difficulty was the presence in these models of longitudinal as well as of transverse waves (see Whittaker 1951, ch. V). Lorentz faced a daunting problem of a different sort: in order to explain certain electromagnetic phenomena, he postulated an ether at rest. He considered a portion of the ether, calculated the resultant $\vec{R}$ of the Maxwellian stresses acting on its surface and found that $\vec{R}$ is generally non-zero. Hence, if he was to assume that the ether was anything like an ordinary substance, he would have also to suppose that it was in constant motion. But this contradicted his original assumption of an ether at rest. He concluded that "the ether is undoubtedly widely different from all ordinary matter" and that "we may make the assumption that this medium, which is the receptacle of electromagnetic energy and the vehicle for many and perhaps for all the forces acting on ponderable matter, is, by its very nature, never put in motion, that it has neither velocity nor acceleration, so that we have no reason to speak of its mass or of forces that are applied to it." (Lorentz 1909, p.30) In other words, Lorentz had reached a point where the behaviour of the electromagnetic field dictated what properties the ether ought to have, no matter how implausible these properties might be: for example, the ether was to be both motionless *and* acted upon by non-zero net forces. Thus, the ether was nothing but the carrier of the field. *This involved a reversal of the heuristic of Lorentz's programme: instead of learning something about the field from a general theory of the ether, he could get at the ether only by way of the field.* In the case of the MFH, for example, Lorentz *first* studied the transformational properties of the electromagnetic field; and only *then* did he extend these properties to other molecular forces. Instead of positing *one medium* endowed with certain properties from which all forces inherit some *common* characteristic, we have an electromagnetic field acting as the archetype which determines the respects in which all forces are similar.

I do not claim that the ether programme was beyond redemption. Of course, there was no immediate reason why the postulation of some non-mechanical properties of the ether should not account for both electromagnetic phenomena and molecular interactions. All I claim is that the heuristic, *as it stood,* had petered out and that the ether programme was in need of a 'creative shift'[8]—a shift which, as a matter of fact, Lorentz did not provide.

Einstein based his heuristic on the requirement that all physical laws should be Lorentz-covariant; i.e. all theories should assume the same form, whether they are expressed in terms of x, y, z, t or in terms of x′, y′, z′, t′. But it would be practically impossible to discover new laws simply by looking out for all Lorentz-covariant equations. A good method is to start

[8]This is a technical term in the methodology of scientific research programmes: cf. Lakatos 1970, p. 137.

from well-tested laws whose past success would anyway have to be explained by any new theory. *Thus the heuristic of Einstein's programme is based on two different requirements: (i) a new law should be Lorentz-covariant, and (ii) it should yield some classical law as a limiting case.* (This is the correspondence principle.)

We have just seen that Lorentz used the ether in order to extend certain properties of the electromagnetic field to molecular forces. His methods were effective in explaining Michelson's and other null results. By requiring, not only that electromagnetic and molecular interactions, but that all forces should obey the same transformation laws, by then taking Maxwell's equations and imposing their transformation properties on the whole of physics, Einstein both strengthened those Lorentzian methods which had proved effective in particular cases and turned them into a heuristic method of general applicability. In this sense, Einstein's programme displayed greater heuristic power than Lorentz's.

Let us give a more formal rendering of these two requirements. Let $R(a_1, a_2, \ldots, a_n) = 0$ be an equation which constitutes a physical law in some inertial frame I. If I′ is any other inertial frame in which the quantities $a_1, a_2, \ldots, a_n$ assume the values $a'_1, a'_2, \ldots, a'_n$, respectively, then, by the relativity principle:

$$\text{(1)}\quad [R(a'_1, a'_2, \ldots, a'_n) = 0] \Leftrightarrow [R(a_1, a_2, \ldots, a_n) = 0]$$

But, as Kretschmann pointed out to Einstein, every empirical law can be given, not only a Lorentz-covariant, but also a generally covariant expression (of course, general covariance implies Lorentz-covariance) (see Kretschmann 1917/Einstein 1918). Thus, on the face of it, the most distinctive requirement of Einstein's heuristic is empty (see §8.1). However, the requirement is trivialised only if one is allowed complete freedom in reformulating the law. If one is restricted to a given number of entities: $a_1, a_2, \ldots, a_n$, then the covariant requirement, far from being empty, becomes a very stringent condition. As we shall see, in each particular case in which the heuristic is applied, the entities occurring in the covariant law are precisely those involved in the corresponding classical law.[9]

We now consider the requirement that a new relativistic law should yield the corresponding classical theory as a limiting case. In the most general case laws will involve the speed of light, the velocities $\vec{v}_1, \ldots, \vec{v}_n$ of a

[9]This problem arises also in the case of general relativity where a different set of restrictions again render the covariance principle non-empty. (e.g.: apart from the energy tensor $T_{\mu\nu}$, only the $g_{\mu\nu}$'s and their first and second order derivatives can occur in the field equations.)

finite number of processes and some other quantities a, b, ... If R = 0 and K = 0 are the relativistic and classical laws respectively, we require that:

$$R \to K, \text{ as } (v_1/c, v_2/c, \ldots, v_n/c) \to (0, 0, \ldots, 0).$$

There are at least two ways of letting v_m/c tend to zero for m = 1, 2, . . . , n. We first take c to be a constant and let $(v_1, \ldots, v_n)$ approach zero. In this case, put $\vec{w}_m = \vec{v}_m/c$ for all m = 1, 2, . . . , n and consider both R and K as functions of c, $\vec{w}_1, \ldots, \vec{w}_n$, a, b, ...

In other words we write:

$$R = R(c, \vec{w}_1, \ldots, \vec{w}_n, a, b, \ldots) \text{ and}$$

$$K = K(c, \vec{w}_1, \ldots, \vec{w}_n, a, b, \ldots)$$

We then make:

$$R(c, \vec{w}_1, \ldots, \vec{w}_n, a, b, \ldots) - K(c, \vec{w}_1, \ldots, \vec{w}_n, a, b, \ldots)$$

approach zero as $w_1, \ldots, w_n$ simultaneously tend to zero. It is of course tacitly assumed that R and K are continuous functions. Hence the difference

$$[R(c, \vec{w}_1, \ldots, \vec{w}_n, a, b, \ldots) - K(c, \vec{w}_1, \ldots, \vec{w}_n, a, b, \ldots)]$$

approaches:

$$[R(c, 0, \ldots, 0, a, b, \ldots) - K(c, 0, \ldots, 0, a, b, \ldots)]$$

as $(w_1, \ldots, w_n)$ approaches (0, . . . , 0). Thus the second requirement reduces to the equation:

$$R(c, 0, \ldots, 0, a, b, \ldots) = K(c, 0, \ldots, 0, a, b, \ldots).$$

Thus, in this first case, the function R which is to be determined will be subjected to the following two conditions:

(1′) $[R(c, \vec{w}_1, \ldots, \vec{w}_n,$

$a, b, \ldots) = 0] \Leftrightarrow [R(c, \vec{w}'_1, \ldots, \vec{w}'_n,$

$a', b', \ldots) = 0]$

(Relativity Principle)

(2) $R(c, 0, \ldots, 0, a, b, \ldots) = K(c, 0, \ldots, 0, a, b, \ldots)$

(Requirement that the classical law be a limiting case of the new law)

We recall that:

$$R(c, \vec{w}_1, \ldots, \vec{w}_n, a, b, \ldots) \underset{\text{Def}}{\equiv} R(c, \vec{v}_1/c, \ldots, \vec{v}_n/c, a, b, \ldots) = 0$$

is the relativistic law which is to replace the classical equation:

$$K(c, \vec{w}_1, \ldots, \vec{w}_n, a, b, \ldots) \underset{\text{Def}}{\equiv} K(c, \vec{v}_1/c, \ldots, \vec{v}_n/c, a, b, \ldots) = 0.$$

If the relativistic law holds good in general, it will in particular be true for $\vec{w}_1 = \ldots = \vec{w}_n = 0$. By (2) it follows that:

(3) $K(c, 0, \ldots, 0, a, b, \ldots) = 0$

This last question means that, when $\vec{v}_1, \ldots, \vec{v}_n$ vanish, the relativistic law collapses into the classical one, which must therefore hold good, at least in this particular case.[10]

There is a second way of making $v_1/c, \ldots, v_n/c$ tend to zero, namely by treating c as a variable parameter, while fixing the velocities $\vec{v}_1, \ldots, \vec{v}_n$, and by then letting c tend to infinity.[11] Putting $c = 1/\alpha$, we can write:

$$R = R_o(\alpha, \vec{v}_1, \ldots, \vec{v}_n, a, b, \ldots) \text{ and}$$
$$K = K_o(\alpha, \vec{v}_1, \ldots \vec{v}_n, a, b, \ldots)$$

We now require that:

$$[R_o(\alpha, \vec{v}_1, \ldots, \vec{v}_n, a, b, \ldots) - K_o(\alpha, \vec{v}_1, \ldots, \vec{v}_n, a, b, \ldots)] \rightarrow 0,$$

as $c \rightarrow \infty$, i.e. as $\alpha = (1/c) \rightarrow 0$.
Assuming that R_o and K_o are continuous, we obtain:

(3′) $R_o(o, \vec{v}_1, \ldots, \vec{v}_n, a, b, \ldots) = K_o(o, \vec{v}_1, \ldots, \vec{v}_n, a, b, \ldots)$[12]

One last way of meeting the requirement that R should tend to K as $(v_1/c, \ldots, v_n/c)$ tends to zero is to assume that, where K is some function

---

[10] Both Einstein and Planck assumed that Newton's second law of motion holds good when the velocity vanishes.

[11] Note that, as $c \rightarrow \infty$, the Lorentz transformation collapses into the Galilean one.

[12] The relativistic law of the conservation of momentum $\Sigma\, m_i \vec{v}_i/\sqrt{1 - v_i^2/c^2} = 0$ is a good illustration of equation (3′). $\Sigma m_i \vec{v}_i \sqrt{1 - v_i^2/c^2} \rightarrow \Sigma m_i \vec{v}_i$, as $c \rightarrow \infty$. Letting $(v_1, \ldots, v_n)$ tend to zero serves no purpose in this case; for if we start from an arbitrary function $f(v_i)$ and consider $\Sigma\, f(v_i)\vec{v}_i$ then: as $(v_1, \ldots, v_n) \rightarrow (o, \ldots, o)$, $\Sigma\, f(v_i)\vec{v}_i \rightarrow o$ = value of $\Sigma m_i \vec{v}_i$ for $v_1 = v_2 = \ldots = v_n = \mathbf{0}$. This does not help us towards determining f. (See below, chapter seven)

of certain classical quantities, R should be the *same* function of the corresponding relativistic quantities. Then, if R and K are continuous and if each relativistic quantity tends to the corresponding classical one, it follows that:

$$R \rightarrow K, \text{ as } (v_1/c, \ldots, v_n/c) \rightarrow (\mathbf{0}, \ldots, \mathbf{0}).$$ [13]

§3.4 The Transformation Laws for the Electromagnetic Field

Einstein's derivation of the transformation laws for the field quantities $\vec{E}$, $\vec{H}$ and ρ illustrates the way in which the relativity postulate (i.e. requirement (i) above) can be heuristically exploited. In section II of his EMB Einstein starts from Maxwell's equations which he regards as expressing genuine laws of nature. They should consequently assume the same form in all inertial frames; i.e. they must be Lorentz-covariant. Note that Maxwell's equations trivially constitute a limiting case of a classical theory; for they are classical and will moreover be left intact by special relativity. Requirement (ii) is thus automatically fulfilled.

Einstein proceeded as follows. He started by considering a special case, namely Maxwell's theory for the vacuum; that is, he considered the case where $\rho = 0$. Let $\xi(\vec{E}, \vec{H}; x, y, z, t)$ stand for the whole set of Maxwell's equations, where $\vec{E}$ and $\vec{H}$ denote the electric and magnetic fields respectively. $\xi(\vec{E}, \vec{H}; x, y, z, t)$ are supposed to hold in some inertial frame K, with respect to which another frame K′ moves with uniform velocity $\vec{v} = (v, 0, 0)$. If we change the independent variables from x, y, z, t (in K) to x′, y′, z′, t′ (in K′), then $\xi(\vec{E}, \vec{H}; x, y, z, t)$ will be transformed into $\zeta(\vec{E}, \vec{H}; x', y', z', t', v)$ say. Thus:

(1) $\xi(\vec{E}, \vec{H}; x, y, z, t) \Leftrightarrow \zeta(\vec{E}, \vec{H}; x', y', z', t', v).$

By the relativity principle, we know that the field quantities $\vec{E}'$ and $\vec{H}'$ satisfy $\xi(\vec{E}', \vec{H}'; x', y', z', t')$ iff $\vec{E}$ and $\vec{H}$ satisfy $\xi(\vec{E}, \vec{H}; x, y, z, t)$. That is:

(2) $\xi(\vec{E}', \vec{H}', x', y', z', t') \Leftrightarrow \xi(\vec{E}, \vec{H}; x, y, z, t).$

It follows from (1) and (2) that:

(3) $\xi(\vec{E}', \vec{H}'; x', y', z', t') \Leftrightarrow \zeta(\vec{E}, \vec{H}; x', y', z', t', v).$

---

[13] Denote by μ_i the relativistic mass $m_i/\sqrt{1 - v_i^2/c^2}$. The relativistic momentum $\Sigma\, \mu_i \vec{v}_i$ and the classical momentum $\Sigma m_i \vec{v}_i$ are the same functions of the masses and of the velocities. Lewis and Tolman (tacitly) assumed that $\mu_i \rightarrow m_i$ as $v_i \rightarrow \mathbf{0}$. (See below chapter seven.)

This last equivalence offers the advantage of containing the same independent variables—namely: x′, y′, z′ and t′—in both of its members. From the requirement that (3) must hold for all values of x′, y′, z′ and t′, Einstein determined $\vec{E}'$ and $\vec{H}'$ in terms of $\vec{E}$, $\vec{H}$ and v.

Let us now give Einstein's derivation in greater detail. Maxwell's equations for empty space can be written as:

$$(4)\quad \nabla \cdot \vec{E} = 0 = \nabla \cdot \vec{H};\ \nabla \times \vec{H} = \frac{1}{c}\frac{\partial \vec{E}}{\partial t};\ \nabla \times \vec{E} = -\frac{1}{c}\frac{\partial \vec{H}}{\partial t}$$

where we put $\vec{E} = (X, Y, Z)$, $\vec{H} = (L, M, N)$. As usual, ∇ stands for the operator $\left(\frac{\partial}{\partial x}, \frac{\partial}{\partial y}, \frac{\partial}{\partial z}\right)$. Since equations (4) are Lorentz-covariant, we can immediately infer that, in the 'moving' frame K′, the following must hold:

$$(5)\ \nabla' \cdot \vec{E}' = 0 = \nabla' \cdot \vec{H}';\ \nabla' \times \vec{H}' = \frac{1}{c}\frac{\partial \vec{E}'}{\partial t'};$$

$$\nabla' \times \vec{E}' = -\frac{1}{c}\frac{\partial \vec{H}'}{\partial t'}$$

where $\vec{E}' = (X', Y', Z')$; $\vec{H}' = (L', M', N')$.

The coordinates x′, y′, z′, t′ are given by the Lorentz transformation:

$$(6)\ x' = \gamma\,(x - vt);\ y' = y;\ z' = z;\ t' = \gamma(t - \frac{v}{c^2}x);$$

where $\gamma \underset{\text{Def}}{=} \left(1 - \frac{v^2}{c^2}\right)^{-1/2}$

It follows from (6) that:

$$(7)\begin{cases} \dfrac{\partial}{\partial x} = \left(\gamma \dfrac{\partial}{\partial x'} - \dfrac{v\gamma}{c^2}\dfrac{\partial}{\partial t'}\right);\ \dfrac{\partial}{\partial y} = \dfrac{\partial}{\partial y'};\ \dfrac{\partial}{\partial z} = \dfrac{\partial}{\partial z'}; \\ \dfrac{\partial}{\partial t} = \left(\gamma\dfrac{\partial}{\partial t'} - v\gamma\dfrac{\partial}{\partial x'}\right). \end{cases}$$

In (4) let us now change the independent variables from x, y, z, t to x′, y′, z′, t′. The first two equations of (4) become:

$$
\textbf{(8)}\begin{cases} 0= \gamma \dfrac{\partial X}{\partial x'} + \dfrac{\partial Y}{\partial y'} + \dfrac{\partial Z}{\partial z'} - \dfrac{v\gamma}{c^2}\dfrac{\partial X}{\partial t'} \text{ and} \\ \\ 0= \gamma \dfrac{\partial L}{\partial x'} + \dfrac{\partial M}{\partial y'} + \dfrac{\partial N}{\partial z'} - \dfrac{v\gamma}{c^2}\dfrac{\partial L}{\partial t'}. \end{cases}
$$

Remember that each vector equation stands for three relations between the scalar components of the vectors. Thus, the first component of the third equation in (4) is:

$$\frac{\partial N}{\partial y} - \frac{\partial M}{\partial z} = \frac{1}{c}\frac{\partial X}{\partial t}$$

which becomes (in virtue of (7)):

$$\textbf{(9)}\quad \frac{\partial N}{\partial y'} - \frac{\partial M}{\partial z'} = \frac{1}{c}\left(\gamma \frac{\partial X}{\partial t'} - v\gamma \frac{\partial X}{\partial x'}\right).$$

The corresponding equation in system (5) is:

$$\frac{\partial N'}{\partial y'} - \frac{\partial M'}{\partial z'} = \frac{1}{c}\frac{\partial X'}{\partial t'}$$

In view of this relation, we propose to put (9) in the form:

$$\frac{\partial A}{\partial y'} - \frac{\partial B}{\partial z'} = \frac{1}{c}\frac{\partial D}{\partial t'}$$

where A, B, D are functions of the coordinates. That is, we propose to eliminate $\dfrac{\partial}{\partial x'}$ from (9). For this we turn to (8) which yields:

$$\frac{\partial X}{\partial x'} = -\frac{1}{\gamma}\left(\frac{\partial Y}{\partial y'} + \frac{\partial Z}{\partial z'} - \frac{v\gamma}{c^2}\frac{\partial X}{\partial t'}\right).$$

Substituting in (9):

$$\frac{\partial}{\partial y'}\left[\gamma\left(N - \frac{v}{c}Y\right)\right] - \frac{\partial}{\partial z'}\left[\gamma\left(M + \frac{v}{c}Z\right)\right] = \frac{1}{c}\frac{\partial X}{\partial t'}.$$

We similarly derive the following:

$$\frac{\partial L}{\partial x'} - \frac{\partial}{\partial z'}\left[\gamma\left(N - \frac{v}{c}Y\right)\right] = \frac{1}{c}\frac{\partial}{\partial t'}\left[\gamma\left(Y - \frac{v}{c}N\right)\right]$$

$$\frac{\partial}{\partial x'}\left[\gamma\left(M + \frac{v}{c}Z\right)\right] - \frac{\partial L}{\partial y'} = \frac{1}{c}\frac{\partial}{\partial t'}\left[\gamma\left(Z + \frac{v}{c}M\right)\right]$$

$$(10)\quad \begin{cases} \dfrac{\partial}{\partial z'}\left[\gamma\left(Y - \dfrac{v}{c}N\right)\right] - \dfrac{\partial}{\partial y'}\left[\gamma\left(Z + \dfrac{v}{c}M\right)\right] = \dfrac{1}{c}\dfrac{\partial L}{\partial t'} \\ \dfrac{\partial}{\partial x'}\left[\gamma\left(Z + \dfrac{v}{c}M\right)\right] - \dfrac{\partial X}{\partial z'} = \dfrac{1}{c}\dfrac{\partial}{\partial t'}\left[\gamma\left(M + \dfrac{v}{c}Z\right)\right] \\ \dfrac{\partial X}{\partial y'} - \dfrac{\partial}{\partial x'}\left[\gamma\left(Y - \dfrac{v}{c}N\right)\right] = \dfrac{1}{c}\dfrac{\partial}{\partial t'}\left[\gamma\left(N - \dfrac{v}{c}Y\right)\right]. \end{cases}$$

Now, it proves useful to rewrite the last two (vector) equations of (5) in terms of scalar components as follows:

$$(11)\quad \begin{cases} \dfrac{\partial N'}{\partial y'} - \dfrac{\partial M'}{\partial z'} = \dfrac{1}{c}\dfrac{\partial X'}{\partial t'} & \dfrac{\partial Y'}{\partial z'} - \dfrac{\partial Z'}{\partial y'} = \dfrac{1}{c}\dfrac{\partial L'}{\partial t'} \\ \dfrac{\partial L'}{\partial z'} - \dfrac{\partial N'}{\partial x'} = \dfrac{1}{c}\dfrac{\partial Y'}{\partial t'} & \dfrac{\partial Z'}{\partial x'} - \dfrac{\partial X'}{\partial z'} = \dfrac{1}{c}\dfrac{\partial M'}{\partial t'} \\ \dfrac{\partial M'}{\partial x'} - \dfrac{\partial L'}{\partial y'} = \dfrac{1}{c}\dfrac{\partial Z'}{\partial t'} & \dfrac{\partial X'}{\partial y'} - \dfrac{\partial Y'}{\partial x'} = \dfrac{1}{c}\dfrac{\partial N'}{\partial t'}. \end{cases}$$

The two systems of equations (10) and (11) must be equivalent or, as Einstein put it, they must express the same thing. ((10) is in fact the proposition $\zeta\,(\vec{E}, \vec{H}; x', y', z', t', v)$ and (11) is $\xi\,(\vec{E}', \vec{H}'; x', y', z', t')$. I have already explained why the equivalence $\zeta\,(\vec{E}, \vec{H}) \Leftrightarrow \xi\,(\vec{E}', \vec{H}')$ obtains.) Hence there exists a quantity $\psi(v)$—which may depend on the global velocity v of the moving frame K′—such that:

$$(12)\quad \begin{cases} X' = \psi(v)X & L' = \psi(v)L \\ Y' = \psi(v)\gamma\left[Y - \dfrac{v}{c}N\right] & M' = \psi(v)\gamma\left[M + \dfrac{v}{c}Z\right] \\ Z' = \psi(v)\gamma\left[Z + \dfrac{v}{c}M\right] & N' = \psi(v)\gamma\left[N - \dfrac{v}{c}Y\right]. \end{cases}$$

At this point, Einstein remarks that the two frames of reference K and K′ are physically equivalent; i.e. K and K′ have symmetric properties. Note that $-\vec{v}$ is the velocity of K with respect to K′. Consequently, had we started from the system K′, then gone over to K, we would have derived the equation:

$$X = \psi\,(-v) \cdot X'.$$

Through comparison with the first equation in (12), we deduce that:

$$X = \psi(-v) \cdot X' = \psi(-v) \cdot \psi(v) \cdot X.$$

Hence:

(13) $\psi(v) \cdot \psi(-v) = 1.$

We propose to show that $\psi(v) = 1$.[14] Let I be any inertial frame in which we choose two coordinate systems $K(x, y, z, t)$ and $K_1(x_1, y_1, z_1, t_1)$, such that K_1 is obtained by rotating K through 180° about the z-axis. Thus:

(14) $x_1 = -x;\ y_1 = -y;\ z_1 = z;\ t_1 = t.$

Now consider a system $K'_1(x'_1, y'_1, z'_1, t'_1)$ moving with velocity $\vec{v} = (v, o, o)$ along the x_1-axis of K_1 and let $K''(x'', y'', z'', t'')$ be obtained by rotating K'_1 through 180° about the z_1-axis. Thus:

(15)
$$x'_1 = \gamma(x_1 - vt_1),\ y'_1 = y_1,\ z'_1 = z_1$$
$$t'_1 = \gamma\left(t_1 - \frac{v}{c^2}x_1\right)$$

and

(16) $x'' = -x'_1;\ y'' = -y'_1;\ z'' = z'_1;\ t'' = t'_1$

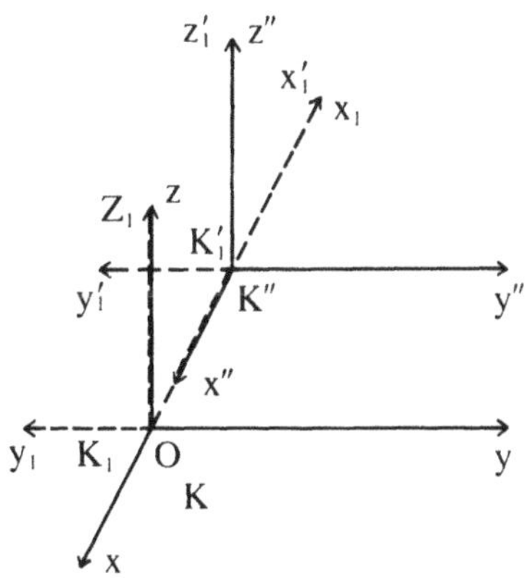

[14]What follows is a reconstruction of Einstein's statement: "Further, from reasons of symmetry, $\psi(v) = \psi(-v)$ and therefore $\psi(v) = 1$" (see Einstein et al. 1923, p. 53).

Obviously K″ moves with velocity $(-v,0,0)$ relatively to K. In K let there be an electromagnetic field given by: $\vec{E} = (0, 0, Z)$ and $\vec{H} = (o, o, o)$, where Z is some constant. This field has exactly the same components in K_1. Hence, by the transformation equation, we have in the system K'_1:

$$\vec{E}'_1 = (0, 0, Z'_1) \text{ and } \vec{H}'_1 = (0, M'_1, 0)$$

where:

$$Z'_1 = \psi(v) \cdot \gamma \cdot Z_1 = \psi(v) \cdot \gamma \cdot Z \text{ and}$$
$$M'_1 = \psi(v) \cdot \frac{v\gamma}{c} Z_1 = \psi(v) \cdot \frac{v\gamma}{c} Z.$$

Remember that K″ is obtained from K'_1 by changing the signs of the x- and y-coordinates, while leaving the other two coordinates unaltered. Thus:

$$\vec{E}'' = (0, 0, Z'') = (0, 0, Z'_1)$$

and

$$\vec{H}'' = (0, M'', 0) = (0, -M'_1, 0).$$

Hence:

$$Z'' = Z'_1 = \psi(v) \cdot \gamma \cdot Z \text{ and } M'' = -M'_1 = -\psi(v) \cdot \frac{v\gamma}{c} Z.$$

We have also remarked that K″ moves with the velocity $-v$ along the x-axis of K. Hence, by the transformation equations for the field:

$$Z'' = \psi(-v)\gamma Z \text{ and } M'' = \psi(-v) \cdot \left(\frac{-v\gamma}{c}\right) \cdot Z.$$

Equating the two expressions of Z″ (or the two expressions of M″), $\psi(v) \cdot \gamma \cdot Z = Z'' = \psi(-v) \cdot \gamma \cdot Z$, we obtain:

$$\psi(v) = \psi(-v).$$

By equation (13), it follows that:

$$[\psi(v)]^2 = 1; \text{ hence:}$$

$$\psi(v) = \pm 1.$$

As v tends to zero we want the system K′ to coincide with K. In particular, we want $X' = \psi(v) \cdot X$ to tend to X as v tends to zero. Therefore:

$\psi(v) = 1.$

Substituting 1 for $\psi(v)$ in (7), we obtain the transformation laws for the electromagnetic field in empty space:

$$(17)^{15} \begin{cases} X' = X & L' = L \\ Y' = \gamma\left(Y - \frac{v}{c} N\right) & M' = \gamma\left(M + \frac{v}{c} Z\right) \\ Z' = \gamma\left(Z + \frac{v}{c} M\right) & N' = \gamma\left(N - \frac{v}{c} Y\right) \end{cases}$$

At this point, it is interesting to examine how Einstein determines the transform ρ' of the charge density ρ. We have seen how Lorentz had made a mistake in transforming ρ: by not realistically interpreting his own concept of local time, he had failed to notice that simultaneity is a three-place relation between two events and a frame of reference. Two events which are simultaneous in one frame need not necessarily be so in another. We shall now see that, having done all the preliminary kinematical work in Part I of the EMB, Einstein did not even need to reconsider the notion of simultaneity in determining ρ.

Einstein turned to the field equations for the general case where convection currents are present:

$$(18) \begin{cases} \nabla . \vec{E} = \rho;\ \nabla . \vec{H} = 0;\ \nabla \times \vec{H} = \frac{1}{c} \cdot \left(\frac{\partial \vec{E}}{\partial t} + \rho \vec{u}\right) \\ \nabla \times \vec{E} = -\frac{1}{c}\frac{\partial \vec{H}}{\partial t} \\ \text{where: } \rho \text{ is the charge density and } \vec{u} = (u_1, u_2, u_3) = \text{charge velocity.} \end{cases}$$

He implicitly assumed that, in this more general case, $\vec{E}$ and $\vec{H}$ transform according to the same rules as those for empty space, i.e. accord-

[15]Note that Lorentz had found exactly the same transformation equations for $\vec{E}$ and $\vec{H}$ in his 1904 paper: "Electromagnetic phenomena in a system moving with any velocity less than that of light."

ing to equations (17). By the relativity postulate, $\nabla \cdot \vec{E} = \rho$ implies that $\nabla' \cdot \vec{E}' = \rho'$ holds in the moving frame. Since $\vec{E}'$ has already been determined, namely through equations (17), the transformation law for ρ' is thereby fixed as follows.

In view of $\nabla' \cdot \vec{E}' = \rho'$ and of $\nabla \cdot \vec{E} = \rho$ we have:

(19) $$\rho' = \frac{\partial X'}{\partial x'} + \frac{\partial Y'}{\partial y'} + \frac{\partial Z'}{\partial z'} \text{ and } \rho = \frac{\partial X}{\partial x} + \frac{\partial Y}{\partial y} + \frac{\partial Z}{\partial z}.$$

By the Lorentz transformation:

(20) $$\begin{cases} \dfrac{\partial}{\partial y'} = \dfrac{\partial}{\partial y};\ \dfrac{\partial}{\partial z'} = \dfrac{\partial}{\partial z};\ \dfrac{\partial}{\partial x'} = \left(\gamma\dfrac{\partial}{\partial x} + \dfrac{v\gamma}{c^2}\dfrac{\partial}{\partial t}\right) \text{ and, by (17):} \\ X' = X;\ Y' = \gamma\left(Y - \dfrac{v}{c}N\right);\ Z' = \gamma\left(Z + \dfrac{v}{c}M\right) \end{cases}$$

Substituting in (19):

(21) $$\begin{cases} \rho' = \left(\dfrac{\partial X'}{\partial x'} + \dfrac{\partial Y'}{\partial y'} + \dfrac{\partial Z'}{\partial z'}\right) = \left[\left(\gamma\dfrac{\partial}{\partial x} + \dfrac{v\gamma}{c^2}\dfrac{\partial}{\partial t}\right)X \right. \\ \qquad \left. + \dfrac{\partial}{\partial y}\left(\gamma\left(Y - \dfrac{v}{c}N\right)\right) + \dfrac{\partial}{\partial z}\left(\gamma\left(Z + \dfrac{v}{c}M\right)\right)\right] \\ \quad = \left[\gamma\left(\dfrac{\partial X}{\partial x} + \dfrac{\partial Y}{\partial y} + \dfrac{\partial Z}{\partial z}\right) + \dfrac{v\gamma}{c}\left(\dfrac{1}{c}\dfrac{\partial X}{\partial t} - \left(\dfrac{\partial N}{\partial y} - \dfrac{\partial M}{\partial z}\right)\right)\right] \\ \quad = \left[\gamma\rho + \dfrac{v\gamma}{c}\left(\dfrac{1}{c}\dfrac{\partial X}{\partial t} - \left(\dfrac{\partial N}{\partial y} - \dfrac{\partial M}{\partial z}\right)\right)\right] \end{cases}$$

By the third equation in (18):

(22) $$\begin{cases} \dfrac{\partial N}{\partial y} - \dfrac{\partial M}{\partial z} = \dfrac{1}{c}\dfrac{\partial X}{\partial t} + \dfrac{\rho}{c}u_1;\ \text{hence:} \\ \dfrac{1}{c}\dfrac{\partial X}{\partial t} - \left(\dfrac{\partial N}{\partial y} - \dfrac{\partial M}{\partial z}\right) = -\dfrac{\rho}{c}u_1 \end{cases}$$

Substituting from (22) into (21), we finally obtain:

(23) $$\rho' = \left[\gamma\rho - \frac{v\gamma\rho}{c^2}u_1\right] = \gamma\rho\left(1 - \frac{vu_1}{c^2}\right)$$

This is the required transformation for the charge density.

With this expression for ρ′ together with the transformation law for the velocity $\vec{u}$, it can be easily verified that Maxwell's theory is Lorentz-covariant, that is:

$$(24)\quad \begin{cases} \nabla' \cdot \vec{E}' = \rho'; \ \nabla' \cdot \vec{H}' = o; \ \nabla' \times H' = \dfrac{1}{c}\left(\dfrac{\partial \vec{E}'}{\partial t'} + \rho' \vec{u}'\right) \\ \nabla' \times \vec{E}' = -\dfrac{1}{c}\dfrac{\partial \vec{H}'}{\partial t'}. \end{cases}$$

This example illustrates the power of Einstein's revolutionary methods. Once the new kinematics had been set up, it cost Einstein very little effort to determine the transformation laws, first for the field and then for the charge density. In other words: by first fixing the space-time structure of STR, Einstein avoided the pitfalls which prevented Lorentz from correctly interpreting the transformed density. There is a striking difference between the efficient simplicity of the heuristic of STR and the laborious constructive methods used by Lorentz in obtaining results similar to Einstein's. It is this heuristic superiority of Einstein's methods which finally persuaded his contemporaries to start working on the relativity programme.

— 4 —

MACH AND EINSTEIN

§4.1 Mach's Alleged Influence on Einstein

The specific problem to which this chapter is addressed is: did Mach's philosophy of science play a significant role in the genesis of relativity theory? The thesis that it did is widely held and was given a sharp formulation by Kenneth Schaffner in an article published in 1974.

This problem is part of a much wider issue: did positivism exert a positive influence on the progress of modern science? Or a baneful influence? Or practically no influence at all? There is a widely held view, voiced mainly by Born and Bridgman, that the modern scientific revolution is the outcome of a positivistic revolution in the philosophy of science. Against this view I propose to show that, except in a restricted and purely negative sense to be explained later (see §4.5), positivism was largely irrelevant to the development of modern physics. While paying lip service to Machian positivism, scientists like Einstein remained old-fashioned realists. Had Einstein really adhered to the tenets of Machism, then special relativity would never have seen the light of day. The operationalist approach to scientific concepts, which supposedly revolutionised modern physics, is nothing more, in Einstein's case, than the proposal to treat certain definitions nominalistically. This vindicates Popper's view (Popper 1945, vol. 2, p. 14) that such nominal-stipulative definitions are empirically empty; they therefore make no difference to the theories in which they are imbedded.

Let me now go back to Schaffner's article, in which he criticised the explanation given above (chapter three) of the emergence of relativity and put forward an alternative operationalist account in the following terms:

> Zahar's account of both the genesis and the reason for the acceptance of Einstein's theory says essentially nothing about the reanalysis of the time concept and the redefinition of simultaneity. This is in marked contrast to how Einstein's contemporaries felt, and it is in part due to this omission of Zahar's that I believe

> his perspective on both the discovery of special relativity and the supersedure of Lorentz's theory of the electrodynamics of moving bodies is fatally flawed . . . A light signal synchronisation procedure could be introduced so that, taking two distant points A and B, we 'establish by *definition* that the time required by light to travel from A to B equals the time it requires to travel from B to A.' (Einstein's italics in his EMB [1905]). A definitional component in the reconstruction of the concept was required because the velocity of light in a *one way* direction cannot be ascertained in any *empirical* manner prior to a determination of distant simultaneity. Since this is an Einsteinian and not a Newtonian world, such a definition is *admissible* . . . I myself believe that the admissibility of Einstein's definition depends on the hypothesis of holding the independence from its source of the velocity of light . . .
>
> Einstein has noted . . . that the philosophical writings of Hume and Mach assisted him in his analysis of the concept of time . . . there is a methodological approach to the reanalysis of fundamental scientific concepts which one finds even in the early editions of Mach's *Science of Mechanics* which could not have failed to make some impression on Einstein, and which may well have served as an unconscious methodological guide for Einstein in his 1905 reanalysis of simultaneity. (Schaffner 1974, pp. 57–60)

For the purpose of my further discussion I should like to separate out two claims which Schaffner makes (both these claims, as we shall see, have been made by other authors like Bridgman):

1. A Machian analysis of the concepts of time and of simultaneity played an important part in the discovery of relativity.
2. Einstein's definition of simultaneity would be inadmissible in a Newtonian world.

These claims seem to be supported by the following passage from Einstein's article "Ernst Mach" which was published in 1916 shortly after Mach's death:

> The reader has already guessed that I am here alluding to certain concepts of the theories of space, time and mechanics which have undergone modifications through the Theory of Relativity. Nobody can take away from the epistemologists the merit of having helped the new developments in these areas; from my own experience I know that I have been, directly or indirectly, aided by epistemologists, especially by Hume and Mach . . .
>
> The quoted lines [from Mach's Mechanics] show that Mach clearly recognised the weak spots in Classical Mechanics and was not far from requiring a General Theory of Relativity, all this about half a century ago! It is not improbable that Mach would have come across Relativity Theory if, at the time when he was in his prime, physicists had concerned themselves with the significance of the constancy of the speed of light.
>
> In the absence of this stimulus which came later from Maxwell's and Lorentz's Electrodynamics, Mach's critical-epistemological needs were insufficient to arouse any feeling for the necessity of a definition of simultaneity for distant events. (Heller 1960, pp. 153-156; my translation)

In view of this quotation, it seems pointless to attack the non-problem of whether Machism exerted an important influence on Einstein's thinking; for Einstein himself admitted that it did and, after all, Einstein ought to know. However, the matter is not as simple as that.

Let me immediately say that I do not intend to examine the question whether, as a matter of psychological fact, Mach's philosophy was present to Einstein's mind when Einstein constructed his theory of relativity. There is absolutely no reason to doubt Einstein's own statement that, at the time he was evolving STR, he was under Mach's direct or indirect influence. We shall in fact see that Mach freed Einstein from certain 'absolutist' prejudices which would have hampered the development of the Relativity Programme (see § 4.5). However, our only concern at this stage is with the question whether there exists any objective connection between Mach's philosophy on the one hand and the theory of relativity as proposed by Einstein on the other. It is well known that scientists often make mistakes when giving accounts of their own intuitive methodologies. Newton may have sincerely believed that he had induced his theories from the facts; but it is obvious that the facts alone—the facts without theoretical admixtures—do not, for example, imply the absolute time hypothesis. Similarly, Einstein may have been mistaken in thinking that Machism helped him to discover relativity. The Machist elements of Einstein's thought may have been irrelevant to this discovery, or—more strikingly still—Einstein may even have used an approach alien to Machism. This latter possibility is clearly indicated by Mach's own rejection of relativity. In the Introduction to his book on the history of optics, Mach wrote:

> I gather from the publications which have reached me, and especially from my correspondence, that I am gradually becoming regarded as the forerunner of relativity. I am able even now to picture approximately what new expositions and interpretations many of the ideas expressed in my book on Mechanics will receive in the future from the point of view of relativity . . . I must, however, as assuredly disclaim to be a forerunner of the relativists as I withhold from the atomistic belief of the present day.
>
> The reason why, and the extent to which, I discredit the present-day relativity theory, which I find to be growing more and more dogmatical, together with the particular reasons which have led me to such a view—considerations based on the psychology of the senses, the theoretical ideas and, above all, the conceptions resulting from my experiments—must remain to be treated in the sequel. (Mach 1913, preface)

Given these contradictory accounts of the relationship between Machism on the one hand and the epistemological foundations of relativity on the other, I have decided to follow Einstein's own advice, namely to investigate what Mach and Einstein actually did and not what they said they did.

Before embarking on this investigation let me point out that the problem tackled in this chapter transcends its connection with a particular historical episode, namely the emergence of relativity theory. Mach's philosophy is strictly positivistic. It has been maintained—by Bridgman and Born among others—that special relativity is the paradigm of a physical theory based on the positivistic requirement that all scientific concepts should be operationally defined. According to Bridgman:

> The important step that Einstein made was that in analysing the connection of the equations with the theory he was led to examine the details of what we do in applying the equations in any specific case. In particular, one of the variables in the equations was the time—what do we do in obtaining the number which replaces the general symbol for time when we apply the equations to a concrete case? As physicists we know that this number is obtained by reading a clock of some sort, that is, it is a number given by a prescribed physical operation. Or the equations may involve the times of two different events in two different places, and to understand completely what is now involved we must analyse what we do in determining the time of two such events. Analysis shows that we read two clocks, one at each place. A new element now enters, because a complete description of all the manipulations involved demands that we set up some method by which we compare the two clocks with which we measure the two events. Out of the examination of what we do in comparing the clocks came, we all know, Einstein's revolutionary recognition that the property of two events which hitherto had been unthinkingly called simultaneity involves in the doing a complicated sequence of physical operations which cannot be uniquely specified unless we specify who it is that is reading the clocks. We know that a consequence of this is that different observers do not always get the same results, so that simultaneity is not an absolute property of two events but is relative to the observing system . . . (Bridgman 1936, pp. 7-8)

Through this operational definition of coordinate time, Einstein is supposed to have set an example for physicists like Heisenberg and Born, who subsequently developed quantum mechanics. Heisenberg allegedly followed Einstein's lead when he decided to construct his new mechanics as a calculus of observables. In a series of letters to Einstein, where he expresses puzzlement over Einstein's rejection of the Copenhagen interpretation of quantum mechanics, Born wrote:

> I was an unconditional follower and apostle of the young Einstein and swore by his theories; I could not imagine that the old Einstein thought differently. Einstein had based Relativity Theory on the principle that concepts which refer to unobservables have no place within physics: a fixed point in empty space and the absolute simultaneity of two distant events are such [unacceptable] concepts. Quantum Theory arose when Heisenberg applied this same theory to the electronic structure of the atom. This was a bold fundamental step, which at once struck me as self-evident and led me to put all my powers at the service of this new idea. Obviously, I found it impossible to grasp that Einstein refused to accept the validity, within Quantum Theory, of this principle which he had very successfully used himself . . . In a previous letter he [Einstein] had expressed

his views by saying that he was averse to the philosophy of the "esse est percipi". (Einstein, Born & Born 1969, pp. 299–300; my translation)

Thus, according to Born, we owe both relativity theory and quantum mechanics to a positivistic approach which was successfully initiated by Einstein. Since relativity and quantum mechanics form the cornerstone of modern physics, it follows from Born's claim that modern physics as a whole is the result of a positivistic revolution in philosophy. Hence if I can show that, on the contrary, special relativity has essentially nothing to do with positivism, then the positivist view of the emergence of modern science will be strongly undermined; this is so, especially because special relativity supposedly acted as a model which was emulated by scientists working in different fields.

Since the thesis of Mach's influence has been so clearly formulated by Schaffner, let me readdress myself to his claims as formulated above. Against Schaffner's claims I shall defend the following two theses:

(1′) If one adopts a Machian critique of scientific *concepts* as opposed for example to a Popperian critique of scientific *propositions*—then both Einstein's definition of coordinate time and special relativity as a whole prove unacceptable on more than one count. Hence, Mach was right in rejecting relativity as incompatible with his own epistemological viewpoint.

(2′) However, if one adopts a propositional approach to scientific hypotheses, then Einstein's definition of time is clearly recognised for what it really is: namely a stipulative nominal definition which, being nearly empty, is compatible with all rival hypotheses. (Popper 1945, vol. 2, p. 14)

It will be shown not only that Einstein's definition of time is compatible with classical physics but also that all classical theories can be reformulated in terms of the time variable as determined by Einstein's synchronisation convention. Needless to say, classical theories, when thus reformulated, assume a form more complicated than the usual one. This result is the obverse of Grünbaum's (Grünbaum 1973, ch. 12). Grünbaum showed that Einstein's definition of time is the most convenient one for special relativity. It is hardly surprising that the same definition should not be the simplest one for classical physics.

It follows from (1′) and (2′) that Einstein's analysis of the time *concept* alone (i.e., independently of certain *propositions* which form part of Einstein's *theory*) could not have led to special relativity. This result supports Popper's general view that propositions and not concepts play a primary role in the progress of science. Moreover, my claim is perfectly compatible with Grünbaum's correct thesis that, in the light of relativity theory, Einstein's particular choice of the time variable is the most convenient one. Taken together with Grünbaum's thesis, my claim entails that the main merit of Einstein's definition is the *purely pragmatic one* of possessing excess symbolic simplicity over rival contentions.

After vindicating these two theses I shall go on to argue that:

(3′) Although the influence of Machism on special relativity was negligible, there is a sense in which the general theory embodies Mach's approach to certain physical hypotheses. We shall see that Mach came very near to requiring a condition of general covariance for certain theories.

This presents us with the following paradox. How is it that an extreme form of positivism like Machism influenced a highly speculative theory like general relativity but not its forerunner, namely special relativity, which appears far more acceptable from an empiricist point of view? Special relativity is a severely tested theory whose concepts seem susceptible of operational definition. As opposed to this, the empirical basis of general relativity is slight and the connection between the coordinates and any process of direct measurement (which is an important feature of special relativity) seems largely lost. In order to resolve this paradox, I shall distinguish between Mach's general positivistic philosophy—henceforth referred to as Machism—and the intuitive methodology which he used in criticising Newtonian mechanics. It turns out that Machism played hardly any part in the genesis either of the special or of the general theory of relativity; however, Mach's intuitive methodology did guide Einstein in the construction of his gravitational theory (i.e., general relativity).

§4.2 A Brief Exposition of Machian Positivism: Operationalism and Mach's Vicious Circle Principle

As a preliminary to showing that it played a negligible role in the genesis of relativity, I shall give a brief exposition of Machism, i.e., of Mach's explicitly articulated brand of positivism.

Mach's ontology consists of the so-called *elements* which are the simple component parts of sensations. Colours, smells and shapes are typical elements of sensations. All elements are interconnected. The task of science is to establish relations between the elements in such a way that economy of thought is maximised. Mach assumed that such elements can be dealt with in a quantitative way (Mach 1906, chs. 1 & 14). Let $a_1, a_2, \ldots, a_m$ be numerical measures of elements of sensation; then science aims at constructing functions $f_1(x_1, \ldots, x_m), f_2(x_1, \ldots, x_m), \ldots, f_n(x_1, \ldots, x_m)$ such that:

(1) $f_i(a_1, \ldots, a_m) = 0$ for all $i = 1, 2, \ldots, n;$

m will in general be larger than n so that, given any (m − n) among the m quantities $a_1, \ldots, a_m$, equations (1) enable us to compute the remaining n ones. This is what Mach means by saying that one important task of science is the completion of facts in thought: if we know (m − n) facts, e.g. the values of $a_1, \ldots, a_{m-n}$ in (1), then the equations enable us to compute $a_{m-n+1}, \ldots, a_m$, i.e. to complete—or extend—the set $\{a_1, \ldots, a_{m-n}\}$ to the set $\{a_1, \ldots, a_m\}$.

Mach uses this conceptual approach in order to replace the asymmetric relation of cause to effect by a symmetric relation: namely the relation of the functional interdependence between the various elements of sensation. Given equations (1), a change in *any* (m − n) among the m elements $a_1, \ldots, a_m$ *causes* a change in the remaining ones. The division into cause and effect is therefore arbitrary. For example: given Boyle's law that, for a fixed mass of gas at constant temperature, $pv = k$, we can say that a variation of pressure causes a corresponding variation of volume and *vice versa*.

Let us note that each of the m quantities $a_1, \ldots, a_m$ refers to an element of sensation, i.e., to an observable. Thus an observable effect must have observable causes: only observables can act on observables. For example, according to Mach, the oblateness of the earth is caused not by the earth's acceleration in absolute space but by its rotation relatively to the observable stars. (This conclusion will play an important part in the genesis of general relativity.)

Let us go back to equations (1) and to the requirement that science should maximise economy of thought. The functions $f_1, f_2, \ldots, f_n$ must therefore be simple in the strictly pragmatic sense of minimising intellectual discomfort. For example, equations (1) should enable us to express $a_{m-n+1}, \ldots, a_m$ in terms of $a_1, \ldots, a_{m-n}$ with the least expenditure of effort. Noting that $a_1, \ldots, a_m$ are measures of elements of sensations, i.e. of observables, we conclude that, according to Mach, only the arguments of the f_i but not the f_i themselves possess ontological status. The functions f_i, i.e. all scientific laws, exist only in our minds and are thus subjective, or at any rate dependent on social and cultural circumstances. It follows that the simplicity of scientific laws has no objective status. This consequence of Machism will play a significant role in connection with the covariance principle. In his diary (published by Dingler) Mach wrote:

> [A law is] Only [the] completion of experience and applicable only as long as experience remains the same. Confirmation and refutation. Completion. Abstraction from present conditions. *That which is constant, the rule* [*is*] *an anchoring point for us* [*and*] *does not exist outside our thought* [*mind*]. (Dingler 1926, p. 100; my translation)

Let us now turn to Mach's operationalist view of scientific concepts. Mach maintained that definitions in science should fulfill the following conditions:

(i) The definition of a scientific concept (or quantity) should provide us with a concrete method for deciding whether or not the concept applies (in the case of physical quantities, the definition should furnish a means of measuring the quantity in question). The meaning of a scientific concept is the sequence of operations (i.e. the set of sensations) which constitutes a decision procedure for the applicability of the concept. Grünbaum correctly pointed out that operationalism in this sense confuses pragmatics with semantics (Grünbaum 1954). However, Mach's approach to definitions is definitely in line with his ontology. A concept or quantity is metaphysical if it is not operationally defined; hence every notion which refers to hidden entities (i.e. to entities inaccessible to direct observation) is metaphysical. One aim of mature science is to purge itself of all metaphysical notions. Mach implicitly assumed that science would thereby become more economical. He did not seriously consider the possibility that the use of metaphysical concepts might, even in the long run, prove indispensable for maximising economy of thought.

According to Mach, theories are metaphysical if they contain notions which are not operationally defined. Thus the status of theories is determined by that of the concepts which occur in them. In this sense, concepts play a primary role and theories a derivative one in Mach's methodology.

This conceptual—as opposed to propositional—approach, had another important effect on Mach's philosophy of science. It led him to try to eliminate the asymmetry between cause and effect in scientific explanation. For example, in examining Boyle's law, Mach concentrated on the *concepts* of pressure and volume. Since changes in p and v co-determine each other, either of them can be regarded as cause or effect. Mach concluded that, in *all* cases, the cause-effect relation can be replaced by the symmetric relation of functional dependence. The propositional approach reinstates the asymmetry between cause and effect as follows. Only states of affairs can be causes of other states of affairs and states of affairs are described, not by concepts, but by propositions. Let T, I, P stand for: scientific theory, initial conditions and prediction respectively. Given T, I will be a cause of P if $\vdash T \Rightarrow (I \Rightarrow P)$. Even though this relation may hold, it does not of course necessarily follow that $\vdash T \Rightarrow (P \Rightarrow I)$. The asymmetry between the cause I and the effect P is (partly) explicated by the asymmetry of the functor $\Rightarrow$, which connects *propositions*. This is not to deny that cause and effect may in some cases be interchangeable, in the sense that in these cases $T \Rightarrow (I \Leftrightarrow P)$ also holds. However, such symmetry *need not* obtain. Although Einstein paid lip service to Mach's view that the only task of science is to introduce order into our sense-perceptions (Einstein 1922, pp. 1-2), i.e. to set up functional relations between elements of sensations, relativity theory is a *causal* theory in the traditional sense of the word. However, in connection with the analogy between classical and relativistic causality, a note of warning should be sounded. Only in classical physics can one properly speak of *initial* conditions obtaining at $t = 0$, say. In GTR, the situation is somewhat different: once the *boundary* conditions on some Cauchy surface are consistently given, then Einstein's field equations uniquely determine the solution, naturally to within an arbitrary choice of some coordinate system. Of course, this system can be so defined that the Cauchy surface represents the equation $x^0 = 0$, i.e. $t = 0$. So, within that frame, one can speak of the initial conditions being given; but this mode of speech, being frame-dependent, has no absolute significance. All the same, GTR remains at least as rigidly deterministic as classical physics (Adler-Bazin-Schiffer 1965, ch. 7).

Apart from condition (i) above, Mach subjects the definitions of scientific terms to the following condition:

(ii) Definitions should be free of theoretical presuppositions; that is, all definitions should be theory-independent.

An immediate consequence of (ii) is:

(iii) The definition of scientific concepts should not depend on a theory containing the concept in question. Let us call this condition Mach's *vicious circle principle*. Let us say of two concepts A and B that *A involves B* if A is defined in the light of a hypothesis in which B occurs. Thus, Mach's vicious circle principle states that no concept should involve itself.

I propose to show that, although Mach did not and indeed could not satisfy (ii), although he could not free his definitions from *all theoretical* assumptions, he did manage to satisfy (iii) and hence avoid vicious circularity, at least in some cases; e.g. in his redefinition of mass. I shall now give both Mach's definition of mass and Einstein's definition of coordinate time, then demonstrate that the two definitions are radically different.

In connection with classical mechanics Mach's intention was to eliminate, without any loss of empirical content, all metaphysical notions from Newtonian theory. The main problem was posed by Newton's second law $\vec{F} = m\vec{a} = md^2\vec{r}/dt^2$, which connects four distinct notions: force, mass, space and time.

Since force, if treated as existing independently of other quantities, is unobservable, Mach's positivism enjoined him to treat the second law *as a definition* of force. Because distance and time are in principle measurable, the remaining problem was to give an independent operational definition of mass. Newton had taken mass to be the measure of the quantity of matter in a given body, but this was precisely the kind of metaphysical definition which Mach abhorred. Mach envisaged measuring mass by means of balance scales. This meant defining mass in terms of weight. However, weight is nothing but gravitational force and force is *mass × acceleration*. We are thus caught in a vicious circle and condition (iii) is violated. This is why Mach abandoned this approach and proposed instead the following solution. Take two bodies A and B; refer them to the frame determined by the stars; remove them from all neighbouring sources of external interference; then define the mass ratio of A and B as the negative inverse ratio of the accelerations which A and B induce in each other. Thus: $m_A/m_B = -\vec{a}_{BA}/\vec{a}_{AB}$, where m_A/m_B = ratio of the mass of A to that of B, $\vec{a}_{BA}$ = acceleration induced in B by A, $\vec{a}_{AB}$ = acceleration induced in A by B.

Mach wrote:

> [I shall] show how I think the conception of mass can be quite scientifically developed. The difficulty of this conception, which is pretty generally felt, lies, it seems to me, in two circumstances: (1) in the unsuitable arrangement of the first conceptions and theorems of mechanics; (2) in the silent passing over important presuppositions lying at the basis of the deduction.
>
> Usually people define $m = p/g$ and again $p = mg$. This is either a *very repugnant circle*, or it is necessary for one to conceive force as "pressure". The latter cannot be avoided if, as is customary, statics precedes dynamics. The difficulty, in this case, of defining the magnitude and direction of a force is well known.
>
> In that principle of Newton, which is usually placed at the head of mechanics, and which runs "Actioni contrariam semper et aequalem esse reactionem: sive corporum duorum actiones in se mutuo semper et aequales et in partes contrarias dirigi", the "actio" is again a pressure, or the principle is again unintelligible unless we already possess the conception of force and mass. But pressure looks very strange at the head of the quite phoronomical mechanics of today. However this can be avoided. (Mach 1909, p. 8a)

Concerning his final definition of the mass-ratio as the negative inverse of the acceleration-ratio Mach maintained:

> A special difficulty seems to be still found in accepting my definition of mass. Streintz . . . has remarked in criticism of it that it is based solely upon gravity, although this was expressly excluded in my first formulation of the definition (1868). Nevertheless, this criticism is again and again put forward, and quite recently even by Volkmann. My definition simply takes note of the fact that bodies in mutual relationship, whether it be of action at a distance, so called, or whether rigid or elastic connections be considered, determine in one another changes in velocities (accelerations). More than this one does not need to know in order to be able to form a definition with perfect assurance and without the fear of building on sand. It is not correct, as Höfler asserts, that this definition tacitly assumes *one and the same force* acting on both masses. It does not even assume the notion of force, since the latter is built up subsequently upon the notion of mass, and gives then the principle of action and reaction quite independently and without falling into Newton's logical error. In this arrangement, one concept is not misplaced on another which threatens to give way under it. (Mach 1893, ch. 2 p. x).

Thus Mach defined mass in terms of length and time, thereby avoiding all vicious circularity. This does *not* imply (as Mach sometimes seems to think it does) that he needed no theoretical assumptions at all. He assumed for example that the ratio $\vec{a}_{BA}/\vec{a}_{AB}$ is independent of the relative positions and velocities of A and B. However, he made use of *no* supposition involving the notion of mass. (Of course, he had to assume that the stars exert no appreciable influence on A and B. This premise involves a theoretical assumption in which the notion of mass might conceivably occur. Had Mach *not* been a positivist, he could have chosen his frame K to be an arbitrary inertial system, *without* bothering to tie K to any observables. In this way, his definition of mass would have become logically watertight; but the philosophical cost would have been too high: Mach would have had to jettison his operationalism.)

§4.3 Einstein's Violation of the Basic Tenets of Machism, in Particular of the Vicious Circle Principle. Einstein's Convention in Classical Physics

My intention now is to examine Einstein's definition of coordinate time. As we shall see, Einstein's approach, far from being Machian, involves a step of precisely the kind which Mach banned. Einstein starts his famous 1905 paper by proposing the following two postulates:

(P1): All laws of nature assume the same form in all inertial frames.

(CL): Light is always propagated in empty space with a definite velocity c which is independent of the state of motion of the emitting body.

An immediate consequence of P1 and (CL) is:

(Q): Let A and B be any two points at rest in any inertial frame. Then the velocity of light from A to B is equal to its velocity from B to A. Hence light takes the same time to travel from A to B as it does from B to A.

On the basis of (Q), Einstein gives his well-known convention for clock-synchronisation. Let C and C′ be two identical clocks respectively at rest at two points O and P of some inertial frame I. Without loss of generality, suppose that O is the origin of I. Let a light signal leave O at C-time t_1 and reach P at C′-time t′; the light beam is instantaneously reflected at P and goes back to O, where it arrives at C-time t_2. Supposing that C and C′ measure the coordinate times in their respective neighbourhoods, we have in view of (Q): $t' - t_1 = t_2 - t'$. i.e.

$$\textbf{(1)}\quad t' = \frac{1}{2}(t_1 + t_2)$$

Note that equation (1) is operationally decidable. The two clocks C and C′ are said to be synchronous if (1) holds for all values of t_1. If C and C′ are thus synchronised, then the coordinate time is determined by means of proposition (Q), which clearly involves the notion of velocity. But velocity is rate of change of position with respect to *time*. We are therefore caught in the sort of vicious circle which offends against condition (iii) above. Einstein thus violated one of Mach's cardinal methodological principles. This is hardly surprising; the two notions of time and space are so primitive that it seems impossible to define either in terms of something more fundamental. Mach thought that he could boil everything down to spatial measurements. He overlooked the fact that, in expressing time as a function of spatial measurement (e.g. of the earth's angle of rotation), he still needed the notion of simultaneity; and that simultaneity is irreducible to purely spatial relations.

There is another respect in which special relativity is unacceptable from a strictly Machian viewpoint. In his EMB Einstein assumes the existence of a set of inertial frames which are privileged over all others. Such frames are abstract entities; they are *not* determined by any relation they might bear to observables. Thus, if we eliminate from Newtonian theory an idle metaphysical component, namely the absolute space hypothesis, and speak instead of the absolute *set* of inertial frames, special relativity turns out to be no less 'absolute' than classical mechanics. Both theories postulate sets of unobservable inertial frames; the main difference between them is the difference between the groups of transformations mapping one inertial frame onto another; these are the Galilean group in the Newtonian case and the Lorentz group in the relativistic one. In STR for example, one method of resolving the twin paradox is by way of the different relations which the twins bear to the set of inertial frames. Hence an observable effect, namely the retardation of *one* of the twins' clocks, is explained in terms of a hidden cause, namely the existence of a set of abstract frames. Einstein himself was later to object to this 'absolute' character of his own theory. It is small wonder that Mach should have disowned relativity.

Let us turn to the circularity inherent in the relativistic definition of coordinate time. Einstein was aware of this circularity when he wrote the following dialogue between himself on the one hand, and an imaginary critic of his popular account of special relativity, on the other.

> I am very pleased with this suggestion [i.e. with the operational definition of simultaneity], but for all that, I cannot regard the matter as settled, because I feel constrained to raise the following objection: "Your definition would certainly be right if only I knew that the light by which the observer at M perceives the lightning flashes travels along the length AM as along the length BM. But an examination of this supposition would only be possible if we already had at our

disposal the means of measuring time. It would thus appear as though we were moving here in a logical circle."

After further consideration you can cast a somewhat disdainful glance at me—and rightly so—and you declare: "I maintain my previous definition nevertheless, because in reality it assumes absolutely nothing about light. There is only *one* demand to be made of the definition of simultaneity, that in every real case it must supply us with an empirical decision as to whether or not the conception that has to be defined is indisputable. That light requires the same time to traverse the path AM as for the path BM is in reality neither a *supposition nor a hypothesis* about the physical nature of light, but a *stipulation* which I can make of my own freewill in order to arrive at a definition of simultaneity." (Einstein 1920, pp. 22–23)

Einstein's answer to the objection he levels at his own definition is clearly correct *provided* we adopt a propositional, as opposed to a conceptual, view of scientific theories. Let the sentence 'C is synchronised with C′ at C-time t_1'—henceforth referred to as S(C, C′, t_1)—stand for equation (1) supplemented by an interpretation of t_1, t_2 and t′. The proposition S (C, C′, t_1) is empirically decidable. Thus Einstein's definition satisfies condition (i) above, but it gives rise to the following problem. Let us weaken Mach's methodological position with regard to definitions by eliminating conditions (ii) and (iii) and requiring only that condition (i) be satisfied. Is it then the case, as Schaffner claims, that, by trying to give a definition of time which fulfils (i), Einstein was led to his theory of relativity?

My answer is again in the negative. This is again hardly surprising, for Einstein himself admits that his definition is stipulative. Such nominal-stipulative definitions are nearly empty, hence compatible with all rival hypotheses. As regards the definition of scientific terms Popper wrote:

> The scientific use of definitions . . . may be called its *nominalist* interpretation as opposed to its Aristotelian or essentialist interpretation. In modern science, only nominalist definitions occur, that is shorthand symbols or labels are introduced in order to cut a long story short. And we can at once *see* from this that definitions do *not* play a very important part in science. For shorthand symbols can always, of course, be replaced by the longer expressions, the defining formula, for which they stand . . . Our scientific knowledge, in the sense in which this term may be properly used, remains entirely unaffected if we eliminate all definitions; the only effect is upon our language, which would lose, not precision, but merely brevity. (Popper 1945, vol. 2 p. 14)

Popper's position sounds somewhat extreme, but it is essentially correct. As regards relativity, I maintain not only that Einstein's definition of time is compatible with classical physics, but also that a classical physicist could adopt Einstein's convention for clock synchronisation—thereby obtaining a time variable t* different from the Galilian variable t—then reformulate all his theories in terms of t*. Evidently classical theories, when expressed in terms of t*, assume a more complicated form than when expressed in terms of t.

Let us assume that we live in a Newtonian universe and let OXYZ be an absolute frame determined, say, by an ether at rest. The ether frame OXYZ is inertial and is such that Maxwell's equations hold in it. Hence, in OXYZ light is propagated with a constant speed c in all directions. Let O'X'Y'Z' be another frame of reference sliding with constant velocity $\vec{v}$ along OX.

We have already established that, if Einstein's convention is adopted, the equations connecting the 'absolute' coordinates with the 'moving' ones are as follows:

$$\textbf{(2)}\ x' = \frac{1}{\mu}(x\text{-}vt);\ y' = \frac{1}{\psi}y;\ z' = \frac{1}{\psi}z;\ t' = \Omega\gamma^2(t - \frac{v}{c_2}x),$$

where: $\gamma \underset{\text{Def}}{=} (1 - v^2/c^2)^{-1/2}$; Ω is the retardation coefficient; μ and ψ are the longitudinal and transverse contraction coefficients respectively. To say that we live in a Newtonian world is to say that there exist neither contraction nor retardation effects; i.e. $\psi = \mu = \Omega = 1$. The last equations thus lead to the special transformation:

$$\textbf{(3)}\ x^* = (x - vt);\ y^* = y;\ z^* = z;\ t^* = \gamma^2(t - \frac{v}{c^2}x).$$

The round trip velocity of light along a direction making the angle θ with O'X' is given by (see chapter two):

$$\textbf{(4)}\ \bar{c}^*(\theta) = c\,[1 - (1 - \gamma^{-2})\,sin^2\theta]^{-1/2}$$

(Note that θ is the angle as measured in the moving frame).

We recall that the value of $\bar{c}^*(\theta)$ is independent of the adopted synchronisation procedure. In a Newtonian world $\bar{c}^*(\theta)$ naturally depends on θ, since the speed of light cannot in such a case be constant in all inertial frames.

Let us compare equations (3) with the Galilean transformation:

$$\textbf{(5)}\ x_G = x - vt;\ y_G = y;\ z_G = z;\ t_G = t.$$

It is hardly surprising that the spatial components of (3) (the first three equations of (3)) should coincide with those of (5). We have already seen that the transformation of the spatial variables does not depend on any convention but flows directly from real physical effects (see eq. (2)). It also comes as no surprise that the fourth equations in (3) and (5) differ widely from each other. This is because Einstein's synchronisation procedure, which yields $t^* = \gamma^2(t - \frac{v}{c^2}x)$, seems unnatural in a Newtonian world where the velocity $\bar{c}^*(\theta)$ is not the same for all θ. Under these circumstances, the choice of absolute time as coordinate time in the moving frame, i.e., the choice of $t_G = t$ instead of $t^* = \gamma^2(t - \frac{v}{c^2} x)$, *appears* to be the only allowable one. We can however eliminate t between (3) and (5) and thus obtain:

$$(6)\ x^* = x_G,\ y^* = y_G,\ z^* = z_G,\ t^* = t_G - \frac{v\gamma^2}{c^2} x_G.$$

Equations (6) establish a one-one linear correspondence between (x^*, y^*, z^*, t^*) and (x_G, y_G, z_G, t_G). All laws expressible in terms of the Galilean coordinates x_G, y_G, z_G, t_G are expressible in terms of x^*, y^*, z^*, t^*; and vice-versa, of course.

For a classical physicist therefore, the time variable t^* determined by Einstein's convention is just as legitimate as the Galilean variable t_G.

We can thus conclude that the role played by Machism in the emergence of special relativity is negligible. If one adopts a strictly Machist view of definitions, then Einstein's convention violates Mach's vicious circle principle. If one adopts a more liberal operationalist attitude, then Einstein's convention fails to distinguish between classical physics on the one hand and STR on the other. Finally special relativity, because it postulates a privileged set of inertial frames, is just as 'absolute' as Newtonian mechanics; the proviso being that Newton's unique absolute frame be replaced by the infinite set of Galilean inertial frames.

§4.4 Length Contraction, Clock-Retardation and the Covariance of Maxwell's Equations in Classical Ether Physics

In this section we shall adopt a synoptic viewpoint, which consists in remaining agnostic about synchronisation procedures while postulating a clock-retardation effect (by some factor $\Omega(v)$) and both longitudinal and transverse contraction effects (by some factors $\mu(v)$ and $\psi(v)$ respectively). We have already seen that the transformed spatial coordinates x', y', z' obey the following equations:

$$(7)\ x' = \frac{1}{\mu}(x - vt);\ y' = \frac{1}{\psi} y;\ z' = \frac{1}{\psi} z.$$

In connection with the transformed time variable t', we assume that t' is linear in x and t; i.e.:

(8) $t' = Ax + Bt$, where A and B may depend on v.

We only require that Ω, μ, ψ, A and B be well-behaved functions of v. Because of the assumed retardation of moving clocks, the two scalars A and B are not independent of each other. Thus, expressing t' in terms of x' and t, we infer from (7) and (8):

(9) $t' = A(\mu x' + vt) + Bt = A\mu x' + (B + Av)t.$

Consider a point (x', y', z') rigidly connected to the moving frame; i.e. let x', y', and z' be constant. Differentiating (9), we have: $dt' = (B + Av)dt$; but, by definition of clock-retardation, $\frac{dt'}{dt} = \Omega$. Hence, $\Omega = B + Av$; i.e. $B = \Omega - Av$. Thus $t' = Ax + Bt = (\Omega - Av)\, t + Ax$, where A is the remaining free scalar.

In conjunction with (7), we obtain the following general transformation:

(10) $x' = \frac{1}{\mu}(x - vt);\ y' = \frac{1}{\psi}y;\ z' = \frac{1}{\psi}z;\ t' = (\Omega - Av)t + Ax.$

As shown in chapter two, the choice of Einstein's convention entails: $t' = \Omega\gamma^2(t - \frac{v}{c^2}x)$. (This holds for all values of μ, ψ and Ω.) Comparing this relation with the last equation in (10), we conclude that in this special case: $A = -\frac{v\gamma^2}{c^2}\Omega$. In this way we are led to a posit, without any loss of generality: $A = -\frac{v\gamma^2}{c^2}\Omega\xi$, where ξ is some function of v. (10) can therefore be rewritten as:

$$\textbf{(11)}\quad \begin{cases} x' = \frac{1}{\mu}(x - vt);\ y' = \frac{1}{\psi}y;\ z' = \frac{1}{\psi}z; \text{ and} \\ t' = \Omega[(1 + (\gamma^2 - 1)\xi)t - \frac{v\gamma^2}{c^2}\xi\, x], \end{cases}$$

where, as usual: $\gamma = (1 - v^2/c^2)^{-1/2}$; and Ω, μ, ψ, ξ are functions of v.

We recall that the round-trip velocity of light along any direction making the angle θ with $\vec{v}$ is given by:

(12) $\bar{c}'(\theta) = c\Omega^{-1}\mu^{-1}\gamma^{-2}[1 - (1 - \psi^2.\mu^{-2}.\gamma^{-2})sin^2\theta]^{-1/2}$

Note that θ is the angle as measured in the moving frame. $\bar{c}'(\theta)$ will be independent of θ if $\psi = \mu\gamma$, i.e. if $\mu = \psi\gamma^{-1}$.

To adopt Einstein's synchronisation procedure is to make (by fiat) the one-way velocity of light identical to its round trip velocity; and we have just seen that this is tantamount to choosing $A = -\frac{v\gamma^2}{c^2}\Omega$; i.e. to taking $\xi = 1$. The form of the function ξ of v thus reflects the choice of a particular synchronisation convention.

Before being involved in further technical details, I should like to list the advantages which are offered by the synoptic viewpoint adopted in this section:

(i) Equations (11) clearly separate the real physical causes described by μ, ψ and Ω from the purely conventional element expressed by ξ. Since μ, ψ, Ω and ξ are all independent of each other, it follows that no adoption of any specific synchronisation procedure can give us a clue as to the underlying physical mechanisms. This should cause no surprise: one can, by means of a convention, i.e. by fiat, neither dictate nor even guess the laws of physical nature.

(ii) Equations (11) also subsume under one scheme all the theories we have so far examined:

We obtain the Newton-Maxwell theory, i.e. Lorentz's theory T_1, by taking $\mu = \psi = \Omega = 1$ and $\xi = 0$. That is: we arrive at T_1 by adopting the Galilean transformation and by thus assuming neither clock-retardation nor dimension-contraction.

The choice $\xi = 0$ is the most convenient one in the Newtonian world. However, as was shown in Section (III) above, there is nothing to prevent us from adopting Einstein's convention in such a world, i.e.. from choosing $\xi = 1$. In conjunction with $\mu = \psi = \Omega = 1$, this yields:

$$x' = x - vt,\ y' = y,\ z' = z,\ t' = \gamma^2\left(t - \frac{v}{c^2}x\right);$$

i.e.. we retrieve the transformation (3) constructed in §4.3.

Lorentz's theory T_2 postulates only one (longitudinal) contraction by the factor $(1 - v^2/c^2)^{1/2} = 1/\gamma$. Hence, in this case: $\mu = 1/\gamma$ and $\psi = \Omega = 1$. Taking $\xi = 1$ means adopting Einstein's convention, i.e.. ensuring that the measured one-way velocity of light along every direction in the moving frame equals the round-trip velocity $\bar{c}'(\theta)$ given by:

$$\bar{c}'(\theta) = c.\Omega.^{-1}\gamma.^{-2}\mu^{-1}[1 - (1 - \psi.^2\mu^{-2}.\gamma^{-2})sin^2\theta]^{-1/2}$$
$$= c\gamma^{-1} = c(1 - v^2/c^2)^{1/2}$$

Note that $\bar{c}'(\theta)$ is independent of θ; that is, the measured speed of light is constant in the moving frame as well as in the stationary one (the two constants are however different). This result was to be expected, since $\mu = 1/\gamma = \psi\gamma^{-1}$.

Lorentz's theory T_3, which is mathematically and observationally equivalent to STR, arises from the choice $\xi = 1$ (Einstein's convention) and from the postulates:

$$\mu = \Omega = 1/\gamma \text{ and } \psi = 1.$$

(iii) Our general approach finally enables us to answer the following question: does the covariance of the Maxwell-Lorentz theory strictly entail STR? We have already given a glib affirmative answer which is certainly correct if, by the covariance of Maxwell's theory, we *also* mean the invariance of the constant c occurring in the field equations. We can however look upon c as a quantity which, though constant within each inertial frame, changes as one goes over from one such frame to another. In an illuminating series of articles, Podlaha and Sjödin (see especially Sjödin 1979) showed that, under this relaxed condition, Maxwell's equations keep their form under a larger set of transformations than the Lorentz group. This is proved as follows.

If the speed of light is to be the same along all directions in the moving frame, then the one-way velocity of light must a fortiori be equal to the round trip speed $\bar{c}'(\theta)$. The latter must, by the same token, be independent of θ. As was shown above, these two conditions are expressed by $\xi = 1$ and by $\mu = \psi\gamma^{-1}$ respectively (the first condition represents a convention and the second a physical law). Substituting in (11):

$$\textbf{(13)}\quad x' = \frac{\gamma}{\psi}(x - vt);\ y' = \frac{1}{\psi}y;\ z' = \frac{1}{\psi}z \text{ and } t' = \Omega\gamma^2\left(t - \frac{v}{c^2}x\right)$$

We immediately derive the inverse transformation:

$$\textbf{(14)}\quad \begin{cases} x = \psi\gamma\left(x' + \dfrac{v}{\psi\Omega\gamma}t'\right);\ y = \psi y';\ z = \psi z'; \\ t = \dfrac{1}{\Omega}\left(t' + \dfrac{\Omega\psi v\gamma}{c^2}x'\right) \end{cases}$$

and

$$\textbf{(15)}\quad \begin{cases} \dfrac{\partial}{\partial x} = \dfrac{\gamma}{\psi}\dfrac{\partial}{\partial x'} - \dfrac{\Omega v\gamma^2}{c^2}\dfrac{\partial}{\partial t'};\ \dfrac{\partial}{\partial y} = \dfrac{1}{\psi}\dfrac{\partial}{\partial y'};\ \dfrac{\partial}{\partial z} = \dfrac{1}{\psi}\dfrac{\partial}{\partial z'} \\ \dfrac{\partial}{\partial t} = \Omega\gamma^2\dfrac{\partial}{\partial t'} - \dfrac{\gamma v}{\psi}\dfrac{\partial}{\partial x'} \end{cases}$$

We shall also need a transformation rule for an arbitrary velocity $\vec{w} = (w_x, w_y, w_z)$:

$$w'_x \underset{\text{Def}}{=} \frac{dx'}{dt'} = \frac{\gamma}{\psi\Omega\gamma^2}\cdot\left(\frac{dx - v.dt}{dt - \frac{v}{c^2}dx}\right) \text{ (by (13))}$$

$$= \frac{1}{\psi\Omega\gamma}\cdot\left(\frac{w_x - v}{1 - \frac{vw_x}{c^2}}\right)$$

Similarly:

$$w'_y \underset{\text{Def}}{=} \frac{dy'}{dt'} = \frac{1}{\psi\Omega\gamma^2}\left(\frac{dy}{dt - \frac{v}{c^2}dx}\right) = \frac{1}{\psi\Omega\gamma^2}\left(\frac{w_y}{1 - \frac{vw_x}{c^2}}\right)$$

Thus the transformation equations for the velocity are:

$$\textbf{(16)}\begin{cases} w'_x = \dfrac{1}{\psi\Omega\gamma}\left(\dfrac{w_x - v}{1 - v.w_x/c^2}\right); \; w'_y = \dfrac{1}{\psi\Omega\gamma^2}\left(\dfrac{w_y}{1 - v.w_x/c^2}\right) \\ \text{and: } w'_z = \dfrac{1}{\psi\Omega\gamma^2}\left(\dfrac{w_z}{1 - v.w_x/c^2}\right) \end{cases}$$

Putting $\xi = 1$ and $\mu = \psi\gamma^{-1}$ in (12) we have:

(17) $\bar{c}'(\theta) = c(\psi\Omega\gamma)^{-1} = c'$ (say).

c′ is thus the measured velocity of light in the moving frame.

By using a simple method introduced by Poincaré, we shall determine the transform ρ′ of the charge density ρ. We thereby presuppose the conservation law of electric charge.

Consider a charge e enclosed in a sphere of radius r, whose centre moves with (uniform) velocity $\vec{w} = (w_x, w_y, w_z)$ in the stationary frame. Thus $\rho = 3e/4\pi r^3$. The equation of the sphere is of the form:

(18) $(x - w_xt + a)^2 + (y - w_yt + b)^2 + (z - w_zt + d)^2 = r^2$

Substituting from (14) into (18), we obtain a relation of the form:

$$\textbf{(19)}\begin{cases} \left[\psi\gamma\left(1 - \dfrac{vw_x}{c^2}\right) x' + a_o \cdot t' + a\right]^2 + [\psi y' + kx' \\ \qquad + b_o \cdot t' + b]^2 + [\Psi z' + hx' + d_o \cdot t' + d]^2 = r^2 \end{cases}$$

This equation represents an ellipsoid whose volume is equal to:

$$\frac{4}{3}\pi r^3 \cdot \frac{1}{\psi^3\gamma\,(1 - v \cdot w_x/c^2)}$$

Assuming conservation of charge, we conclude:

$$\textbf{(20)}\ \rho' = e \cdot \frac{3\,\psi^3\gamma\,(1 - v \cdot w_x/c^2)}{4\pi r^3} = \rho\psi^3\gamma\left(1 - \frac{v \cdot w_x}{c^2}\right)$$

As usual, we write Maxwell's equations in the form:

$$\textbf{(21)}\quad \begin{cases} \nabla \cdot \vec{E} = \rho;\ \nabla \cdot \vec{H} = 0; \\ \nabla \times \vec{H} = \dfrac{1}{c}\left(\dfrac{\partial \vec{E}}{\partial t} + \rho\vec{w}\right);\ \nabla \times \vec{E} = -\dfrac{1}{c}\dfrac{\partial \vec{H}}{\partial t}. \end{cases}$$

We know how to transform x, y, z, t, ρ and $\vec{w}$. It can thus be easily (but tediously) verified that Maxwell's equations keep the same form, provided we put:

$$\textbf{(22)}\quad \begin{cases} \vec{E}' = (E'_x, E'_y, E'_z) = \left(\psi^2 E_x,\ \psi^2\gamma\left(E_y - \dfrac{v}{c}H_z\right),\ \psi^2\gamma\left(E_z + \dfrac{v}{c}H_y\right)\right) \\ \text{and:} \\ \vec{H}' = (H'_x, H'_y, H'_z) = \left(\psi^2 H_x,\ \psi^2\gamma\left(H_y + \dfrac{v}{c}E_z\right),\ \psi^2\gamma\left(H_z - \dfrac{v}{c}E_y\right)\right) \end{cases}$$

In other words:

$$\textbf{(23)}\quad \begin{cases} \nabla' \cdot \vec{E}' = \rho';\ \nabla' \cdot \vec{H}' = 0 \\ \nabla' \times \vec{H}' = \dfrac{1}{c'}\left(\dfrac{\partial \vec{E}'}{\partial t'} + \rho'\vec{w}'\right);\ \nabla' \times \vec{E}' = -\dfrac{1}{c'}\dfrac{\partial \vec{H}'}{\partial t'} \end{cases}$$

Therefore: in moving the frame all electromagnetic disturbances are propagated with the constant velocity $c' = c/(\psi\Omega\gamma)$ which generally differs from c. This result holds for two arbitrary functions Ω and ψ of v. In the case of Lorentz's theory T_3 (and also of STR), $\Psi = 1$ and $\Omega = 1/\gamma$; so $c' = c$. As is well-known, c is in this particular instance a universal invariant. This is more generally the case whenever $\Omega\gamma\psi = 1$.

§4.5 Reassessment of the Role of Positivism in the Emergence of Relativity Theory

Though the theses put forward in the last two sections are, on the whole, correct, the anti-positivist case was somewhat overstated. In this closing section I shall reconsider the role of positivism in the progress of science. For this I need to distinguish between two empiricist theses which have often been conflated. Of these theses, the first can be recast into a form which is both sound and fruitful while the second remains totally unacceptable.

(i) As explained above, a basic tenet of Mach's operationalism was that scientific concepts should be both empirically decidable and definable in a theory-independent way. As it stands, this operationalist principle proves untenable, first because empirical decidability cannot be extended to *all* concepts occurring in a scientific hypothesis, secondly because freedom from *all* theoretical presuppositions is an unattainable 'ideal'. It is however a (contingent) fact that all the basic statements physical science needs can be made to depend exclusively on a minimal observational theory, which remains unaffected by scientific revolutions (see chapter five). *Let us note that the observational theory in question is both low-level and (contingently) incorrigible.* Observation-statements are synthetic propositions; hence they need not be true, at least from the strictly logical point of view. It is however a fact that, unless one questions the sincerity of the reporter, observation reports about coincidences for example are not doubted. Such reports may be shelved pending some satisfactory explanation; such an explanation may never become available, but the report in question is not therefore dubbed false; it may forever remain a puzzle, but the last thing one doubts is its veracity. It also seems to be a contingent biological fact that our perceptual categories are not altered by the introduction of new revolutionary hypotheses. One can of course imagine species which, through discovering theories about waves of high frequencies say, would start 'seeing' or 'hearing' such frequencies. It would however appear that the human race is not this kind of quickly evolving species. There seems to be complete stability at the human phenomenal level, i.e., the level at which scientific hypotheses come into contact with perceptual experience.

To sum up: this first empiricist principle can be restated as follows: *the observational level is both low and contingently infallible.* This is to my mind a true, indeed trivially true, proposition.

(ii) Mach and some of his followers, like Bridgman, adopted the further principle that scientific hypotheses should, in the last analysis, refer exclusively to the phenomenal level. In other words: phenomena should constitute the whole domain of discourse of theories: theories should be interpretable as sentences about sense-data. Their only function should be to organise and connect the so-called elements of sensation.

It follows from this principle that, provided we hold the problem of induction in abeyance, a phenomenological analysis of what we perceive and of what we actually do in experimental situations will lead us to the discovery of new laws. This was the conclusion drawn by Bridgman, who also attributed this method of phenomenological analysis to Einstein. An analysis of what we do (or of what we should do) in synchronising distant clocks allegedly led to the discovery of frame-dependent simultaneity.

One motive for the adoption of this second principle is the quest for certainty: by restricting science to the phenomenal level one hopes that the infallibility which obtains at this level would flood the whole of our knowledge. This attitude is again illustrated by Bridgman, who thought that operationalism would enable us to arrive at absolutely unalterable concepts and laws:

> For, in so restricting the possible operations, our theories reduce in the last analysis to descriptions of operations actually carried out in actual situations, and so cannot involve us in inconsistency or contradiction, since these do not occur in actual physical situations. Thus is solved at one stroke the problem of so constructing our fundamental physical concepts that we shall never have to revise them in the light of new experience. (Bridgman 1936, p. 9)

This quest for certainty also explains why so many philosophers opt for some form or other of idealism: we can be sure only of our thoughts and sensations; hence, if the world consisted of nothing but mind-stuff, we could at least hope to have *certain* knowledge of it. For example, in his "Cartesian Meditations" Husserl clearly states that one object of his phenomenological reduction is the attainment of indubitable knowledge; and he inevitably ends up by adopting a version of transcendental idealism. Another motive behind the recourse to phenomenalism is the desire to democratise science. If the most abstruse physical theory is in the last analysis nothing but a long sentence about sense-data, one could, at least in principle, make it accessible to every layman.

Let us note that the two principles described above are independent of each other and that, in a certain sense, they pull in opposite directions. According to the first principle, observational theories, if they exist at all, are infallible but very weak. In fact, they are infallible precisely because they are weak, i.e. because they presuppose very little. They therefore leave the door open to unbridled speculation about possible entities behind, or alongside, the phenomena; the only proviso is that all scientific hypotheses obtained in this way be compatible with the basic statements. This is Popper's well-known conclusion: very little is *positively* dictated by the facts; facts perform a mainly *negative* task, the task of weeding out false theories. The second positivist principle however forbids any reference to entities which are not *constituted* by the phenomena, i.e., to irreducibly theoretical entities. Our hypotheses should, in some sense, be co-extensive with our observables.

My general claim is that Einstein accepted principle (i) which he may well have extracted from, or read into, Mach's *mechanics*; but that he paid only lip-service to principle (ii). Such lip-service he *did* however pay in his *Meaning of Relativity* where he claimed that:

> The object of all science, whether natural science or psychology, is to coordinate our experiences and to bring them into a logical system. (Einstein 1922, Introduction)

Needless to say, Einstein gives no suggestion as to how theoretical statements should be effectively reinterpreted as sentences about 'experiences'.

In order to appreciate Mach's contributions to epistemology, it ought to be recalled that the mechanics adopted in the nineteenth century was not the original Newtonian system but the so-called classical framework as entrenched by Kant. This framework supposedly consisted of a set of unassailable propositions. Even a philosopher as recent as Poincaré wondered whether classical mechanics is an empirical or a necessary a priori science (Poincaré 1902a, ch. 6). Kant had allegedly proved that the very possibility of objective experience presupposes absolute time, absolute space and Euclidean geometry. Mach, who after all took his starting point from Kantianism, had the great merit of showing that experience as such does not require the full strength of the absolute space and time hypotheses. Mach went of course much further than this: he tried to show that even the mere *notions* of absolute space and time were monstrous impossibilities and that experience was independent of any theory whatsoever; in these extreme claims he failed. However, he unwittingly showed that the theories presupposed by experience are so low-level as to be neutral, say, between Newtonian mechanics and a new physics which would set out to explain inertia in terms of rotation with respect to the stars. He showed, contra Kant, that experience does not presuppose absolute time but only clock-time. He similarly demonstrated that the notion of absolute space is not involved in the perception of moving extended bodies. However, what Mach achieved *vis-à-vis* Einstein was contrary to his avowed intention of banning speculation from science. By analysing what we actually do when we observe, Mach liberated physicists from the Kantian straightjacket; he thereby set Einstein free to speculate at the most untouchable level and thus alter dogmas about space and time which had hitherto been considered sacrosanct. In "Physics and Reality" Einstein wrote:

> The lack of definiteness which, from the point of view of empirical importance, adheres to the notion of time in classical mechanics, was veiled by the axiomatic representation of space and time as things given independently of our senses. Such a use of notions—independent of the empirical basis to which they owe their existence—does not necessarily damage science. *One may however be easily led into the error of believing that these notions, whose origin is forgotten, are necessary and unalterable accompaniments of our thinking, and this error may constitute a serious danger to the progress of science.* (Einstein 1950, p. 68; my italics)

Thus, when Einstein faced the incompatibility of the following three principles:

(α) The relativity postulate,
(β) Maxwell's equations (*viz.* the constancy of c), and
(γ) The Galilean transformation,

he already knew from Mach's teaching that he *could modify* the Galilean transformation, hence the absoluteness of time, without *necessarily* contradicting experience.[1] However, the way in which such a modification should be carried out was determined, not by a phenomenological analysis of experience, but by examining the transformational properties of Maxwell's equations, hence the possible invariance of c. In other words: Mach *negatively* helped Einstein by showing him that absolute time could be dispensed with; but Machism offered *no positive assistance* when it came to the actual construction of a new transformation. The Lorentz transformation was determined as one under which the law of inertia, Maxwell's equations and the velocity of light, are preserved. As explained in the last section, Podlaha and Sjödin showed, under very general assumptions, that the transformation taking a stationary frame into an inertial one moving with velocity v along the x-axis is given by:

$$x' = \frac{1}{\mu}(x - vt)\ ;\ y' = \frac{1}{\psi}y\ ;\ z' = \frac{1}{\psi}z\ ;\ t' = Ax + (\Omega - vA)\,t.$$

μ and ψ are respectively the longitudinal and transverse contraction factors; Ω is the clock-retardation coefficient; A is an independent parameter which reflects the Synchronisation procedure chosen in the moving frame. *All* phenomena can be described with *any* value of this parameter, which is thus the result of a stipulative convention. Since A does not determine μ, ψ or Ω, it follows that no amount of reflecton on the 'correct' synchronisation procedure can yield a new physical principle. This supports my thesis that

[1]This view is somewhat similar to the one expressed by Mary Hesse: "Absence of any possible operational definition *permits* but does not *compel* elimination of a concept, and it sometimes happens that a concept has been retained unnecessarily because it has been assumed to have a direct operational meaning, although this is later found not to be the case . . . The fact that . . . the aether was in some senses unobservable meant that it could be abandoned without conflict with other empirical facts" (1961, p. 8). In the present paper I am more concerned with propositions than with concepts. Anyway, it seems to me that one can decide not to use certain concepts, even though they may have a direct operational meaning. There is however an obvious parallel between propositions *not* presupposed by experience and concepts *not* susceptible of operational definition: both *can*, but of course *need* not, be jettisoned, without necessarily falling foul of experience.

metaphysical and mathematical speculation, not conceptual analysis, played the dominant role in the discovery of relativity.[2]

Against its avowed intentions, the Machian type of empricism underlying principle I effected a Popperian revolution. By showing how *little* is dictated by empirical results, operationalism did not ban 'metaphysics' from science. On the contrary, it opened the floodgates to imaginative speculation at the most fundamental levels of scientific theorising (space and time for example). The only constraint on speculation is the well-known Popperian one: no matter how wild the conjectures, they must still be capable of contact (more precisely of conflict) with possible experience.

In connection with Mach's influence on the genesis of general relativity the following quotation from his work on the *Conservation of Energy* will be found very revealing:

> Obviously it does not matter whether we think of the earth as turning round on its axis, or at rest while the celestial bodies revolve around it. Geometrically these are exactly the same case of a relative rotation of the earth and of the celestial bodies with respect to one another. Only, the first representation is astronomically more convenient and simpler.
>
> But if we think of the earth as at rest and the other celestial bodies revolving round it, there is no flattening of the earth, no Foucault's experiment and so on, at least according to our usual conception of the law of inertia. Now, one can solve the difficulty in two ways: either all motion is absolute, or our law of inertia is wrongly expressed. Neumann preferred the first supposition, I, the second. The law of inertia must be so conceived that exactly the same thing results from the second supposition as from the first. By this it will be evident that in its expression, regard must be paid to the masses of the universe. (Mach 1909, pp. 76–77)

[2]In 1977 Malament established the following deep result: let O be a time-like line representing some inertial observer and let K be the causal connectibility relation; i.e., for any 2 events in a Lorentzian frame, $p = (x_1,y_1,z_1,t_1)$ and $q = (x_2,y_2,z_2,t_2)$, we have:

$$(pKq) \equiv [c^2 (t_2 - t_1)^2 - (x_2 - x_1)^2 - (y_2 - y_1)^2 - (z_2 - z_1)^2 \geqslant O].$$

57 is well known that K is frame-independent. Let S_o (p,q) be the usual simultaneity relation with respect to O; that is, the relation S_o (p,q) is based on Einstein's synchronisation convention, so that $S_o (p,q) \equiv (t_2 = t_1)$. Malament proved that S_o (p,q) is the only (non-trivial) equivalence relation definable exclusively in terms of K and of membership of O. Malament regards this theorem as having removed the conventional character of Einstein's synchronisation procedure. I beg to disagree; for, on the one hand, nothing compels us to define synchronism in terms of K; all that can reasonably be required is that, whatever convention we adopt, we be in a position to express K. This condition is clearly met for every choice of A; on the other hand, as pointed out by Michael Redhead, the condition that simultaneity should be an equivalence and in particular a transitive relation is far from innocuous or self-evident. It can even be said that Einstein's great insight consisted in recognising that simultaneity is, generally, speaking, intransitive. There is therefore no a priori reason for imposing a transitivity condition on the simultaneity relation. See Malament 1977, and Redhead 1983, p. 184, note 38.

Mach can be taken as requiring that a condition of general covariance be imposed on certain theories. He maintains that we should be able to explain the oblateness of the earth whether we adopt the stars, or the earth itself, as our frame of reference. He does not explicitly assert that the explanations must be the same in both cases, but some such assertion is implicit in the above passage. Mach must have known that classical theories can be referred to rotating frames, provided one introduces an inertial field determined, for example, by the relative motion of the earth and the stars. It was also known that physical laws assume a different (indeed a more complicated) form in a rotating frame. Thus, in order to make sense of the above quotation, in order to make it non-vacuous, we have to assume that Mach was suggesting that certain laws should assume the same form in all frames, i.e. that they should be generally covariant. Thus Einstein was right in claiming that Mach had identified the weak spots in classical mechanics and was not far from requiring a general theory of relativity, all of this about half a century before Einstein.

Einstein's arguments in favour of general covariance are well-known (see chapter eight) and can be briefly formulated as follows. Laws have an ontological status over and above the observables which they correlate with one another. Moreover, simplicity is an objective property of scientific laws. Hence, if the laws assume a particularly simple form in some frames of reference, this would indicate that these frames are privileged over others. But, according to the relativity postulate, there exist no privileged frames. All laws must therefore assume the same form in *all* frames of reference. Note that this argument makes essential use of the assumption that laws and their form possess an objective status. We have already seen that Machism contradicts this assumption. According to Machism, a law is nothing more than a computational device; both the device and its degree of convenience have a subjective function. Hence the simplicity of a scientific law does not indicate that the chosen frame of reference is privileged in any ontological (objective) sense; in fact the frame as such, being unobservable, has no place within Machian ontology. From a Machian viewpoint, there is no good reason why the choice of a certain coordinate system should not prove more convenient, more economical, than another. There is in fact every reason, from a Machian point of view, to congratulate a scientist on finding the most convenient coordinate system for the simplest formulation of physical laws. In proposing general covariance, Mach was therefore violating a tenet of his own brand of positivism; he was instinctively following a philosophy of science with strong Platonistic undertones; apart from the elements of sensations, scientific theories express the structure of the world and thus have ontological status; hence, a particularly simple form assumed by our theories would reveal an objective property of the frames to which they are referred.

— 5 —

POINCARÉ'S INDEPENDENT DISCOVERY OF THE RELATIVITY PRINCIPLE

§5.1 Whittaker's Contention

Speaking about Kant, Schopenhauer (1818, vol. 1 appendix) said that it is much easier to criticise a genius than to praise him. His errors, always finite in number, are easily circumscribed while his contribution to truth remains both unfathomable and inexhaustible. I say this by way of an excuse for the criticisms that I am about to level at Jules Henri Poincaré's philosophy. We all know that Pioncaré was one of the greatest mathematicians of all time. In a recent work Popper describes him, quite simply, as the greatest of all philosophers of science. In *Men of Mathematics* E.T. Bell (1937, ch. 28) refers to Poincaré as to the "last universalist". Poincaré has in fact left his mark on practically all branches of mathematics and of theoretical physics. He had a prodigious knowledge of complex analysis; he founded the field of analysis situs, or topology; he studied the theory of groups, the probability calculus, classical mechanics, electricity, optics and astronomy; he attacked the three-body problem and proved a fundamental theorem relating to the second law of thermodynamics. This list can be prolonged almost indefinitely. In this chapter, I propose to examine Poincaré's contribution to special relativity with a view to deciding whether his so-called conventionalism helped him towards the discovery of this theory. Edmund Whittaker (1953, vol. II ch. 2) is to my knowledge the only historian of science to attribute the discovery of relativity exclusively to Poincaré; he does this in a chapter of his *History of the Theories of Aether and Electricity* in which Einstein is barely mentioned. Though Whittaker was unjust towards Einstein, his positive account of Poincaré's actual achievement contains much more than a simple grain of truth.

In his excellent account of Poincaré's philosophy, *Science and Convention,* Giedymin adopts a position analogous to the one which will be presented and defended in this paper. Giedymin is to my mind one of the few philosophers of science who do justice to Poincaré's contribution to relativity without being misled by the word 'conventionalism' into holding that Poincaré was an instrumentalist (Giedymin 1982, ch. 5).

In what follows I propose to defend a thesis as brutally simple as Whittaker's, namely that Poincaré did discover special relativity, that his philosophy of science provided him with heuristic guidelines, but that certain ambiguities within that same philosophy prevented both his contemporaries and many historians from appreciating the true value of his contribution. This account in no way tarnishes the merit of Einstein who, following a path parallel to Poincaré's, finally transcended special relativity by constructing a generally covariant theory of gravitation.

In connection with Poincaré, we shall find it useful to reinterpret his philosophy by adopting a unified standpoint which will give us a synoptic view of his whole position. In other words, we shall carry out a rational reconstruction which is not strictly compatible with the detail of all the theses advanced by Poincaré (the conjunction of all these theses is inconsistent) but which affords, in return, a better insight into his conception of the foundation of geometry and of relativity. We shall see that, in 1905, Poincaré had gone far beyond the results obtained by Einstein; that, in 1900, he had already given the operational definition of clock synchronisation which is usually, but incorrectly, attributed to Einstein (see §4.3); that he had enunciated the Lorentz-covariance principle and founded the relativity programme on the structure of the Lorentz-group; that he had corrected the transformation rule for the electric density proposed by Lorentz in 1904 (see §2.6), then used this new rule in order to transform the whole electromagnetic field. He had also reflected on the parallel between gravity and inertia and considered the possibility of a variation of gravitational attraction with the velocity of the mobile. He had finally constructed a Lorentz-covariant theory of gravitation. Poincaré had thus pushed special relativity to the utmost limits of which it seemed capable. Thus, with regard to Lorentz, Poincaré and Einstein, there is for the historian the duty both of acknowledging Einstein's unsurpassed contribution to relativity and of rectifying a certain imbalance. Between 1912 and 1915 (after Poincaré's death), Einstein perceived the necessity of going beyond special relativity towards a generally covariant theory of gravitation. Because of the mathematical beauty of the general theory and because it was the exclusive product of a single man's genius, there has been a natural tendency to play down, even to forget, Poincaré's achievement and thus consider Einstein as the sole founder of relativity in both its special and its general phases. I hope the present chapter will redress this injustice.

§5.2 Three Different Kinds of Hypothesis

Proceeding analytically, I shall isolate within Poincaré's philosophy various strands which at first sight appear contradictory. I shall distinguish between his so-called conventionalism on the one hand and a marked tendency towards an inductivist form of empiricism on the other. By adopting the unified stance mentioned above, I shall then attempt to resolve these contradictions. It turns out that Poincaré tacitly subscribed to a Kantian brand of realism and that his conventionalism was, in the last analysis, a vast *petitio principii*. He asserted *both* that certain hypotheses are neither true nor false but more or less convenient *and*, in a different context, that these same hypotheses are convenient to the extent that they are true. Thus he treated the degree of convenience, the level of aesthetic-mathematical perfection, as an index of verisimilitude. It should be immediately added that Poincaré used verisimilitude as an intuitive concept having nothing to do with the correspondence theory of truth.

For Poincaré as for Popper, it is impossible to pursue science without preconceived ideas, i.e., without provisional conjectures.

> "It is often said that experiments should be made without preconceived ideas. That is impossible. Not only would it make every experiment fruitless, but even if we wished to do so, it could not be done. Every man has his own conception of the world, and this he cannot so easily lay aside. We must, for example, use language, and our language is necessarily steeped in preconceived ideas. Only they are unconscious preconceived ideas, which are a thousand times the most dangerous of all." (SH p. 143).

Poincaré divides scientific hypotheses into three separate categories. The first consists of certain assumptions "which are quite natural and necessary. It is difficult not to suppose that the influence of very distant bodies is quite negligible, that small movements obey a linerar law, and that effect is a continuous function of its cause." (SH, p. 152) Among these assumptions, which should be the last to be abandoned, are certain so called conventions, like the postulates of Euclidean geometry and those of rational mechanics.

The second category contains theories which are termed indifferent because they perform a purely psychological function. Such are certain metaphysical-interpretative hypotheses, which assist our understanding and simplify certain calculations but make no difference as to the mathematical form of our theories and to their empirical predictions. "For example in most questions the analyst assumes, at the beginning of his calculations, either that matter is continuous, or the reverse, that it is formed of atoms. In either case his results would have been the same. On the atomic supposition he has a little more difficulty in obtaining them—that is all." (SH, p. 152)

The third and last category consists of generalisations obtained by induction from scientific facts. Let us start by examining the nature and status of the latter.

§5.3 Crude Facts and Scientific Facts

Against Edouard Le Roy's nominalism Poincaré maintains that the scientist does not arbitrarily create, or fabricate, scientific facts; these are, according to Poincaré, nothing but the translation into a more convenient language, i.e., into the language of science, of so-called crude facts. Crude facts are expressed by propositions which, for want of a better word, I shall call crude propositions; these are decidably true or false. Here, we encounter the first ambiguity in Poincaré's position: where does the crude fact belong? Is it purely phenomenological or does it have an objective aspect? Is it purely subjective or does it also refer to a reality located outside the mind? In certain passages of SH and VS Poincaré says that propositions like 'the current is on,' which certainly possess objective import, describe crude facts. In other parts of the same works, he maintains that each crude fact is unique, that it consists in certain subjective perceptions and that no proposition can completely characterise, hence individuate it; this is because our language has at its command "only a finite number of terms to express the shades, in number infinite, that my impressions might cover." (VS, p. 117) Crude facts can nonetheless definitively falsify, or verify, the crude propositions which fix on certain of their aspects. Given the individual character of each of our sensations, it seems to me that an objective basic statement like 'the current is on' cannot by right, i.e., by means of logic alone, be the mere translation into a convenient language of the observation report 'I see the spot of the galvanometer move'. Only the crude visual fact, as distinct from the alleged passage of the electric current, is indubitable. Yet, Poincaré needs the objective basic statements themselves in order to arrive at the experimental laws, like 'all metals are good conductors', which enable us to predict and hence to act. We shall therefore rationally reconstruct Poincaré's position by modifying it a little. We shall say that every basic statement can, in the last analysis, be reduced to a proposition which is exclusively about our perceptions and can be decided by them. We should immediately add the rider that in this context 'can be reduced to' does *not* mean 'is logically equivalent to'. More precisely: let p be a statement describing some physical process, e.g. the 'current is on'; let p′ be the statement, expressed in phenomenological terms, that a certain spot is seen to move. p' thus signifies a perception, or a group of perceptions, which correspond to p. It can be said that p', which is sometimes called a protocol sentence, is the result of a phenomenological reduction (*epoché*) which consists in eliminating from a proposition all references to a reality transcending the observer's consciousness. According to the phenomenological view, a laboratory experiment can always be described by propositions of the same form as p' and can thus verify or falsify such propositions. Moreover there exists a physico-psychological hypothesis A such that $(A \Rightarrow (p \Leftrightarrow p'))$. p is thus, not logically, but materially or contingently equivalent to p'. Note that A contains (implicit) clauses about the reliability of the instruments used in an experiment and about the mental and physical health of the observer.

As long as we accept A we can assert $p \Leftrightarrow p'$, hence also $p' \Rightarrow p$. If the experiment is carried out n times and the (phenomenological) results $p'_1, \ldots, p'_n$ are found, then, in view of $p'_i \Rightarrow p_i$ for all $i = 1, 2, \ldots, n$, we obtain: $p_1, \ldots, p_n$. These objective singular statements can then be generalised and yield some experimental law, P say. Induction is thus fallible on two counts. First, if A is false, then we can no longer infer p_i from p'_i; so $p_1, \ldots, p_n$ need not hold good and p may therefore be false. Secondly, even if A were true, P may still break down, because P goes beyond $p_1, \ldots, p_n$.

Let us briefly examine the testability of theories from this phenomenological viewpoint. In order to test a complex theory T, one of whose conjuncts is A, one extracts form T and from some (objective) initial condition q_1 an (objective) prediction q_2. Thus:

1. $T \Rightarrow A$ and $T \Rightarrow (q_1 \Rightarrow q_2)$

To q_1 and q_2 correspond two protocol sentences q'_1 and q'_2 such that

2. $A \Rightarrow (q_1 \Leftrightarrow q'_1)$ and $A \Rightarrow (q_2 \Leftrightarrow q'_2)$

From (1) and (2) it follows that:

3. $T \Rightarrow (q'_1 \Rightarrow q'_2)$

In fact, in order to obtain (3), it suffices to assume, instead of (2), the weaker implications: $A \Rightarrow (q'_1 \Rightarrow q_1)$ and $A \Rightarrow (q_2 \Rightarrow q'_2)$. T is thus falsifiable by means of propositions describing events which can be effectively ascertained, e.g. $q'_1 \wedge \neg q'_2$.

Let us compare Poincaré's position, as thus reinterpreted, with Duhem's and with Popper's. For Duhem (1906, part II ch. 4), crude facts and propositions exist but they belong exclusively to the domain of commonsense knowledge; they have no right of entry into science, which makes use only of scientific facts; these are both more precise and much less certain than their crude counterparts and there exists no theory, such as A above, which could map the ones onto the others. There is therefore a hiatus between science and commonsense. According to Popper (1935, ch. 5), there are no crude propositions at all since no synthetic proposition is indubitable; science is continuous with commonsense; it can even be said that science is commonsense writ large. Consequently, for Poincaré as for Popper, there is continuity between science and commonsense but, contrary to Popper, Poincaré maintains that certain basic statements are inaccessible to doubt.

§5.4 Unity and Induction

According to Poincaré, the possibility of physical science rests on the two principles of the unity and of the uniformity of nature. The first principle, which says that all physical entities are interdependent, is absolutely indispensable; it constitutes a transcendental truth. Hence the methodological primacy of the unity principle over all others.

> Let us first of all observe that every generalisation supposes in a certain measure a belief in the unity and simplicity of Nature. As far as the unity is concerned, there can be no difficulty. If the different parts of the universe were not as the organs of the same body, they would not react one upon the other; they would mutually ignore each other, and we in particular should only know one part. We need not, therefore, ask if Nature is one, but how she is one. (SH, p. 145)

Let us note that for Poincaré the degree of unity of a given hypothesis is very different from that of its simplicity, as the following passage from SH clearly indicates:

> With this tendency there is no doubt a loss of simplicity. Such and such an effect was represented by straight lines; it is now necessary to connect these lines by more or less complicated curves. On the other hand unity is gained. Separate categories quieted but did not satisfy the mind. (SH, p. 182)

What constitutes the unity of a theory is a sort of organic compactness arising from the interconnectedness of its parts. Such a theory is not, from a pragmatic point of view, necessarily simpler than its rivals.

We shall now examine the inductive principle which enables us to proceed from the particular to the general. Like all classical empiricists Poincaré believed that, without induction, science would not have come into existence. This principle plays an important role in the selection of facts which we choose, not at random, but in terms of the possible generalisations they might allow. This is why we concentrate on simple facts having a great chance of recurring. Biologists for example prefer to study cells rather than whole organisms, because the former present simple recurrent patterns which easily lead to possible generalisations. Since however any—necessarily finite—number of experimental results can be generalised in an infinite number of ways, we need a rule for restricting the range of possibilities.

The uniformity principle is such a rule which enjoins us to connect points given in space, not by an angular line but by a differentiable curve possessing a bounded derivative. Induction understood in this way constitutes an indispensable tool for the researcher. It might be thought that Poincaré looked upon induction as an instrument of discovery having no metaphysical implications. This is however not the case. He repeatedly said that science ought to aim at the truth.

Consequently, a scientist who pursues such an aim while generalising from a limited number of facts implicitly believes in the simplicity or uniformity of nature. Poincaré asks the question, meaningful only to a realist, whether this simplicity is profound or only superficial. He remarks that the two principles of unity and of simplicity pull in opposite directions: if nature is one, then every entity in it, being connected to every other entity, must be a function of very many variables and will thus obey complex laws. Given the primacy of the unity principle, we thus have to ask whether we can proceed *as if* nature were simple. Carnot's principle, that is the second law of thermodynamics, assures us that we can do so without great risk; for nature tends towards more (apparent) uniformity arising from the intimate blending of many complex factors.

> The laws which govern visible bodies would then be merely consequences of the molecular laws. Their simplicity would then be merely apparent and would conceal an extremely complex reality since its complexity would be measured by the very number of molecules. But it is precisely because this number is very large that the divergences in details would be mutually compensatory and we would therefore believe that there was harmony. (LE, p. 10)

§5.5 The Testability of Laws

We shall now examine the very important question of the logical status of empirical laws. Poincaré's attitude turns out once more to be ambiguous. In certain passages he maintains that generalisations are both verifiable and refutable, in others that, for logical reasons, experience can only disconfirm experimental laws. Let us start with the problem of empirical refutability which remains relatively simple. Poincaré's falsificationism proves at least as intransigent, not to say as extremist, as Popper's. The following two pasages could easily have been written by the author of the *Logic of Scientific Discovery*:

> Thus when a rule has been established, we have first to look for the cases in which the rule stands the best chance of being found in fault. This is one of many reasons for the interest of astronomical facts and of geological ages. By making long excursions in space or in time, we may find our ordinary rules completely upset . . . (SM, pp. 20-21)

And

> Ah well! this impatience is not justified. The physicist who has just given up one of his hypotheses should, on the contrary, rejoice, for he found an unexpected opportunity of discovery. (SH, p. 150)

Let us note that, for Poincaré, the descriptive terms occurring in empirical laws possess meanings fixed in advance; he contemptuously rejects the artifice which consists, for instance, in altering the meaning of the word 'swan' in case the proposition 'all swans are white' were refuted. Long before Popper, Poincaré had put a ban on so-called conventionalist stratagems. Arguing once more against Edouard Le Roy, he wrote:

> Recall first the examples he [Le Roy] has given. When I say: Phosphorus melts at 44°, I think I am enunciating a law; in reality it is just the definition of phosphorus; if one should discover a body which, possessing otherwise all the properties of phosphorus, did not melt at 44°, we should give it another name, that is all, and the law would remain valid . . .
>
> It is clear that, if laws were reduced to that, they could not serve in prediction; then they would be good for nothing, either as means of knowledge or as principle of action. (VS, pp. 122-123)

And

> There is no escape from this dilemma: either science does not enable us to foresee, and then it is valueless as a rule of action; or else it enables us to foresee in a fashion more or less imperfect, and then it is not without value as means of knowledge. (VS, p. 115)

Let us go back to the question of the alleged verifiability of empirical laws. In SH Poincaré claims that "the hypotheses of the third category are real generalisations. They must be confirmed or invalidated by experience. Whether *verified* or condemned, they will always be fruitful . . . ' (p. 153; my italics). However, in both VS and LE he clearly realises that experiments which can refute a law are logically incapable of definitively establishing it. Commenting on the possibility of an evolution of laws, he wrote:

> Thus there is not a single law which we can enunciate with the certainty that it has always been true in the past with the same approximation as today; in fact not even with the certainty that we will never be able to demonstrate that it has been false in the past. And yet, there is nothing in this to prevent the scientist from maintaining his belief in the principle of immutability, since no law will ever be relegated to the rank of being transitory without being replaced by another law more general and more comprehensive. (LE, p. 13)

We thus have, once again, to rectify Poincaré's position in order to render it more coherent. We retain the view that the primitive terms contained in empirical laws have unalterable meanings, which are therefore fixed in advance. Such laws are by right either true or false. However, since generalisations are universal propositions, experience can effectively only falsify them. Not being verifiable, generalisations are only partially decidable.

As regards empirical laws, Poincaré thus instinctively adheres to the correspondence theory of truth.

§5.6 Mathematical Induction and Physical Induction

In order to underline the importance which inductive principles assume for Poincaré, I shall allow myself a short digression into his philosophy of mathematics. I propose to show that his belief in scientific induction led him to assess the nature and role of mathematical induction in a way which remains highly questionable.

Poincaré takes for granted that mathematics constitutes a creative discipline, and this in two distinct senses of the word 'creative': on the one hand, all mathematical postulates are not logical truths; on the other hand, all the rules of inference used in mathematics are not those of the syllogism. In more modern parlance we can say that, for Poincaré, the mathematical rules of inference cannot all be deductive; otherwise, all theorems would be implicitly contained in the axioms and would thus provide no genuinely new information, a consequence which he finds intuitively unacceptable. Among the mathematical principles there must therefore be some which are synthetic a priori. Mathematical induction, which can be formalised by the following inference scheme

$$\frac{P(O),\ P(n) \Rightarrow P(n+1)}{(\forall n)P(n)},$$

was chosen as the candidate par excellence for this elevated status.

Poincaré's evaluation of this principle is a little ambiguous. Since nobody doubted its infallibility, it had to be somewhat different from physical induction. According to Poincaré, this difference lies in that mathematical induction refers to no external reality but describes the process by which the mind infallibly knows itself capable of generating the indefinite sequence of natural numbers. (Note: 'indefinite' and not 'infinite'). As a first step towards understanding the role of this principle, Poincaré regards it as an indefinite sequence of syllogisms all condensed into one scheme. That is: the rule of inference written above is shorthand for

$$\frac{P(O),\ P(O) \Rightarrow P(1),\ \ldots\ ,\ P(n-1) \Rightarrow P(n)}{P(n)}$$

for any (given) n.

It is however clear that, if interpreted in this way, mathematical induction can effect only economy of thought, not any genuine increase in content; for, by modus ponens alone, i.e., by a *deductive* rule of inference, we can pass from $P(O),(P(O) \Rightarrow P(1)), \ldots$, and $(P(n-1) \Rightarrow P(n))$ to $P(n)$, for any given n. Poincaré finally sees the key to the creativity problem in his thesis that mathematical induction proceeds from the particular to the general, thereby increasing the content of its premises. On this view, there should be a profound analogy between mathematical induction on the one hand and physical induction on the other, the only difference being that the former is infallible while the latter is not.

In his review of SH Russell (1910) objected on the one hand that mathematical induction in no way enables us *to generalise*; for it is understood that the premise $P(n) \rightarrow P(n+1)$ has the same meaning as $(\forall n)$ $(P(n)$ $P(n+1))$; hence we go from one universal proposition to another, namely from $[P(O) \land (\forall n)(P(n) \Rightarrow P(n+1))]$ to $(\forall n)P(n)$. Thus, any similarity between the two principles of mathematical and of physical induction vanishes. On the other hand, Russell maintains that the set ω of natural numbers is *defined* by means of the inductive principle. ω could for example be said to be the smallest inductive class, i.e. the intersection of all classes X such that $[(O \in X) \land (\forall t)((t \in X) \Rightarrow (t \cup \{t\} \in X))]$; ω would then, by definition, obey the principle of mathematical induction.

Russell's first argument seems impossible to contradict. As for the second, it proved far less decisive, especially after the breakdown of the logicist programme. In his reply, Poincaré seems not to have understood, or rather not to have wanted to understand, Russell's first objection which he interprets as asserting that the inductive principle is more general than its conclusion; i.e. that the scheme

$$\frac{P(O),\ P(n) \Rightarrow P(n+1)}{(\forall n)P(n)}$$

is more 'general' than the formula $(\forall n)P(n)$. Such a reply constitutes either a truism or a category mistake and, anyway, it distorts Russell's position. It is surprising that Poincaré did not choose instead to attack Russell's second argument. He could for instance have said that Russell only shifts the difficulty: it is not enough to define ω in logical terms; it has, moreover to be demonstrated that, in proving the theorems of arithmetic, only logical axioms are invoked. We know that Russell subsequently resorted to the postulates of infinity and reducibility which, on his own admission, are not logical in character. Some mathematical axioms are thus synthetic after all and, if they are to remain independent of experience, then they must be declared synthetic a priori. But I have the impression that, in 1902, Poincaré would have disliked such a line of criticism; or rather, that he would have considered it insufficient since it implicitly concedes that the whole content of mathematics lies in its axioms. No matter what additional assumptions Russell might have needed, he would still have insisted on using strictly deductive rules of inference; and this Poincaré found unacceptable. So, despite Russell's justified criticism, he repeated, in VS for example, that mathematical induction is fruitful because it proceeds from the particular to the general. The inductive process is thus viewed as the mechanism of progress in mathematics as well as in the physical sciences. In physics conventions provide mere classificatory schemes and, in mathematics, both the logical axioms and the deductive rules of inference are vacuous. Only generalisations can, in both cases, increase the content of our knowledge.

§5.7 Poincaré's Empiricist Meaning Criterion

In order to understand the rôle and status which Poincaré ascribes to scientific conventions, it proves useful to examine his critique of classical mechanics. In SH he wonders whether dynamics constitutes a rational discipline or an experimental science. He asserts that the principle of inertia is not a proposition susceptible of being true of false because it cannot be directly confronted with experience. There exist no bodies which are either isolated or else entirely removed from the action of all external forces; therefore, the principle of inertia cannot be tested independently of certain auxiliary hypotheses. Consider next Newton's law of motion $\vec{f} = m\vec{a}$. Even if one wanted to treat this equation as a definition of force, one would have to define the mass m, i.e. provide a means of measuring m. Poincaré resorts to Mach's definition (above §4.2), according to which

$$\frac{m_A}{m_B} = -\frac{\vec{a}_B}{\vec{a}_A},$$

where m_A, m_B and $\vec{a}_A$, $\vec{a}_B$ respectively denote the masses and the accelerations of two bodies, A and B, supposed to be isolated from the rest of the universe. Poincaré observes first that such a definition brings into play Newton's third law which says that every action provokes an equal and opposite reaction; whence $m_A \vec{a}_A = - m_B \vec{a}_B$ for all initial positions and velocities of A and B. This principle of reaction is therefore taken for granted: in Poincaré's words it is treated as a definition. Secondly, knowing that A and B cannot be completely isolated, we have to take account of external factors affecting the system; we have, for example to consider the vector sum of all the external forces and of the interactions between A and B. We would be implicitly relying on Newton's fourth law, according to which the force exerted by a third body C does not affect the action of A on B (or that of B on A) but gets linearly added to it. Hence this law will also be regarded as a definition.

From the above considerations Poincaré concludes that the principles of mechanics are nothing but conventions which have nothing to fear from any fresh experiments we might be tempted to perform.

One way of coherently accounting for Poincaré's position is to say that he tacitly adopts the following empiricist criterion of meaning. Let us call a 'simple proposition' any well-formed formula possessing no proper sentential components. Such a formula will thus either be atomic or else start with some quantifier. It will be considered meaningful if it is either verifiable, or falsifiable, or both. In other words, a simple proposition will be susceptible of truth or falsity if and only if it is at least partially decidable. As for complex propositions, they will be regarded as meaningful if and only if all their sentential components have the same property. (I am aware that this criterion is not only language–, but also formulation–dependent. I have the feeling however that such a defect would not have worried Poincaré; he would have regarded the syntactical formulation of a given theory as part and parcel of that theory). It would be a mistake to think that Poincaré instinctively adhered to the verificationist principle, later proposed by the Vienna Circle, which identifies the meaning of a statement with the method of its verification. Far from it. Since most propositions of which theoretical science treats are universal and therefore unverifiable, it can be claimed, without too great a risk of error, that Poincaré effectively adopted a falsificationist definition of meaning. Such a definition should not be confused with Popper's demarcation criterion. Following Tarski, Popper maintains that every syntactically well-formed formula can be semantically interpreted and is therefore susceptible of truth or falsity. Since science and (good) metaphysics are both meaningful, the demarcation criterion delineates their common frontier. Not so for Poincaré: for him crude propositions are fully decidable and hence pose no problem. Unverifiable experimental laws can be experimentally refuted and are therefore, by right, either true or false. And as regards purely existential statements which are verifiable but not falsifiable (metaphysical in Popper's sense), Poincaré hardly concerns himself with them, no doubt because they play a negligible role in theoretical science. The conventions however, which occupy a central position in mathematical physics, can be neither verified nor refuted and are therefore neither true nor false, at least in Poincaré's sense of these words. Let us recall that Tarski's truth-definition consists, first in fixing a domain of individuals, secondly in analysing a given proposition into primitive terms each of which receives, *separately*, some given meaning, finally in assigning a truth value to the proposition as then recomposed. For Poincaré, such an atomising process ceases to have any legitimacy when one comes to deal with totally undecidable hypotheses. When considered in isolation, the descriptive terms occurring in such propositions signify nothing; but a proposition connecting them functions like a metaphor, which is adequate to the extent that it proves convenient. Commentating on the value of atomic theories Poincaré wrote:

> Such a philosopher claims that all physics can be explained by the mutual impact of atoms. If he simply means that the same relations obtain between physical phenomena as between the mutual impact of a large number of billiard balls—well and good! This is verifiable, and perhaps it is true. But he means something more, and we think we understand him, because we think we know what an impact is. Why? Simply because we have often watched a game of billiards. Are we to understand that God experiences the same sensations in the contemplation of his work that we do in watching a game of billiards? If it is not our intention to give his assertion this fantastic meaning and if we do not wish to give it the more restricted meaning I have already mentioned, which is the sound meaning, then it has no meaning at all. Hypotheses of this kind have therefore only a metaphorical sense. (SH, pp. 163–164)

Thus a good convention reflects, by means of its syntactical structure, some deep reality which it cannot directly signify. Convenience, and convenience alone, operates like an index of verisimilitude. A hypothesis proves the more convenient, the nearer it is to the truth; but, to repeat, Poincaré uses an intuitive notion of verisimilitude which does not depend on the correspondence theory. It is a purely syntactical notion based on the mathematical structure of a given proposition, not on any semantic relation between its descriptive terms and some external reality. Poincaré subscribes to a quasi-Kantian thesis which allows us to simulate the noumenal world, but not to refer to it directly. This form of Kantianism has considerable methodological implications. For example, Poincaré is a thoroughgoing anti-essentialist. A convention is fundamentally incapable of expressing the essence of things, since it does not directly express *anything* at all. It is illegitimate to adduce in favour of a given hypothesis that it posits a particularly intelligible ontology, for no convention has any metaphysical presuppositions or pretensions. To repeat: only the degree of convenience of a hypothesis reveals its truthlikeness. Poincaré will therefore be ruthless in rejecting a theory which has become inconvenient, no matter how 'intelligible' it might initially have seemed; for he attributes a purely psychological status to the notion of intelligibility.

Attacking once again the mechanistic programme, he wrote:

> For a mechanical explanation to be good it must be simple; to choose it from among all the explanations that are possible there must be other reasons than

> the necessity of making a choice. Well, we have no theory as yet which will satisfy this condition and consequently be of any use. Are we then to complain? That would be to forget the end we seek, which is not mechanism; the true and only aim is unity. (SH, p. 177)

As regards the relationship between convenience and simplicity on the one hand and verisimilitide on the other, he said:

> The imposing simplicity of Mayer's principle equally contributes to strengthen our faith. In a law immediately deduced from experiments, such as Mattiotte's law, this simplicity would rather appear to us a reason for distrust; but here this is no longer the case. We take elements which at first glance are unconnected; these arrange themselves in an unexpected order and form a harmonious whole. We cannot believe that this unexpected harmony is a mere result of chance. (SH, p. 131)

As contrasted with Lorentz, Poincaré does not believe in the ontology of classical physics because dynamics, being a convention, can be true of no ontology at all. Although he initially claimed that classical mechanics both is and will always be the most convenient system, he also conceded that absolute time does not constitute a necessary a priori notion. According to Poincaré, we have no direct intuition of the simultaneity of two events which are spatially separated or which take place in two distinct consciousnesses. Only the temporal order within one consciousness is given, i.e., does not depend on our volition. However, physical (coordinate) time requires a stipulation and Poincaré accepts *the absolute time convention* only provisionally: "Provisionally, then we shall admit absolute time and Euclidean geometry" (SH, p. 91). As for the Galilean transformation which, until the very end, Lorentz regarded as the only 'true' one, Poincaré did not hesitate in throwing it overboard as soon as it proved inconvenient; he replaced it by the Lorentz transformation, which alone was operative.

§5.8 Geometry in the Light of the Empiricist Criterion

Poincaré's empiricist criterion of meaning enables us to understand better his conception of the foundations of geometry. He claims that geometry is not about physical space but about ideal figures immersed in an abstract manifold. A geometrical figure A′ is obtained from a concrete object A through an idealising process which consists in simplifying A so as to make it accessible to mathematical reasoning. The discrepancy between A′ and A will be attributed to the action of certain forces whose resultant is denoted by $\vec{D}$. Let P be the physical law governing these forces, which, in Reichenbach's terminology (Reichenbach 1958, first part §6), are said to be differential if they either depend on the physical constitution of A or else can be shielded against; such forces thus generally vary from one substance to another. Letting G denote underlying geometry, it can be asserted that only the conjunction (G Λ P) taken as a whole can be pitted against experience. Taken in isolation, G is thus irrefutable. Since G moreover contains universal propositions, it cannot be verified either; G is thus a convention placed beyond the reach of experience. In this sense and in conformity with Poincaré's empiricist meaning criterion, a geometry is neither true nor false but more or less convenient.

In affirming that G can be tested only in conjunction with some physical theory, we have so far done nothing but pose a Duhem-Quine problem. In the case of the geometry G, Poincaré solves this problem, so to speak negatively, by determining a translation mapping G into another arbitrary geometry G_o. Let A″ be the ideal body of which G_o treats. Poincaré effectively constructs a physical theory P_o which explains the transition from A″ to A in two steps: P_o postulates, on the one hand, a universal field $\vec{U}$ which uniformly affects all substances; the presence of $\vec{U}$ entails the alteration of A″ into A′. On the other hand, like P, P_o asserts the existence of $\vec{D}$ which effects the transition from A′ to A. Putting it more directly, if a little crudely, it can be said that P_o posits ($\vec{U} + \vec{D}$), while P posits only $\vec{D}$.

Since the two systems (G Λ P) and (G_o Λ P_o) obviously have the same empirical content, the choice between them is, at any rate in Poincaré's view, a matter of pure convention. Taking stock of this conclusion, Reichenbach stipulated in 1927 that G must be so chosen as to make all universal fields vanish; which allows Reichenbach to give general relativity a sort of methodological priority. To conclude this short survey of Poincaré's philosophy of geometry, let us mention one important difference which separates it from Reichenbach's. Poincaré readily envisages ideal objects such as A′ which, though fictitious, are nonetheless perfectly legitimate. Reichenbach's more purist form of empiricism forbids the use of such devices. By means of so called coordinative definitions, Reichenbach tries to set up a correspondence between basic geometrical notions on the one hand and concrete objects such as A on the other; but, since geometry does not directly apply to any real objects, he has to correct A by means of ($-\vec{D}$) while avoiding the construction of ideal entities like A′. Needless to say, such methods are both less elegant and less clear than those used by Poincaré.

§5.9 Convenience and the Correspondence Principle

Let us try to clarify the notion of convenience which plays a central role in Poincaré's philosophy of science. This notion possesses two distinct aspects. At the theoretical level, the degree of convenience of a hypothesis is identical with that of its unity; it expresses itself, as was said above, by the multiple connections between the various components of the hypothesis in question.

It has also been noted that theoretical unity differs from pragmatic simplicity. At the empirical level, the hypothesis will be judged convenient if it entails certain experimental laws and either makes novel predictions or else establishes unexpected connections between known facts; all this without resorting to any auxiliary assumptions incoherent with the central hypothesis. These two aspects are thus complementary: a theory is convenient to the extent that it accounts for experience while preserving its unity. Under the impact of facts, a theory can disintegrate; in such a case, one should avoid local manoeuvres whose net effect is to fragment the theory while trying to bring it into line with the facts. It is preferable to rebuild anew, but on condition of satisfying the correspondence principle which Poincaré justifies as follows: if an old hypothesis H turns out to have been systematically 'convenient' throughout some domain Δ, it is improbable that this should be due to pure chance; it can be assumed that H reveals true relations which ought to reappear, perhaps in a slightly modified form, within a new theory T. In other words, T must tend to H whenever certain parameters, by tending to zero, restrict T to the domain Δ. Realising that classical thermodynamics is probably false and should therefore be replaced by an atomic hypothesis, Poincaré wrote:

> These conceptions up to now have always been confirmed by experiment and the proofs today are numerous enough that they cannot be attributed to chance. It will therefore be necessary if new experiments reveal exceptions, not to abandon the theory but to modify it, to make it more comprehensive so as to permit it to include new facts. (LE, p. 77)

The following quotation, taken from a paper presented at the 1904 Saint Louis Conference, summarises Poincaré's conception of a scientific revolution and his views on conventions, experiment and future mathematical physics:

> Suppose now that all these efforts fail and, after all I do not believe they will, what must be done? Will it be necessary to seek to mend the broken principles by giving what we French call a *coup de pouce?* That evidently is always possible and I retract nothing of what I have said above . . .
>
> But then, what have we gained by this stroke? The principle is intact, but thenceforth of what use is it? It enabled us to foresee that in such or such circumstance we could count on such a total quantity of energy; it limited us; but now that this indefinite provision of energy is placed at our disposal, we are no longer limited by anything and, as I have written in 'Science and Hypothesis' if a principle ceases to be fecund, experiment, without contradicting it directly, will nevertheless have condemned it.
>
> This, therefore, is not what would have to be done; it would be necessary to rebuild anew . . .
>
> As I said we have already passed through a similar crisis. I have shown you that in the second mathematical physics, that of the principles, we find traces of the first, that of central forces; it will be just the same if we must know a

> third. Just so with the animal that exuviates, that breaks its too narrow carapace and makes itself a fresh one, under the new envelope one will recognise the essential traits of the organism which have persisted. (VS, pp. 109–110)

Thus, scientific progress consists not in carrying out ad hoc modifications designed to rescue an existing theory but in starting afresh, making sure however that the old superseded theory is a limiting case of the new one.

Having given a global description of Poincaré's philosophy of science, we can now compare it with Popper's. Such a comparison helps towards a better understanding both of Poincaré's and of Popper's positions. It appears at first sight that we are here dealing with two completely antagonistic viewpoints. Whereas Popper is a thoroughgoing anti-inductivist, Poincaré looks upon induction as the basis of the whole scientific—and mathematical—enterprise. Poincaré takes the incorrigibility of 'crude' propositions for granted; Popper is prepared to contest the truth and reliability of basic statements. Finally, with regard to the most abstract theories of mathematical physics, Popper adopts a realist attitude incompatible with Poincaré's conventionalism. However, it can be meaningfully asked whether such philosophical divergences ever surface at the strictly methodological level. The answer I propose to vindicate is: maybe, but very little.

Let us first note that Poincaré regards induction as an instrument of research based on a metaphysical thesis which is loosely connected with Carnot's principle. Thus physical induction enables us to *discover* new laws, but not to *validate* them. Only novel facts can either provisionally confirm a law or else definitively refute it. In this sense Poincaré distinguishes between the context of discovery and that of justification. Moreover, with respect to all generalisations, Poincaré rejects all conventionalist stratagems whose sole purpose is to immunise a law against empirical refutations. That, contrary to Popper, Poincaré believes in the absolute reliability of crude facts, means in effect that his brand of falsificationism is more uncompromising than Popper's. In Lakatos's terminology, Poincaré turns out to be a naive falsificationist.

As for Popper's realism concerning fundamental hypotheses, it proves far less operative at the methodological level than his anti-essentialism. Let us not forget that the *Logic of Scientific Discovery* does not make use of Tarski's correspondence theory; neither does it contain any notion of verisimilitude, which shows that Popperian methodology is largely independent of metaphysical realism.

Exactly like Poincaré, Popper judges a theory to be acceptable only if it provides a non ad hoc explanation of the facts, no matter how the latter are captured. He appraises a theory in terms only of its experimental consequences and of what he calls its organic compactness. He categorically refuses to attribute a privileged status to any ontology, and this irrespective of any intelligibility it might be claimed to possess. It has already been shown that the same attitude flows from Poincaré's conventionalism.

Such convergences between Popper's and Poincaré's methodological views should cause no surprise: although the Popperian *demarcation criterion* is different from the falsificationist *criterion of meaning*, there exists between these two criteria an undeniable parallel. To assert that a scientific law can only enter into *conflict* with experience is methodologically equivalent to saying that the law has meaning only if it is empirically *refutable*.

§5.10 Prelude to Relativity

With regard to conventions, we have already seen that Poincaré put forward two theses and made two prognoses. He maintained that we should always be in a position to keep both classical mechanics and Euclidean geometry, provided we were ready to alter the rest of physics. These two theses, more particularly the second one, have a logical character and can therefore not be contradicted. Poincaré also predicted that classical mechanics and Euclidean geometry would always prove the most convenient systems and would thus have nothing to fear from fresh experiments. Ironically, by criticising Lorentz's theories and then launching the relativity programme, Poincaré refuted the first prognosis. Unfortunately he did not live long enough to accept, as he no doubt would have done, Einstein's refutation of his second prediction.

I shall now present a survey of the most important results obtained by Lorentz, then reconstruct Poincaré's position from the criticisms which he levelled at these results.

Taking the velocity of light to be equal to 1, we can write the Maxwell-Lorentz equations in the form:

$$(1)\ \nabla . \vec{E} = \rho;\ \nabla . \vec{H} = 0;\ \nabla \times \vec{E} = -\frac{\partial \vec{H}}{\partial t};\ \nabla \times \vec{H} = \frac{\partial \vec{E}}{\partial t} + \rho \vec{v}$$

To these relations Lorentz adds an equation for the ponderomotive force $\vec{L}$ which acts on a particle moving with the velocity $\vec{v}$ and carrying charge e:

$$(2)\ \vec{L} = e(\vec{E} + \vec{v} \times \vec{H})$$

Lorentz then proposes to determine the field from the movement of the electrons or, more generally, from the distributions of the velocity $\vec{v}$ and of the electric density ρ: he is thus led to consider a partial differential equation of the form (see §2.2 above)

(3) $\Box\varphi = \psi(t,x,y,z) = \psi(t, \vec{r}),$

where is φ the unknown function of t, x, y, z; $\vec{r} = (x,y,z)$ and $\Box$ is the operator defined by:

(4) $\Box = \frac{\partial^2}{\partial x^2} + \frac{\partial^2}{\partial y^2} + \frac{\partial^2}{\partial z^2} - \frac{\partial^2}{\partial t^2}$ (d'Alembertian).

Lorentz finds the following solution:

(5) $$\varphi(t, \vec{r}) = -\frac{1}{4\pi}\iiint \frac{1}{|\vec{r} - \vec{r}\,'|} \psi(t - |\vec{r} - \vec{r}\,'|, x', y', z').dx'.dy'.dz',$$

where $\vec{r}\,'$ is the vector (x′,y′,z′) with respect to which the integration is carried out.

Basing himself on equations (1) and (2), Lorentz had also calculated the resultant $\vec{R}$ of all the forces which act on a system on charged particles occupying a bounded region of space:

(6) $$\vec{R} = -\frac{d}{dt}\left[\iiint (\vec{E} \times \vec{H})dw\right],$$

dw being the differential element dx.dy.dz and the integral being extended to the whole of space.

Starting from equation (6), Lorentz had determined the electromagnetic mass of the electron; he had examined the latter's self-field and found that this field exerts on its source a braking force equal to (see §2.7 & §7.2):

(7) $$-\frac{d}{dt}\left[\frac{\mu_o \vec{v}}{\sqrt{1 - v^2}}\right],$$

where μ_o is, to within a constant factor, the electrostatic energy of the electron. Thus $\mu_o = k\epsilon/c^2 = k\epsilon$, where: ϵ denotes the electrostatic energy, k is a constant and c the velocity of light, which we have taken to be 1. This expression clearly shows that the energy ϵ possesses the inertial equivalent $k\epsilon/c^2$. $\mu_o/\sqrt{1 - v^2}$ thus represents electromagnetic inertia, which varies with the speed v of the particle and is added to its material mass m. Lorentz assumed m to be constant. Thanks to certain experiments carried out by Kaufmann, he had been able to conclude that the material mass m of the negatively charged electron must vanish. The electromagnetic mass thus accounts for the total inertia of the electron.

Let us now go back to solution (5) which holds only in a frame at rest in the ether. Since all measurements are carried out in a coordinate system fixed in the earth, it is important to determine the form assumed by Maxwell's equations in a moving frame. Lorentz naturally used the Galilean transformation which, until the very end, he regarded as the only legitimate one. The following relations constitute a Galilean transformation:

(8) $\bar{x} = x - ut;\ \bar{y} = y;\ \bar{z} = z;\ \bar{t} = t.$

From (8) it follows that:

(9) $\dfrac{\partial}{\partial x} = \dfrac{\partial}{\partial \bar{x}};\ \dfrac{\partial}{\partial y} = \dfrac{\partial}{\partial \bar{y}};\ \dfrac{\partial}{\partial z} = \dfrac{\partial}{\partial \bar{z}};\ \dfrac{\partial}{\partial t} = \dfrac{\partial}{\partial \bar{t}} - u\dfrac{\partial}{\partial \bar{x}}$

The operator $\square$ becomes:

(10) $$\square = \frac{\partial^2}{\partial x^2} + \frac{\partial^2}{\partial y^2} + \frac{\partial^2}{\partial z^2} - \frac{\partial^2}{\partial t^2}$$

$$= (1 - u^2)\frac{\partial^2}{\partial \bar{x}^2} + \frac{\partial^2}{\partial \bar{y}^2} + \frac{\partial^2}{\partial \bar{z}^2} - \left(\frac{\partial^2}{\partial \bar{t}^2} - 2u\frac{\partial^2}{\partial \bar{x}\,.\,\partial \bar{t}}\right)$$

Given the availability of solution (5) for equation (3), it proves *mathematically* convenient to find a transformation:

$$(\bar{x},\bar{y},\bar{z},\bar{t}) \mapsto (x',y',z',t'),$$

which brings the operator

$$(1 - u^2)\frac{\partial^2}{\partial \bar{x}^2} + \frac{\partial^2}{\partial \bar{y}^2} + \frac{\partial^2}{\partial \bar{z}^2} - \left(\frac{\partial^2}{\partial \bar{t}^2} - 2u\frac{\partial^2}{\partial \bar{x}.\partial \bar{t}}\right)$$

back to the form

$$\left(\frac{\partial^2}{\partial x'^2} + \frac{\partial^2}{\partial y'^2} + \frac{\partial^2}{\partial z'^2} - \frac{\partial^2}{\partial t'^2}\right),$$

or at any rate to the form

$$\left(\frac{\partial^2}{\partial x'^2} + \frac{\partial^2}{\partial y'^2} + \frac{\partial^2}{\partial z'^2} - h^2\frac{\partial^2}{\partial t'^2}\right),$$

where h is some constant.

As already shown, strictly mathematical considerations led Lorentz to envisage the following transformation:

(11) $x' = \gamma\bar{x};\ y' = \bar{y};\ z' = \bar{z};\ t' = \bar{t} - u\gamma^2\bar{x};$

where γ is the constant $1/\sqrt{1 - u^2}$

Equations (11) imply:

$$(11') \quad \Box = \nabla'^2 - \gamma^2\frac{\partial^2}{\partial t'^2} = \frac{\partial^2}{\partial x'^2} + \frac{\partial^2}{\partial y'^2} + \frac{\partial^2}{\partial z'^2} - \gamma^2\frac{\partial^2}{\partial t'^2}$$

The product of the two transformations (8) and (11) is:

$$(12) \quad x' = \gamma\bar{x} = \gamma(x - ut);\ y' = \bar{y} = y;\ z' = \bar{z} = z;$$
$$t' = \bar{t} - u\gamma^2\bar{x} = t - u\gamma^2(x - ut) = \gamma^2(t - ux).$$

Note: to within an extra factor γ in the expression of t', equations (12) constitute what Poincaré subsequently called the 'Lorentz transformation.'[1] It can be said that special relativity was born, in 1892, on the day when Lorentz decided to provide equations (12) with a physical interpretation. Since $x' = \gamma\bar{x} = \bar{x}/\sqrt{1 - u^2}$, it follows that $\bar{x} = x'\sqrt{1 - u^2}$; i.e. $\bar{x}$ is obtained through contracting x' by the factor $\sqrt{1 - u^2}$. As shown above, Lorentz proved that a translation through the ether alters the form and the intensity of the ponderomotive force. A charge in motion is equivalent to a current giving rise both to an electric and to a magnetic field; together, these exert on the electron a force different from a purely electrostatic action. Assuming that the molecular forces also represent a state of the ether and are therefore affected by motion in this medium in the same way as are their electromagnetic counterparts, Lorentz showed that all rigid bodies must be contracted by the factor $\sqrt{1 - u^2}$ along the direction of their velocity $\vec{u}$. In this way he accounted for the null result of Michelson's experiment (see §2.4). In his *Versuch* of 1895 he also pursued a parallel approach which consisted, on the one hand, in neglecting all second order quantities in u and, on the other, in jointly transforming the coordinates and the field in such a way that Maxwell's equations keep their form.[2] If one neglects all second order terms, equations (12) become

$$(13) \quad \begin{array}{l} x' = \gamma\bar{x} \doteqdot (x - ut), \text{ since } \gamma = (1 - u^2)^{-1/2} \doteqdot 1;\ y' = \bar{y} = y; \\ z' = \bar{z} = z \text{ and finally: } t' = \gamma^2(t - ux) \doteqdot t - ux \end{array}$$

Let us call t' the local time and put:

$$(14) \quad \vec{E}' = \vec{E} + \vec{u} \times \vec{H};\ \vec{H}' = \vec{H} - \vec{u} \times \vec{E}.$$

It is easily established that in the free ether:

$$(15) \quad \nabla'.\vec{E}' = 0 = \nabla'.\vec{H}';\ \nabla' \times \vec{E}';\ -\frac{\partial \vec{H}'}{\partial t'};\ \nabla' \times \vec{H}' = \frac{\partial \vec{E}'}{\partial t'};$$

where ∇' is the operator $(\partial/\partial x', \partial/\partial y', \partial/\partial z')$

1. For more details, see Chapter Two, Section (II).
2. See, above, Chapter Two, Section (V).

To repeat: throughout these operations, all second order quantities are neglected. Equations (15) obviously have the same form as those holding in the absolute frame. From this result Lorentz concluded that no uniform translation in the ether gives rise to any first order effect. In other words, the first order covariance of Maxwell's theory entails the absence of all effects of that same order. Special relativity can be considered as a generalisation of this so-called theorem of corresponding states (see §2.6). It is worth mentioning that Lorentz had taken a first step towards this generalisation by showing how the contraction hypothesis explains the absence of second-order effects, at any rate in the case of Michelson's result.

No doubt because of his commitment to the concept of absolute time, Lorentz never managed, or was never eager, to interpret t′ realistically. In most applications, he maintained that the difference between t and t′ was so negligible that these two quantities could be safely identified. In fact, not only did he omit to give a physical interpretation of t′, he also continued to regard the Galilean transformation as the only legitimate one; he called the quantities x′, y′, z′ and t′, *effective* coordinates; i.e. he looked upon them as mere devices enabling him to carry out certain calculations and make empirical predictions. Thus, in a moving frame, $\bar{x}$ remains the 'true' abscissa but, due to the contraction of all measuring instruments, the measured coordinate will be equal, not to $\bar{x}$, but to $x' = \bar{x}/\sqrt{(1 - u^2)}$.

Let us go back to Poincaré who, with regard to space, claimed that we observe only relative motions, not the movement of matter with respect to the ether. This *crude fact* leads us instinctively to subscribe to a universal principle of relativity. This instinctive belief is prior to all articulated hypotheses. As for the existence of the ether, it is to be treated as an acceptable convention only as long as it proves convenient. Poincaré formulated the relativity principle as follows: all natural laws assume the same form in all inertial frames. This is precisely Einstein's formulation but, in SH, Poincaré linked the relativity principle to the Galilean group. He added however that, because of our general adherence to this principle, we would have wished to extend it to all frames without exception. We are stopped only by the crude fact that accelerated motions give rise to easily detectable effects.

Concerning the notion of time, there exists, according to Poincaré, only one crude fact which cannot be ignored; namely the temporal order obtaining within *one* consciousness. We have no intuitive idea of the equality of two durations or of the simultaneity of two events which are either spatially separated or take place in the minds of two distinct individuals. Thus physical time presupposes a convention, which ought to be judged only by the degree of convenience possessed by the hypotheses using this convention. We already understand why Poincaré, not being committed to the notion of absolute time, will readily abandon it in favour of some more convenient concept, e.g. that of local time.

In his 1895 article "Concerning a Theory of M Larmor", Poincaré maintains that Lorentz's hypothesis, whilst remaining superior to its rivals, violates the two principles of relativity and of reaction. It is obvious, on the one hand, that Maxwell's equations are not covariant with respect to the Galilean group, i.e. with respect to transformation (8). On the other hand, since the integral $\iiint(\vec{E} \times \vec{H})dw$ is not generally constant, equation (6) entails that the resultant $\vec{R}$ of all pondermotive forces acting on an isolated system of charged particles does not necessarily vanish. Thus total momentum is not conserved, at any rate if we restrict ourselves to matter alone; i.e. if we take no account of the ether. Poincaré concedes that the action of ponderable matter could be regarded as being instantaneously compensated by a reaction of the neighbouring ether; but this would imply that the ether is in motion, which contradicts one of Lorentz's initial assumptions. Towards the end of the article, Poincaré makes a profound remark which, however, remains unclarified; namely that the violation of the relativity principle is intimately connected with that of Newton's third law (principle of reaction).

In a chapter of his *Électricité et Optique* (1901), then again in his paper "Lorentz's Theory and the Principle of Reaction" (1900), Poincaré offered a tentative solution to this problem. He knew that Lorentz had used an uninterpreted notion of 'local time' in order to explain the absence of all first-order effects attributed to the earth's motion. As already explained, Poincaré's methodological position enjoined him to make all scientific theories testable by means of crude facts. The operations constituting such facts generally involve time measurements which, as noted above, are made possible by certain stipulations. We have also seen that the local time t' plays a central role in the empirical predictions of Lorentz's theory. Poincaré was thus led to give an operational definition of t' by means of a convention, which was later attributed to Einstein: two clocks, A and B, at rest with respect to each other, are said to be synchronous if a light ray, having left A at A-time 0, reaches B at B-time t, is instantaneously reflected at B and goes back to A, where it arrives at A-time 2t. The time which is thus determined is nothing but Lorentz's local time t'. Simultaneity will therefore depend on the state of motion of the system AB, i.e. on the chosen frame of reference; but it is impossible for observers at rest within this frame to realise that their local time is not the absolute one.

Let us now show in what way the principle of reaction, i.e., Newton's third law, involves the notion of simultaneity. Consider one electron A which acts on another electron B; where A and B are separated by a distance l. An electromagnetic disturbance leaving A will be propagated towards B with the speed of light c; A will have resumed a state of rest when, after time l/c, the disturbance reaches B. Thus the action on B is *not instantaneously* compensated by a reaction, which will reach A only after a second interval of time l/c. Maxwell's equations therefore violate Newton's third law, because they entail the propagation of waves at a finite velocity incompatible with the simultaneity of action and reaction. We also begin to understand how Newton's third law connects with classical relativity: classical (Galilean) relativity rests on the notion of absolute simultaneity, which is also presupposed by the principle of reaction. Thus it is hardly surprising that Lorentz's hypothesis, which is incompatible with Galilean relativity, should also transgress Newton's third law. According to Poincaré, we would have jettisoned the relativity principle, had it proved irreconcilable both with experience and with our accepted theories: but the crude facts themselves, in the form of Michelson's results, had stubbornly confirmed the relativity of optical phenomena. Poincaré accused Lorentz of having accounted for these results through a whole cluster of hypotheses. This criticism is, to my mind, unjustified: after all, Lorentz had made only one additional assumption, namely the molecular forces hypothesis (MFH), which roughly says that all forces transform like the Lorentz-force (see §2.4 & §2.6). This natural assumption entails both the Lorentz-Fitzgerald contraction and, as subsequently realised, the retardation of moving clocks. Poincaré would have been right to say, not that Lorentz had 'accumulated hypotheses', but that he had failed to connect the MFH with his concept of local time. In fact, Lorentz was in no position to establish such a connection since he had simply omitted to interpret t'. The synchronisation convention proposed by Poincaré filled this gap. Poincaré however expected, both of himself and of others, much more than that: already in SH he regarded the absence of first and second order optical effects, not as *accidental* phenomena, but as special cases of a universal principle. In conformity with his philosophy and by virtue of the relativity principle governing classical mechanics, Poincaré had *generalised* this principle by asserting that uniform motion gives rise to no detectable effects whatsoever. He elevated relativity to the rank of a postulate which applies not only to mechanics but also to electromagnetism. At the 1904 Saint Louis conference, he underlined the necessity of creating a unified theory in order to account for certain electromagnetic phenomena which seemed disparate and accidental. It was in fact Poincaré himself who later constructed such a theory, in 1905. We recall that a convention is convenient, in Poincaré's sense, if it is unified and if it accounts for experience without resorting to ad hoc hypotheses. It should be noted that in Lorentz's system all central experimental results were explained by recourse to effective coordinates, i.e. to quantities which, being *effectively measured,* describe crude facts. The Galilean transformation, though taken to be the only 'true' one, breaks the transition from the 'absolute' frame to the effective coordinate system. Since he denied scientific hypotheses any *direct* ontological reference, Poincaré, unlike Lorentz, was ready to sweep the Galilean coordinates out of existence and base relativity on the Lorentz transformation, i.e., on the effective coordinates x′, y′, z′, t′. (Cf. eqs. (12).) Thus he obtained the first component of the relativity programme, namely the requirement of Lorentz-covariance. As for the second component, the correspondence principle, we have already seen that Poincaré regarded it as indispensable for any research programme. In order for the relativity principle to be satisfied, the material mass of a particle, in case it exists at all, must vary with the velocity in the same way as does its electromagnetic inertia; which means: material inertia $= km/\sqrt{(1 - v^2)}$, where m is the rest mass and k a constant. Finally, the contraction of moving bodies could be explained in terms not of any physical forces which uniformly affect all substances, but of the kinematical properties of the ether itself. Let us not forget that, for Poincaré, all references to the ether are mere *façons de parler,* the ether being in fact described by the structure of space-time.

Poincaré's new relativistic hypothesis appeared in 1905 under the title "On the Dynamics of the Electron".

§5.11 The Dynamics of the Electron

Let us summarise the conclusions reached above. Together with Einstein but independently of him, Poincaré created the relativity programme. On the one hand, Poincaré's empiricist conception of meaning led him towards an operational definition of Lorentz's local time t′: the local time is that which is measured by a clock synchronised according to the so-called Einstein convention: a light signal leaves some point A at A-time t_1 and arrives, at B-time t_2, at another point B, where the signal is reflected; it then goes back and reaches A at A-time t_3. The two clocks at A and B are said to be synchronous if and only if: $t_2 = ½(t_1 + t_3)$. On the other hand, Poincaré definitively rid physics of the Galilean transformation which Lorentz had always interposed between the ether frame and the effective coordinates; i.e. between x, y, z, t and: $x' = \gamma(x - ut)$, $y' = y$, $z' = z$, $t' = \gamma(t - \frac{u}{c^2}x)$, where $\gamma \underset{\text{Def}}{=} 1/\sqrt{(1 - u^2/c^2)}$. Poincaré considers the effective coordinates as the only relevant ones. In opposition to Lorentz, he is not wedded to the classical ontology, to the doctrine of the ether and of absolute time which makes the Galilean transformation the only intelligible one. This is because, according both to Poincaré's conventionalism and to his empiricist meaning criterion, hypotheses are (in the sense of the correspondence theory) neither true nor false but more or less convenient. The degree of convenience, or of mathematical perfection, operates like an *index of verisimilitude*. Through its global structure, a convenient theory can simulate an objective reality which it cannot directly describe. Though 'meaningless', theories can nonetheless be verisimilar.

Poincaré is thus a Kantian of sorts: scientific hypotheses can mirror the noumena without being able to apprehend them. In the last analysis, absolute time and absolute space (the ether) are devices whose only function is to assist our imagination; they do not reveal the deep nature of things as they really are and should therefore be rejected as soon as they prove cumbersome. Since the Galilean coordinates had become inconvenient, Poincaré eliminated them in favour of the Lorentz-transformation, a transformation which goes directly from the rest-frame to the effective coordinates.

Poincaré chose his units so as to make the speed of light equal to 1; he then put forward a slightly more general transformation than the one proposed by Lorentz:

$$x' = k\ell(x + t),\ y = \ell y,\ z' = \ell z,\ t' = k\ell(t + x),$$

where: $k \underset{\text{Def}}{=} 1/\sqrt{(1 - \epsilon^2)}$, $(-\epsilon)$ = speed of the moving frame, and ℓ is a parameter provisionally taken to be independent of ϵ.

Poincaré also set up the heuristic on which special relativity was to be based. Every physical law is subject to two conditions: it must, on the one hand, be covariant with respect to the Lorentz group, i.e., it must assume the same form in all inertial frames; on the other hand, it should, whenever possible, tend towards some corresponding classical law when the envisaged velocities are small in comparison with c, i.e., in comparison with 1. Thus, the two pillars of the relativity programme are: the covariance requirement and the correspondence principle.

Poincaré used the new heuristic in order both to reconstruct and to correct Lorentz's transformation rules for the electromagnetic field (see §2.6). In this way, he developed a new kinematics and then used this kinematics, together with the conservation law for the charge, in order to modify Lorentz's transformation of the electric density. In 1905 he went well beyond Lorentz's and Einstein's achievements: he proposed a Lorentz-covariant theory of gravitation predicting the propagation of gravitational waves at the speed of light. Poincaré also asked a question which must have occurred to Einstein, namely: why should the speed of light constitute *the* basic invariant and thus play a fundamental role, not only in electrodynamics, but in all other branches of physics? To this question he imagined two possible answers: either all forces have, in the last analysis, some common electromagnetic origin, or it is the observer himself who inserts the velocity of light into nature through using light signals in order to measure distance and time intervals. Poincaré thus envisages two answers of very different philosophical character; the first is realistic and the second partly conventionalistic. He is aware of a constant tension between the scientist who creates conventions in order to interpret nature and nature, which sets definite limits to the otherwise free activity of the scientist.

In order to vindicate the claims made on Poincaré's behalf in the last paragraph, I shall examine in some detail the most important sections of "On the Dynamics of the Electron". The most fascinating aspect of this text is that in it one can grasp, as it were live, the creative thought processes of a genius; Poincaré can be observed constructing, step by step, the various components of a new system. Given his mathematical intuition, it often proves difficult to follow his reasoning; it is necessary to reconstruct, in the form of a proof, what he regards as *one* obvious logical step. However, the two constant regulative ideas of his relativity programmes remain those already mentioned, namely the principles of covariance and of correspondence.

There is another aspect of "On the Dynamics of the Electron" which makes Poincaré's arguments difficult to follow: he uses no vector notation. It is often difficult to identify a divergence, a curl, or even a simple vector product. I shall therefore attempt to simplify the proofs by the systematic use of three- and four-vectors.

§5.11.1 Kinematics, Electromagnetism and Dynamics

We have already said that Poincaré takes the speed of light to be equal to 1 and that he writes the Lorentz transformation in the form:

(1) $x' = k\ell(x + \epsilon t)$; $y' = \ell y$; $z' = \ell z$; $t' = k\ell(t + \epsilon x)$; where ϵ and ℓ are parameters initially taken to be independent of each other; and $k \underset{\text{Def}}{=} 1/\sqrt{1 - \epsilon^2}$ ($-\epsilon$ is in effect the velocity of the moving frame).

The inverse transformation is obtained by solving (1) for x, y, z, t in terms of x′, y′, z′ and t′. Thus:

$$\textbf{(2)}\ x = \frac{k}{\ell}(x' - \epsilon t');\ y = \frac{1}{\ell}y';\ z = \frac{1}{\ell}z';\ t = \frac{k}{\ell}(t' - \epsilon x').$$

It also proves useful to determine the transformation equations of the operators: $\partial/\partial x$, $\partial/\partial y$, $\partial/\partial z$, $\partial/\partial t$, which occur in Maxwell's equations.

$$\frac{\partial}{\partial x} = \frac{\partial x'}{\partial x}\cdot\frac{\partial}{\partial x'} + \frac{\partial t'}{\partial x}\cdot\frac{\partial}{\partial t'} = k\ell\frac{\partial}{\partial x'} + k\ell\epsilon\frac{\partial}{\partial t'}.$$

Hence:

$$\textbf{(3)}\ \frac{\partial}{\partial x} = k\ell\left(\frac{\partial}{\partial x'} + \epsilon\frac{\partial}{\partial t'}\right).\ \text{Similarly: } \frac{\partial}{\partial t} = k\ell\left(\frac{\partial}{\partial t'} + \epsilon\frac{\partial}{\partial x'}\right);\ \frac{\partial}{\partial y} = \ell\frac{\partial}{\partial y'}$$

$$\text{and } \frac{\partial}{\partial z} = \ell\frac{\partial}{\partial z'}.$$

The inverse transformations are:

$$\textbf{(4)}\ \frac{\partial}{\partial x'} = \frac{k}{\ell}\left(\frac{\partial}{\partial x} - \epsilon\frac{\partial}{\partial t}\right); \frac{\partial}{\partial t'} = \frac{k}{\ell}\left(\frac{\partial}{\partial t} - \epsilon\frac{\partial}{\partial x}\right); \frac{\partial}{\partial y'} = \frac{1}{\ell}\frac{\partial}{\partial y}; \frac{\partial}{\partial z'} = \frac{1}{\ell}\frac{\partial}{\partial z}.$$

We need to transform the velocities, which also occur in Maxwell's equation. Let $\vec{v} = (v_1, v_2, v_3)$ and $\vec{v}' = (v_1', v_2', v_3')$ be two corresponding velocities in the stationary and mobile frames respectively.

$$v_1' = \frac{dx'}{dt'} = \frac{d[k\ell(x + \epsilon t)]}{d[k\ell(t + \epsilon x)]} = \frac{k\ell(dx + \epsilon dt)}{k\ell(dt + \epsilon dx)} = \frac{(dx/dt) + \epsilon}{1 + \epsilon(dx/dt)};$$

i.e.

$$\textbf{(5)}\ v_1' = \frac{v_1 + \epsilon}{1 + \epsilon v_1}.\ \text{Similarly: } v_2' = \frac{v_2}{k(1 + \epsilon v_1)}; v_3' = \frac{v_3}{k(1 + \epsilon v_1))}.$$

We can also express v_1, v_2, v_3 in terms of v_1', v_2', v_3'. We have:

$$\textbf{(6)}\ v_1 = \frac{v_1' - \epsilon}{1 - \epsilon v_1'}; v_2 = \frac{v_2'}{k(1 - \epsilon v_1')}; v_3 = \frac{v_3'}{k(1 - \epsilon v_1')}.$$

We have already see that Poincaré writes Maxwell's equations in the form:

(7) $\nabla . \vec{H} = 0; \nabla . \vec{E} = \rho$, where ρ is the electric density;

$$\textbf{(8)}\ \nabla \times \vec{H} = \frac{\partial \vec{E}}{\partial t} + \rho\vec{v};$$

$$\textbf{(9)}\ \nabla \times \vec{E} = -\frac{\partial \vec{H}}{\partial t};$$

(10) $\vec{L} = e(\vec{E} + \vec{v} \times \vec{H})$, where $\vec{L}$ is the Lorentz force

acting on a particle whose charge is e and velocity $\vec{v}$.

Poincaré attacks the problem of transforming the electromagnetic field through considering the vector and scalar potentials: $\vec{A}$ and ψ. The equation $\nabla . \vec{H} = 0$ entails the existence of a vector $\vec{A}$ such that:

(11) $\vec{H} = \nabla \times \vec{A}$. $\vec{A}$ is called the vector potential.

Substituting from (11) into (9):

$$\nabla \times \vec{E} = -\frac{\partial}{\partial t}(\nabla \times \vec{A}) = -\nabla \times \left(\frac{\partial \vec{A}}{\partial t}\right);$$

i.e.

$$\nabla \times \left(\vec{E} + \frac{\partial \vec{A}}{\partial t}\right) = 0.$$

It follows that:

(12) $\vec{E} + \dfrac{\partial \vec{A}}{\partial t} = -\nabla\psi$ for some scalar potential ψ.

i.e. $\vec{E} = -\dfrac{\partial \vec{A}}{\partial t} - \nabla\psi.$

Since equations (11) and (12) do not uniquely determine $\vec{A}$ and ψ, we can impose the following constraint, also called Lorentz's condition:

(13) $\nabla . \vec{A} + \dfrac{\partial\psi}{\partial t} = 0$. i.e. $\nabla . \vec{A} = -\dfrac{\partial\psi}{\partial t}.$

The two equations (7) and (12) imply:

$$\begin{aligned}\rho &= \nabla . \vec{E} \\ &= -\nabla . \left(\frac{\partial \vec{A}}{\partial t} + \nabla\psi\right) = -\left(\frac{\partial}{\partial t}(\nabla . \vec{A}) + \nabla^2\psi\right) \\ &= -\left(\frac{\partial}{\partial t}\left(-\frac{\partial\psi}{\partial t}\right) + \nabla^2\psi\right),\end{aligned}$$

by virtue of (13).

(14) Hence: $-\rho = \left[\nabla^2 - \dfrac{\partial^2}{\partial t}\right]\psi \equiv \Box\psi$, where $\Box$ is the operator (called d'Alembertian) defined as follows:

(15) $\Box = \nabla^2 - \dfrac{\partial^2}{\partial t^2} = \dfrac{\partial^2}{\partial x^2} + \dfrac{\partial^2}{\partial y^2} + \dfrac{\partial^2}{\partial z^2} - \dfrac{\partial^2}{\partial t^2}.$

Similarly, equations (8), (11) and (12) imply:

$$\nabla \times (\nabla \times \vec{A}) = \frac{\partial}{\partial t}\left(-\frac{\partial \vec{A}}{\partial t} - \nabla\psi\right) + \rho\vec{v}\text{; i.e. } \nabla(\nabla . \vec{A}) - \nabla^2\vec{A}$$

$$= \rho\vec{v} - \frac{\partial^2 \vec{A}}{\partial t^2} - \nabla\left(\frac{\partial\psi}{\partial t}\right).$$

By virtue of (13), we obtain:

$$\textbf{(16)}\ \Box\vec{A} \equiv \left[\nabla^2 - \frac{\partial^2}{\partial t^2}\right]\vec{A} = -\rho\vec{v}$$

Equations (13), (14) and (16) enable us to calculate $\vec{A}$ and ψ in terms of p, $\vec{v}$ and the boundary conditions. Since these relations involve the d'Alembertian $\Box$, we are led to determine the transformation rule governing this operator. It follows from (3) and (15) that:

$$\Box \equiv \frac{\partial^2}{\partial x^2} + \frac{\partial^2}{\partial y^2} + \frac{\partial^2}{\partial z^2} - \frac{\partial^2}{\partial t^2} = \left[k^2\ell^2\left(\frac{\partial}{\partial x'} + \epsilon\frac{\partial}{\partial t'}\right)^2 + \ell^2\frac{\partial^2}{\partial y'^2} +\right.$$

$$\left.\ell^2\frac{\partial^2}{\partial z'^2} - k^2\ell^2\left(\frac{\partial}{\partial t'} + \epsilon\frac{\partial}{\partial x'}\right)^2\right] = \ell^2\left[\frac{\partial^2}{\partial x'^2} + \frac{\partial^2}{\partial y'^2} + \frac{\partial^2}{\partial z'^2} - \frac{\partial^2}{\partial t'^2}\right],$$

since $k \underset{\text{Def}}{=} 1/\sqrt{1 - \epsilon^2}$. i.e.

$$\textbf{(17)}\ \Box = \ell^2\left[\nabla'^2 - \frac{\partial^2}{\partial t'^2}\right] = \ell^2\,\Box'\text{. That is: } \Box' = \ell^{-2}\,\Box$$

According to the relativity principle, the density ρ' and the potential: $\vec{A}'$, ψ' in a moving inertial frame obey the following equations:

$$\textbf{(18)}\ \Box'\psi' = -\rho' \text{ and } \Box'\vec{A}' = -\rho'\vec{v}'.$$

Substituting from (17) and (5) into (18), we obtain:

$$\textbf{(19)}\ \Box\psi' = -\ell^2\rho' \text{ and } \Box\vec{A}'$$

$$= -\ell^2\rho'\left(\frac{v_1 + \epsilon}{1 + \epsilon v_1}, \frac{v_2}{k(1 + \epsilon v_1)}, \frac{v_3}{k(1 + \epsilon v_1)}\right)$$

Should we succeed in determining the transformation law for the charge density ρ, equations (19) will then enable us to transform $\vec{A}$ and ψ, hence also $\vec{E}$ and $\vec{H}$, since $\vec{E} = -\nabla\psi - \frac{\partial\vec{A}}{\partial t}$ and $\vec{H} = \nabla \times \vec{A}$. At this point

Poincaré applies the principles of the conservation and invariance of the electric charge. He considers a charged particle placed at the centre of a mobile sphere having a constant radius r and moving with the constant velocity $\vec{v} = (v_1, v_2, v_3)$. The equation of the sphere is:

(20)
$$(x - v_1t)^2 + (y - v_2t)^2 + (z - v_3t)^2 = r^2.$$

The charge density ρ at the centre of the sphere is thus given by:

$$\rho = e\Big/\frac{4}{3}\pi r^3 = 3e/4\pi r^3,$$ where e is the charge carried by the particle.

In virtue of transformation (2), the equation of the sphere referred to a moving frame is:

$$\frac{k^2}{\ell^2}[(x' - \epsilon t') - v_1(t' - \epsilon x')]^2 + \left[\frac{1}{\ell}y' - v_2\frac{k}{\ell}(t' - \epsilon x')\right]^2 + \left[\frac{1}{\ell}z' - v_3\frac{k}{\ell}(t' - \epsilon x')\right]^2 = r^2;\text{ i.e.}$$

(21)
$$k^2(1 + \epsilon v_1)^2\left[x' - \frac{v_1 + \epsilon}{1 + \epsilon v_1}t'\right]^2 + [y' + kv_2(\epsilon x' - t')]^2 + [z' + kv_3(\epsilon x' - t')]^2 = \ell^2r^2$$

We thus obtain an ellipsoid whose volume is equal to $4\pi\ell^3r^3/3k(1 + \epsilon v_1)$. Since the charge e is an invariant, the density ρ′ in the moving frame equals

$$e\Big/\left(\frac{4\pi\ell^3r^3}{3k(1 + \epsilon v_1}\Big/\right),\text{ i.e.}$$

(22)
$$\rho' = \frac{3e}{4\pi r^3}\cdot\frac{k(1 + \epsilon v_1)}{\ell^3}$$

Comparing (22) with the second equation of (20), we obtain the following transformation law for ρ:

(23)
$$\rho' = \frac{k\rho}{\ell^3}(1 + \epsilon v_1) = \frac{\rho(1 + \epsilon v_1)}{\ell^3\sqrt{1 - \epsilon^2}},$$ by definition of k.

Let us now go back to equations (19) which, by virtue of (23), become:

(24)
$$\Box\psi' = -\ell^2\rho' = -\frac{k}{\ell}\rho(1 + \epsilon v_1) = -\frac{k}{\ell}(\rho + \epsilon\rho v_1)\text{ and:}$$

$$\Box\vec{A}' = -\ell^2\rho'\vec{v}'$$
$$= -\frac{k\rho}{\ell}(1+\epsilon v_1)\left(\frac{v_1+\epsilon}{1+\epsilon v_1}, \frac{v_2}{k(1+\epsilon v_1)}, \frac{v_3}{k(1+\epsilon v_1)}\right)$$

Putting $\vec{A}' = (A_1', A_2', A_3')$, we have:

(25) $$\Box A_1' = -\frac{k}{\ell}(\rho v_1 + \epsilon\rho)\,;\ \Box A_2' = -\frac{1}{\ell}\rho v_2\,;\ \Box A_3' = -\frac{1}{\ell}\rho v_3.$$

We recall that $\Box\psi = -\rho$ and $\Box\vec{A} = -\rho\vec{v}$ (see (14) and (16)). Putting $\vec{A} = (A_1, A_2, A_3)$, we obtain:

(26) $$\Box\psi = -\rho;\ \Box A_1 = -\rho v_1;\ \Box A_2 = -\rho v_2;\ \Box A_3 = -\rho v_3.$$

Equations (24) and (26) entail:

$$\Box\psi' = -\frac{k}{\ell}(\rho + \epsilon\rho v_1) = \frac{k}{\ell}(\Box\psi + \epsilon\Box A_1) = \Box\left[\frac{k}{\ell}(\psi + \epsilon A_1)\right].$$

Similarly, (25) and (26) imply:

$$\Box A'_1 = \frac{k}{\ell}(\Box A_1 + \epsilon\Box\psi) = \Box\left[\frac{k}{\ell}(A_1 + \epsilon\psi)\right] \text{ and}$$
$$\Box A'_2 = \frac{1}{\ell}\Box A_2 = \Box\left(\frac{1}{\ell}A_2\right),\ \Box A'_3 = \Box\left(\frac{1}{\ell}A_3\right).$$

We conclude:

(27) $$\begin{cases} \Box A'_1 = \Box\left[\frac{k}{\ell}(A_1 + \epsilon\psi)\right];\ \Box A'_2 = \Box\left(\frac{1}{\ell}A_2\right];\ \Box A'_3 = \\ \qquad \Box\left(\frac{1}{\ell}A_3\right) \text{ and } \Box\psi' = \Box\left[\frac{k}{\ell}(\psi + \epsilon A_1)\right] \end{cases}$$

Equations (27) will thus be satisfied if we take:

(28) $$\begin{cases} A'_1 = \frac{k}{\ell}(A_1 + \epsilon\psi);\ A'_2 = \frac{1}{\ell}A_2; \\ \qquad A'_3 = \frac{1}{\ell}A_3 \text{ and } \psi' = \frac{k}{\ell}(\psi + \epsilon A_1). \end{cases}$$

In order for $\vec{A}'$ and ψ' to represent the vector and scalar potentials in the moving frame, it remains to be shown that Lorentz's condition is satisfied; i.e. we have to establish that

$$\nabla'.\vec{A}' + \frac{\partial\varphi'}{\partial t'} = 0.$$

We have:

$$\nabla'.\vec{A}' + \frac{\partial\psi'}{\partial t'} = \frac{\partial A'_1}{\partial x'} + \frac{\partial A'_2}{\partial y'} + \frac{\partial A'_3}{\partial z'} + \frac{\partial\psi'}{\partial t'}$$

$$= \frac{1}{\ell^2}\left[k^2\left(\frac{\partial}{\partial x} - \epsilon\frac{\partial}{\partial t}\right)(A_1 + \epsilon\psi) + \frac{\partial A_2}{\partial y} + \frac{\partial A_3}{\partial z} + k^2\left(\frac{\partial}{\partial t} - \epsilon\frac{\partial}{\partial x}\right)(\psi + \epsilon A_1)\right] \text{ (by (4) and (28))}$$

$$= \frac{1}{\ell^2}\left[k^2(1 - \epsilon^2)\frac{\partial A_1}{\partial x} + \frac{\partial A_2}{\partial y} + \frac{\partial A_3}{\partial z} + k^2(1 - \epsilon^2)\frac{\partial\psi}{\partial t}\right]$$

$$= \frac{1}{\ell^2}\left[\frac{\partial A_1}{\partial x} + \frac{\partial A_2}{\partial y} + \frac{\partial A_3}{\partial z} + \frac{\partial\psi}{\partial t}\right]$$

(since $k = (1 - \epsilon^2)^{-1/2}$

$$= \frac{1}{\ell^2}\left[\nabla.\vec{A} + \frac{\partial\psi}{\partial t}\right] = 0 \text{ (by (13)). Consequently:}$$

(29) $\nabla'.\vec{A}' + \dfrac{\partial\psi'}{\partial t'} = 0.$

In other words: Lorentz's condition is satisfied. Thus, A′ and ψ' can be taken to be the electromagnetic potentials in the moving frame; which allows us to write:

(30) $\nabla' \times \vec{A}' = \vec{H}'$ and $-\dfrac{\partial\vec{A}'}{\partial t'} - \nabla'\psi' = \vec{E}'.$

It remains for us to express $\vec{E}'$ and $\vec{H}'$ directly in terms of $\vec{E}$ and $\vec{H}$. From the second equation in (30) we deduce:

$$E'_1 = -\frac{\partial A'_1}{\partial t'} - \frac{\partial \psi'}{\partial x'} = -\frac{k^2}{\ell^2}\left[\left(\frac{\partial}{\partial t} - \epsilon\frac{\partial}{\partial x}\right)(A_1 + \epsilon\psi)\right.$$
$$\left. + \left(\frac{\partial}{\partial x} - \epsilon\frac{\partial}{\partial t}\right)(\psi + \epsilon A_1)\right] \text{ (by (4) and (28))}$$
$$= -\frac{k^2(1-\epsilon^2)}{\ell^2}\left[\frac{\partial A_1}{\partial t} + \frac{\partial \psi}{\partial x}\right] =$$
$$-\frac{1}{\ell^2}\left[\frac{\partial A_1}{\partial t} + \frac{\partial \psi}{\partial x}\right] \text{(since } k = (1-\epsilon^2)^{-1/2}\text{)}$$
$$= \frac{1}{\ell^2} E_1 \text{ (by (12), which entails } E_1 = -\frac{\partial A_1}{\partial t} - \frac{\partial \psi}{\partial x}\text{)}$$

Similarly:

$$E'_2 = -\frac{\partial A'_2}{\partial t'} - \frac{\partial \psi'}{\partial y'} = -\frac{k}{\ell}\left(\frac{\partial}{\partial t} - \epsilon\frac{\partial}{\partial x}\right)\left(\frac{A_2}{\ell}\right) -$$
$$\frac{1}{\ell}\frac{\partial}{\partial y}\left(\frac{k}{\ell}(\psi + \epsilon A_1)\right)$$
$$= \frac{k}{\ell^2}\left[-\frac{\partial A_2}{\partial t} + \epsilon\frac{\partial A_2}{\partial x} - \frac{\partial \psi}{\partial y} - \epsilon\frac{\partial A_1}{\partial y}\right]$$
$$= \frac{k}{\ell^2}\left[-\left(\frac{\partial A_2}{\partial t} + \frac{\partial \psi}{\partial y}\right) + \epsilon\left(\frac{\partial A_2}{\partial x} - \frac{\partial A_1}{\partial y}\right)\right]$$

Hence

$$E'_2 = \frac{k}{\ell^2}[E_2 + \epsilon H_3],$$

since:

$$\vec{E} = -\left(\frac{\partial \vec{A}}{\partial t} + \nabla\psi\right) \text{ and } \vec{H} = \nabla \times \vec{A}$$

In the same way we obtain:

$$E'_3 = \frac{k}{\ell^2}\left[E_3 - \epsilon H_2\right].$$

Thus:

(31) $E'_1 = \frac{1}{\ell^2}E_1;\ E'_2 = \frac{k}{\ell^2}(E_2 + \epsilon H_3);\ E'_3 = \frac{k}{\ell^2}(E_3 - \epsilon H_2).$

As for the transformation of $\vec{H}$, it results from the equation $\vec{H}' = \nabla' \times \vec{A}'$. Thus:

$$H'_1 = \frac{\partial A'_3}{\partial y'} - \frac{\partial A'_2}{\partial z'} = \left[\frac{1}{\ell}\frac{\partial}{\partial y}\left(\frac{1}{\ell}A_3\right) - \frac{1}{\ell}\frac{\partial}{\partial z}\left(\frac{1}{\ell}A_2\right)\right] \text{ (by (4) and (28))}$$

$$= \frac{1}{\ell^2}\left[\frac{\partial A_3}{\partial y} - \frac{\partial A_2}{\partial z}\right] = \frac{1}{\ell^2}H_1 \text{ (by (11)).}$$

Also:

$$H'_2 = \frac{\partial A'_1}{\partial z'} - \frac{\partial A'_3}{\partial x'} = \frac{1}{\ell}\frac{\partial}{\partial z}\left(\frac{k}{\ell}(A_1 + \epsilon\psi)\right) - \frac{k}{\ell}\left(\frac{\partial}{\partial x} - \epsilon\frac{\partial}{\partial t}\right)\left(\frac{1}{\ell}A_3\right)$$

$$= \frac{k}{\ell^2}\left[\left(\frac{\partial A_1}{\partial z} - \frac{\partial A_3}{\partial x}\right) + \epsilon\left(\frac{\partial A_3}{\partial t} + \frac{\partial \psi}{\partial z}\right)\right]$$

$$= \frac{k}{\ell^2}[H_2 - \epsilon E_3] \text{ (by (11) and (12)).}$$

Similarly:

$$H'_3 = \frac{k}{\ell^2}[H_3 + \epsilon E_2].$$

Thus:

$$\textbf{(32)}\quad H'_1 = \frac{1}{\ell^2}H_1;\ H'_2 = \frac{k}{\ell^2}(H_2 - \epsilon E_3);\ H'_3 = \frac{k}{\ell^2}(H_3 + \epsilon E_2).$$

Thus, we have obtained the transformation equations (cf §2-7) of the field $(\vec{E}, \vec{H})$ via those of the potentials $\vec{A}$ and ψ. In view of (28), if $\ell = 1$, then the quadruple $(\vec{A}, \psi)$ obeys the same transformation rule as the coordinates x,y,z and t; i.e., if $\ell = 1$, $(\vec{A}, \psi)$ is a 4-vector.

Since the force $\vec{L} = e(\vec{E} + \vec{v} \times \vec{H})$ plays a central rôle in the whole of Lorentz's physics, it is essential that we should be able to transform it too. We shall make use of the equations (6), (31) and (32). Without loss of generality let us take the charge e to be equal to 1. Since now $\vec{L} = \vec{E} + \vec{v} \times \vec{H}$, we can write:

$$L_1 = E_1 + (v_2H_3 - v_3H_2);\ L_2 = E_2 + (v_3H_1 - v_1H_3);$$
$$L_3 = E_3 + (v_1H_2 - v_2H_1)$$

By virtue of the covariance principle, we must also have:

$$L'_1 = E'_1 + (v'_2H'_3 - v'_3H'_2); L'_2 = E'_2 + (v'_3H'_1 - v'_1H'_3);$$
$$L'_3 = E'_3 + (v'_1H'_2 - v'_2H'_1)$$

Equations (5), (31) and (32) entail:

$$L'_1 = \frac{1}{\ell^2} E_1 + \left[\frac{v_2}{k(1 + \epsilon v_1)} \frac{k}{\ell^2} (H_3 + \epsilon E_2) - \frac{v_3}{k(1 + \epsilon v_1)} \cdot \frac{k}{\ell^2} \cdot (H_2 - \epsilon E_3)\right]$$

$$= \frac{1}{\ell^2}\left[E_1 + \frac{(v_2H_3 - v_3H_2) + \epsilon(v_2E_2 + v_3E_3)}{(1 + \epsilon v_1)}\right]$$

$$= \frac{1}{\ell^2(1 + \epsilon v_1)}\left[E_1 + (v_2H_3 - v_3H_2) + \epsilon(E_1v_1 + E_2v_2 + E_3v_3)\right],$$

$$= \frac{1}{\ell^2(1 + \epsilon v_1)}\left[L_1 + \epsilon \vec{E}.\vec{v}\right],$$

since, by the above, $L_1 = E_1 + (v_2H_3 - v_3H_2)$. Noting that $\vec{L} = \vec{E} + \vec{v} \times \vec{H}$ implies $\vec{L}.\vec{v} = (\vec{E} + \vec{v} \times \vec{H}).\vec{v} = \vec{E}.\vec{v} + (\vec{v} \times \vec{H}).\vec{v} = \vec{E}.\vec{v}$, we conclude:

$$\textbf{(33)}\ L'_1 = \frac{1}{\ell^2(1 + \epsilon v_1)}\left[L_1 + \epsilon \vec{L}.\vec{v}\right].$$

Similarly:

$$L'_2 = E'_2 + (v'_3H'_1 - v'_1H'_3 =$$
$$\left[\frac{k}{\ell^2} \cdot (E_2 + \epsilon H_3) + \frac{v_3}{k\,(1 + \epsilon v_1)} \cdot \frac{H_1}{\ell^2} - \frac{(v_1 + \epsilon)}{(1 + \epsilon v_1)} \frac{k}{\ell^2} \cdot (H_3 + \epsilon E_2)\right]$$

$$= \frac{k}{\ell^2(1 + \epsilon v_1)}\left[(E_2 + \epsilon H_3)(1 + \epsilon v_1) + \frac{v_3H_1}{k^2} - (v_1 + \epsilon)(H_3 + \epsilon E_2)\right] = \frac{k}{\ell^2(1 + \epsilon v_1)}\left[(1 - \epsilon^2)(E_2 - v_1H_3) + \frac{v_3H_1}{k^2}\right] = \frac{1}{k\ell^2(1 + \epsilon v_1)}\left[E_2 + (v_3H_1 - v_1H_3)\right]$$

since $k \underset{\text{Def}}{=} (1 - \epsilon^2)^{-1/2}$. Remembering that $L_2 = E_2 + (v_3H_1 - v_1H_3)$, we finally obtain:

$$(34)\ L'_2 = \frac{L_2}{k\ell^2(1 + \epsilon v_1)}\,.$$

Similarly:

$$(35)\ L'_3 = \frac{L_3}{k\ell^2(1 + \epsilon v_1)}\,.$$

Thus, the transformation law for the Lorentz-force $\vec{L}$ is:

$$(36)\ L'_1 = \frac{L_1 + \epsilon \vec{L}\,.\,\vec{v}}{\ell^2(1 + \epsilon v_1)}\ ;\ L'_2 = \frac{L_2}{k\ell^2(1 + \epsilon v_1)}\ ;\ L'_3 + = \frac{L_3}{k\ell^2(1 + \epsilon v_1)}\,.$$

Taking $\ell = 1$ and $\vec{v} = 0$, we retrieve the relations: $L'_1 = L'_1$, $L'_2 = L_2/k$ and $L'_3 = L_3/k$, which Lorentz had established between the electrostatic force in a moving frame and the corresponding force in the stationary system (see §2.4). Poincaré thus generalised Lorentz's results and was about to exploit these results in constructing a Lorentz-covariant theory of gravitation.

Equations (36) resemble the transformation laws both of the coordinates and of the velocities. We shall push this analogy further by considering the differential quantity:

$$dt^2 - (dx^2 + dy^2 + dz^2) = dt^2(1 - v^2).$$

It follows from equations (1) that:

$$\begin{aligned} dt'^2(1 - v'^2) &= dt'^2 - (dx'^2 + dy'^2 + dz'^2) \\ &= \ell^2\,[k^2(dt + \epsilon dx)^2 - dy^2 - dz^2 - k^2\,(dx + \epsilon dt)^2] \\ &= \ell^2\,[k^2(1 - \epsilon^2).dt^2 - k^2(1 - \epsilon^2).dx^2 - dy^2 - dz^2] \\ &= \ell^2\,[dt^2 - (dx^2 + dy^2 + dz^2)]\ ; \end{aligned}$$

since $k \underset{\text{Def}}{=} (1 - \epsilon^2)^{-1/2}$.

Therefore:

$$\begin{aligned} (37)\ dt'^2(1 - v'^2) &= dt'^2 - (dx'^2 + dy'^2 + dz'^2) \\ &= \ell^2\,[dt^2 - (dx^2 + dy^2 + dz^2)] = \ell^2.dt^2.\,(1 - v^2). \end{aligned}$$

Thus:

(38) $dt'\sqrt{1 - v'^2} = \ell dt\sqrt{1 - v^2}$

The third equation of (1) entails: dt′ = kℓ(dt + ϵdx). Hence:

(39) $dt' = k\ell(1 + \epsilon v_1)\, dt$, since $v_1 \underset{\text{Def}}{=} dx/dt$.

Comparing (38) with (39), we conclude:

$$\textbf{(40)}\ \sqrt{1 - v'^2} = \frac{\sqrt{1 - v^2}}{k(1 + \epsilon v_1)} = \frac{\sqrt{1 - v^2}.\sqrt{1 - \epsilon^2}}{(1 + \epsilon v_1)}$$

Similarly, by using the third equation of (2), we obtain:

$$\textbf{(41)}\ \sqrt{1 - v^2} = \frac{\sqrt{1 - v'^2}.\sqrt{1 - \epsilon^2}}{(1 - \epsilon v_1)}$$

Dividing (36) by (40), we have:

$$\textbf{(42)}\ \frac{L'_1.}{\sqrt{1 - v'^2}} = \frac{k}{\ell^2}\left(\frac{L_1}{\sqrt{1 - v^2}} + \epsilon\frac{\vec{L}.\vec{v}}{\sqrt{1 - v^2}}\right);\ \frac{L'_2}{\sqrt{1 - v'^2}} = \frac{1}{\ell^2}\frac{L_2}{\sqrt{1 - v^2}};\ \frac{L'_3}{\sqrt{1 - v'^2}} = \frac{1}{\ell^2}\frac{L_3}{\sqrt{1 - v^2}}$$

$\left(\frac{\vec{L}}{\sqrt{1 - v^2}}, \frac{\vec{L}.\vec{v}}{\sqrt{1 - v^2}}\right)$ thus seems to obey a transformation law analogous to that of the coordinates (x,y,z,t) = ($\vec{r}$,t). This leads us to try and determine

$$\frac{\vec{L}'.\vec{v}'}{\sqrt{1 - v'^2}}$$

with a view to completing the system of equations (42). Multiplying (42) by (5), then adding the equations thus obtained, we have:

$$\frac{\vec{L}'.\vec{v}'}{\sqrt{1 - v'^2}} = \frac{1}{k\ell^2(1 + \epsilon v_1)\sqrt{1 - v^2}}\Big[k^2 (L_1 + \epsilon\vec{L}.\vec{v})(v_1 + \epsilon) + L_2v_2 + L_3v_3\Big]$$

$$= \frac{1}{k\ell^2 (1 + \epsilon v_1)\sqrt{1 - v^2}}\Big[k^2 (L_1 +$$

$$\epsilon\vec{L}.\vec{v})(v_1 + \epsilon) + (\vec{L}.\vec{v} - L_1v_1)\Big]$$

$$= \frac{1}{k\ell^2(1 + \epsilon v_1)\sqrt{1 - v^2}}\left[\left(\frac{1 + \epsilon v_1}{1 - \epsilon^2}\right)\vec{L}.\vec{v} + \right.$$

$$\left.\left(\frac{1 + \epsilon v_1}{1 - \epsilon^2}\right)\epsilon L_1\right]$$

(by substituting $(1 - \epsilon^2)^{-1/2}$ for k)

$$= \frac{1}{k\ell^2\sqrt{1 - v^2}(1 - \epsilon^2)}\left[\vec{L}.\vec{v} + \epsilon L_1\right] =$$

$$\frac{k}{\ell^2}\left[\frac{\vec{L}.\vec{v}}{\sqrt{1 - v^2}} + \epsilon\frac{L_1}{\sqrt{1 - v^2}}\right]$$

In conjunction with (42) we obtain:

$$\frac{L'_1}{\sqrt{1 - v'^2}} = \frac{k}{\ell^2}\left(\frac{L_1}{\sqrt{1 - v^2}} + \epsilon\frac{\vec{L}.\vec{v}}{\sqrt{1 - v^2}}\right); \frac{L'_2}{\sqrt{1 - v'^2}} =$$

(43)

$$\frac{1}{\ell^2}\frac{L_2}{\sqrt{1 - v^2}}; \frac{L'_3}{\sqrt{1 - v'^2}} = \frac{1}{\ell^2}\frac{L_3}{\sqrt{1 - v^2}}; \frac{\vec{L}'.\vec{v}'}{\sqrt{1 - v'^2}} =$$

$$\frac{k}{\ell^2}\left(\frac{\vec{L}.\vec{v}}{\sqrt{1 - v^2}} + \epsilon\frac{L_1}{\sqrt{1 - v^2}}\right)$$

Thus the transformation of

$$\left(\frac{\vec{L}}{\sqrt{1 - v^2}}, \frac{\vec{L}.\vec{v}}{\sqrt{1 - v^2}}\right)$$

is, to within the factor $1/\ell^3$, the same as that of $(x,y,z,t) = (\vec{r},t)$. Should we decide to take: $\ell = 1$, then

$$\left(\frac{\vec{L}}{\sqrt{1 - v^2}}, \frac{\vec{L}.\vec{v}}{\sqrt{1 - v^2}}\right)$$

will be a 4-vector.

Poincaré realised that Lorentz had managed to explain Michelson's null results by extending to molecular interactions the law governing the transformation of the ponderomative force $\vec{L}$. Poincaré consequently decided to treat $\vec{L}$, or rather the 4-vector

$$\left(\frac{\vec{L}}{\sqrt{1 - v^2}}, \frac{\vec{L}.\vec{v}}{\sqrt{1 - v^2}}\right),$$

as the paradigm of all physical forces, in particular of the gravitational attraction. However, in "On the Dynamics of the Electron" there is no attempt to derive gravitation from electromagnetism. Poincaré makes use only of the theorem that all 4-forces transform alike—which holds by the very definition of a 4-vector—and of the assumption that gravitation can be represented by a 4-force. This is why I do not share Arthur Miller's view (1975) that Poincaré could not have discovered relativity because he was wedded to the electromagnetic world image. It is true that Poincaré, like many other scientists, entertained the possibility of reducing physics to electromagnetism; after all Einstein himself was to flirt (very briefly) with this idea when constructing his field equations in 1915. As for Poincaré, he took care not to let his work in theoretical physics hinge on the possibility of such a reduction. Though he was confident that electromagnetism accounted for the total inertia of the electron, he remained agnostic about the mass of positive ions. We have already mentioned his view that, for the relativity principle to hold good, the material mass of a particle must functionally depend on its velocity in the same way as does its electromagnetic inertia. This clearly shows that, for Poincaré, the material mass need not necessarily vanish. We have also seen that he regarded the existence of the ether as highly problematic, that his talk about a universal medium is often a mere *façon de parler* about the structure of space-time. To my mind, this suffices to establish that Poincaré was the (co-)author of a universal relativity principle which applies to the whole of physics and which, as will presently be shown, is firmly based on the structure of the Lorentz group.

§5.11.2 The Lorentz Group

We have already seen that, unlike Lorentz, Poincaré based his approach to relativity on the theory of groups.

Let $\mathcal{H}$ be the set of all transformations of the form:

(1) $x' = k\ell\,(x + \epsilon t);\ y' = \ell y;\ z' = \ell z;\ t' = k\ell\,(t + \epsilon x);\ k \underset{\text{Def}}{=} (1 - \epsilon^2)^{-1/2}$

Every element (1) of $\mathcal{H}$ is thus determined by two independent parameters, ϵ and ℓ; it can therefore be denoted by $T(\epsilon,\ell)$. Solving (1) for x, y, z, t in terms of x′, y′, z′, t′, we obtain the inverse transformation:

$$\text{(2)}\ x = \frac{k}{\ell}(x' - \epsilon t');\ y = \frac{1}{\ell}y';\ z = \frac{1}{\ell}z';\ t = \frac{k}{\ell}(t' - \epsilon x')$$

The inverse of $T(\epsilon, \ell)$ is thus identical with $T(-\epsilon, 1/\ell)$ which, by definition of $\mathcal{H}$, must lie in $\mathcal{H}$.

Let us show that $\mathcal{H}$ is a group. We have just established that the inverse of every member of $\mathcal{H}$ exists and belongs to $\mathcal{H}$; i.e.

$$\text{(3)}\ [T(\epsilon, \ell)]^{-1} = T(-\epsilon, 1/\ell) \in \mathcal{H}$$

It remains to be shown that $\mathcal{H}$ is closed with respect to multiplication. Consider another element $T(\epsilon', \ell')$ of $\mathcal{H}$ which sends the point (x′, y′, z′, t′) into the point (x″, y″, z″, t″); i.e.

$$\text{(4)}\ x'' = k'\ell'(x' + \epsilon' t');\ y'' = \ell' y';\ z'' = \ell' z';\ t'' = k'\ell'(t' + \epsilon' x'),\ \text{where } k'' \underset{\text{Def}}{=} (1 - \epsilon'^2)^{-1/2}.$$

In order to find $T(\epsilon', \ell').T(\epsilon, \ell)$ it suffices to express x″, y″, z″, t″ in terms of x, y, z, t. Eliminating x′, y′, z′, t′ between (1) and (4), we obtain:

$$y'' = \ell' y' = \ell'\ell y;\ z'' = \ell'\ell z;$$

similarly:

$$x'' = k'\ell'(x' + \epsilon' t') = k'\ell'[k\ell(x + \epsilon t) + \epsilon' k\ell(t + \epsilon x)] = k'k\ell'\ell(1 + \epsilon'\epsilon)\left[x + \left(\frac{\epsilon' + \epsilon}{1 + \epsilon'\epsilon}\right)t\right]$$

Also:

$$t'' = k'k\ell'\ell(1 + \epsilon'\epsilon)\left[t + \left(\frac{\epsilon' + \epsilon}{1 + \epsilon'\epsilon}\right)x\right]$$

Let us put $\ell'\ell = \ell''$, $(\epsilon' + \epsilon)/(1 + \epsilon'\epsilon) = \epsilon''$ and let us show that $k'k(1 + \epsilon'\epsilon) = (1 - \epsilon''^2)^{-1/2}$; i.e. let us prove that

$$\frac{1 + \epsilon'\epsilon}{\sqrt{(1 - \epsilon^2)(1 - \epsilon'^2)}} = \frac{1}{\sqrt{1 - [(\epsilon' + \epsilon)/(1 + \epsilon'\epsilon)]^2}}$$

This can be easily verified by squaring both sides. Therefore:

$$(5)\begin{cases} x'' = k''\ell''(x + \epsilon''t);\ y'' = \ell''y;\ z'' = \ell''z;\ t'' = k''\ell''(t + \epsilon''x); \\ \qquad \text{where } \epsilon'' = \dfrac{\epsilon' + \epsilon}{1 + \epsilon'\epsilon};\ \ell'' = \ell'\ell;\ k'' = \dfrac{1}{\sqrt{(1 - \epsilon''^2)}} \end{cases}$$

in other words:

$$(6)\ T(\epsilon', \ell').\, T(\epsilon, \ell) = T\left(\frac{\epsilon' + \epsilon}{1 + \epsilon'\epsilon}, \ell'\ell\right)$$

Since, by definition of $\mathcal{H}$, T $\left(\dfrac{\epsilon' + \epsilon}{1 + \epsilon'\epsilon}, \ell'\ell\right) \in \mathcal{H}$, it follows that $\mathcal{H}$ is closed with respect to multiplication (i.e. to transformation-composition) and thus constitutes a group.

Poincaré defines the Lorentz group $\mathcal{L}$ in two ways which, though equivalent, remain conceptually distinct. According to one of the definitions, $\mathcal{L}$ is the set of all transformations each of which can be expressed as the product of a spatial rotation and a member of $\mathcal{H}$ followed by a second spatial rotation. According to the other definition, every member of $\mathcal{L}$ is the product of a transformation of the form: $x_o = \ell x$, $y_o = \ell y$, $z_o = \ell z$, $t_o = \ell t$ and of a linear transformation which leaves the quadratic form $(x_o^2 + y_o^2 + z_o^2 - t_o^2)$ unaltered. Thus, denoting the final coordinates by x', y', z', t', we have:

$$(7)\ x'^2 + y'^2 + z'^2 - t'^2 = x_o^2 + y_o^2 + z_o^2 - t_o^2 = \ell^2(x^2 + y^2 + z^2 - t^2).$$

This means that each member of $\mathcal{L}$ multiplies the quadratic expression $(x^2 + y^2 + z^2 - t^2)$ by ℓ^2; it will therefore preserve the unit velocity, i.e. the velocity of light.

By means of *physical* considerations, Lorentz had determined the value of ℓ as being necessarily equal to 1. He had reasoned as follows: if all substances are subject only to forces of electromagnetic origin, then moving bodies will be contracted by the factor $\sqrt{1 - v^2}$ along the direction of their velocity $\vec{v}$; clocks will also be affected and will, in fact, be retarded by the same factor $\sqrt{1 - v^2}$. The effective, or measured, coordinates are:

(7') $x' = k(x + \epsilon t);\ y' = y;\ z' = z;\ t' = k(t + \epsilon x)$, where $k \underset{\text{Def}}{=} (1 - \epsilon^2)^{-1/2}$ and $(-\epsilon)$ is the velocity of the moving frame with respect to the ether. This means that, in equation (1), we should equate ℓ to 1.

Lorentz also draws the same conclusion from a more general, i.e. from a weaker, assumption than that of the exclusively electromagnetic origin of all forces. He proposed the following molecular forces hypothesis (see §2.4):

since all molecular interactions express states of the electromagnetic ether, they are affected by motion through this medium in the same way as are the electromagnetic forces. In other words: all forces transform like $\vec{L} = e(\vec{E} + \vec{v} \times \vec{H})$.

Let us finally note, in connection with Lorentz, that he 'explains' the phenomena of contraction and retardation against an absolute kinematical background which is fixed a priori.

By virtue of his conventionalism, Poincaré does not believe in the intrinsic intelligibility of classical kinematics and so proceeds very differently from Lorentz. Being deeply opposed to essentialism, he rids himself of absolute time, i.e. of the Galilean transformation, which had proved inconvenient. He deduces the same result as Lorentz's, namely $\ell = 1$, from the relativity principle together with the assumption that the velocity $(-\epsilon)$ of the moving frame uniquely determines the parameter ℓ. Thus, let us put $\ell = \varphi(\epsilon)$. Let $\mathcal{P}$ be the subset of $\mathcal{L}$ consisting of all the transformations for which $\ell = \varphi(\epsilon)$. In view of the relativity principle, all allowable frames are equivalent, which entails that $\mathcal{P}$ must form a group. Let $T(\epsilon, \ell)\epsilon\mathcal{P}$. By definition of $\mathcal{P}$, $\ell = \varphi(\epsilon)$. Since $\mathcal{P}$ is a group, $T(\epsilon, \ell)\epsilon\mathcal{P}$ implies $[T(\epsilon, \ell)]^{-1}\epsilon\mathcal{P}$. By (3):

$$T(-\epsilon, 1/\ell) = [T(\epsilon, \ell)]^{-1}\epsilon\mathcal{P}.$$

Hence:

$$1/\ell = \varphi(-\epsilon);\ \text{i.e. } 1/\varphi(\epsilon) = \varphi(-\epsilon).$$

We can thus write:

(8) $\varphi(\epsilon).\ \varphi(-\epsilon) = 1.$

Let us now go back to equation (1) and let us suppose once more that $T(\epsilon, \ell)\epsilon\mathcal{P}$; i.e. $\ell = \varphi(\epsilon)$. Let us rotate each of the two systems (x, y, z, t) and (x′, y′, z′, t′) through 180° about the y-axis. Let $\bar{x}, \bar{y}, \bar{z}, \bar{t}$) and ($\bar{x}', \bar{y}', \bar{z}', \bar{t}'$) be the coordinates obtained in this way. We have:

(9) $\bar{x} = -x;\ \bar{y} = +y;\ \bar{z} = -z;\ \bar{t} = +t;$

$$\bar{x}' = -x';\ \bar{y}' = +y';\ \bar{z}' = -z';\ \bar{t}' = +t'.$$

The equations (1) and (9) entail:

$$\bar{x}' = -x' = -k\ell(x + \epsilon t) = -k\ell(-\bar{x} + \epsilon\bar{t})$$
$$= k\ell(\bar{x} - \epsilon\bar{t});\ \bar{y}' = y' = \ell y = \ell\bar{y};\ \bar{z}' = -z' = -\ell z = \ell\bar{z};$$
$$\bar{t}' = t' = k\ell(t + \epsilon x) = k\ell(\bar{t} - \epsilon\bar{x}).$$

Therefore:

(10) $\bar{x}' = k\ell(\bar{x} - \epsilon\bar{t}); \bar{y}' = \ell\bar{y}; \bar{z}' = \ell\bar{z}; \bar{t}' = k\ell(\bar{t} - \epsilon\bar{x}).$

The transformation which sends $(\bar{x}, \bar{y}, \bar{z}, \bar{t})$ to $(\bar{x}', \bar{y}', \bar{z}', \bar{t}')$ is therefore $T(-\epsilon, \ell)$. Since $T(-\epsilon, \ell)$ must belong to $\mathcal{P}$, we have: $\ell = \varphi(-\epsilon)$. But we also have: $\ell = \varphi(\epsilon)$. Hence:

(11) $\varphi(\epsilon) = \varphi(-\epsilon).$

In view of (8) and (11): $1 = \varphi(\epsilon) . \varphi(-\epsilon) = [\varphi(\epsilon)]^2$. Thus:

(12) $\varphi(\epsilon) = \pm 1$

Poincaré chooses $\varphi(\epsilon) = +1$. (Note that $\varphi(\epsilon) = -1$ would contradict the group structure of $\mathcal{P}$; for:

$$T(\epsilon', -1) . T(\epsilon, -1) = T\left(\frac{\epsilon' + \epsilon}{1 + \epsilon'\epsilon}, +1\right);$$

$\mathcal{P}$ would thus not be closed with respect to multiplication)

§5.11.3 A Theory of Gravitation

Poincaré bases his approach to gravitation on an *analogy* with electromagnetism. Let us recall that the Lorentz-force $\vec{L}$ which acts on a particle carrying charge e and moving with velocity $\vec{v}$ is given by:

(1) $\vec{L} = e(\vec{E} + \vec{v} \times \vec{H}),$

where $\vec{E}$ and $\vec{H}$ denote the electric and magnetic fields respectively.

Let us also remember that Poincaré chooses his units so as to make the speed of light equal to 1. He then goes on to determine the transformation law for $\vec{L}$ and is thus led to consider, long before Minkowski, a force possessing 4 components, namely

$$\left(\frac{\vec{L}}{\sqrt{1 - v^2}}, \frac{\vec{L} . \vec{V}}{\sqrt{1 - v^2}}\right).$$

We have already seen that this force constitutes a 4-vector, i.e. it obeys the same transformation law as do the differentials of the coordinates.

Since gravitation treats of the mutual attractions of material bodies, Poincaré tries to determine the simplest Lorentz-invariants which can be constructed from the velocities and from the relative position of two point masses. He conceives the Lorentz group as the set of all rotations of a 4-dimensional space whose points are given the coordinates (x,y,z,it), where $i = \sqrt{-1}$. Let us denote (x,y,z) by $\vec{r}$.

Let (x_o,y_o,z_o,t_o) and $(x_o + x, y_o + y, z_o + z, t_o + t)$ be the coordinates of the attracted body P and of the attracting body P_1 respectively. P_1 acts on P, not at the considered instant to, but at $t_o + t$. In order for the principle of causality to be satisfied, $t_o + t$ must precede t_o, i.e.

(2) $t < O$.

Denote by $\vec{v} = (\xi,\eta,\zeta)$ and $\vec{v}_1 = (\xi_1,\eta_1,\zeta_1)$ the velocities of P and P_1 at the instants t_o and $t_o + t$ respectively. In accordance with the Relativity Principle, Poincaré requires that t be determined by a covariant equation of the form:

(3) $\varphi(t, \vec{r}, \vec{v}, \vec{v}_1) = 0.$

He also expects his gravitational law to assume the following general form:

(4) $\dfrac{1}{\text{mass}} \cdot \text{force} = \text{some kinematical quantity.}$

As a candidate for the right-hand side of (4), he considers the simplest Lorentz-invariants constructed from the following three 4-vectors:

(5) (x,y,z,it); $(\Delta x,\Delta y,\Delta z,i\Delta t)$; $(\Delta_1 x,\Delta_1 y,\Delta_1 z,i\Delta_1 t)$

The following diagramme indicates what these vectors represent

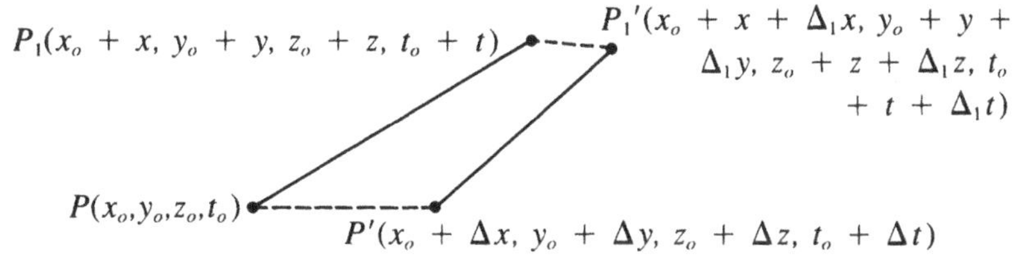

(Note that the suffix 1 refers to the attracting body)

Through forming the scalar products and the modules of the vectors in (5), we obtain the Lorentz-invariants:

(6)
$$\begin{cases} x^2 + y^2 + z^2 - t^2 = r^2 - t^2; \\ \qquad [t.\ \Delta t - (x.\ \Delta x + y.\ \Delta y + z.\ \Delta z)]; \\ [(\Delta t)^2 - (\Delta x^2 + \Delta y^2 + \Delta z^2)]^{1/2}; \\ \qquad [t.\ \Delta_1 t - (x.\ \Delta_1 x + y.\ \Delta_1 y + z.\ \Delta_1 z)]; \\ [(\Delta_1 t)^2 - (\Delta_1 x^2 + \Delta_1 y^2 + \Delta_1 z^2)]^{1/2}; \\ \qquad [\Delta t.\ \Delta_1 t - (\Delta x.\ \Delta_1 x + \Delta y.\ \Delta_1 y + \Delta z.\ \Delta_1 z)] \end{cases}$$

Through dividing these quantities, by one another, in order to eliminate differentials, we obtain the following:

$$\left[t - \left(x \cdot \frac{dx}{dt} + y \cdot \frac{dy}{dt} + z \frac{dz}{dt}\right)\right] \Big/ \left[1 - \left(\left(\frac{dx}{dt}\right)^2 + \left(\frac{dy}{dt}\right)^2 + \left(\frac{dz}{dt}\right)^2\right]^{1/2} = \frac{t - \vec{r} \cdot \vec{v}}{\sqrt{1 - v^2}}$$

We similarly construct the invariants

$$\frac{t - \vec{r} \cdot \vec{v}_1}{\sqrt{1 - v_1^2}} \text{ and } \frac{1 - \vec{v} \cdot \vec{v}_1}{\sqrt{(1 - v^2)(1 - v_1^2)}}$$

(through dividing the last quantity in (6) by the product of the third and fifth).

(7) Put $A = \dfrac{t - \vec{r}\,.\,\vec{v}}{\sqrt{1 - v^2}}$; $B = \dfrac{t - \vec{r}\,.\,\vec{v}_1}{\sqrt{1 - v_1^2}}$; $C = \dfrac{1 - \vec{v}\,.\,\vec{v}_1}{\sqrt{(1 - v^2)(1 - v_1^2)}}$

Poincaré thus considers the following 4 invariants:

(8) $(r^2 - t^2)$; A; B; C.

Note that, though this was not the method actually adopted by Poincaré, we could have avoided all recourse to differentials by constructing, right from the start, the following three 4-vectors:

(9) $(\vec{r}, it)$; $\left(\dfrac{\vec{v}}{\sqrt{1 - v^2}}, \dfrac{i}{\sqrt{1 - v^2}}\right)$; $\left(\dfrac{\vec{v}_1}{\sqrt{1 - v_1^2}}, \dfrac{i}{\sqrt{1 - v_1^2}}\right)$.

We would then have considered the modules and scalar products of these vectors, thus retrieving the invariants (8).

In the limiting cases where the new hypothesis should tend to the Newtonian law of gravitation, we have to consider velocities which are negligible with respect to the speed of light; i.e. we shall have cases where $\vec{v} = (\xi, \eta, \zeta)$ and $\vec{v}_1 = (\xi_1, \eta_1, \zeta_1)$ are much smaller than 1. Let us immediately note that, by (7), the quantities:

(10) $(C - 1)$ and $(A - B)^2$ are of the 2nd order in v and v_1 and can therefore be neglected when we come to apply the Correspondence Principle.

Let us now focus on the left-hand side of (4), i.e. on force/mass. Let $\vec{G}$ be the gravitational field at P; $\vec{G}$ will be taken to depend on $\vec{r}$, $\vec{v}$ and

$\vec{v}_1$. By virtue of the analogy with the Lorentz-force $\vec{L}$, we shall require that

(11) $\left(\frac{\vec{G}}{\sqrt{1-v^2}}, \frac{i\vec{G}.\vec{v}}{\sqrt{1-v^2}}\right)$

be a 4-vector, i.e. that it should be subject to the same transformation law as the coordinates (x,y,z,it). Consider the following 4-vectors:

(12) $\left(\frac{\vec{G}}{\sqrt{1-v^2}}, \frac{i.\vec{G}.\vec{v}}{\sqrt{1-v^2}}\right); (\vec{r},it); \left(\frac{\vec{v}}{\sqrt{1-v^2}}, \frac{i}{\sqrt{1-v^2}}\right);$
$\left(\frac{\vec{v}_1}{\sqrt{1-v_1^2}}, \frac{i}{\sqrt{1-v_1^2}}\right).$

(13) Put $T = \vec{G}.\vec{v}$, $k_o = (1-v^2)^{-1/2}$ and $k_1 = (1-v_1^2)^{-1/2}$.

The expressions (12) become:

(14) $(k_o\vec{G}, ik_oT); (\vec{r}, it); (k_o\vec{v}, ik_o); (k_1\vec{v}_1, ik_1)$

Let us now try to construct a Lorentz-covariant theory by expressing the first vector in (14) as a linear combination of the remaining three. The coefficients of this combination must naturally be Lorentz-invariant. Thus:

$$k_o\vec{G} = \alpha\vec{r} + \beta k_o\vec{v} + \gamma k_1\vec{v}_1 \text{ and } k_oT = \alpha t + \beta k_o + \gamma k_1, \text{ i.e.}$$

(15) $\begin{cases} \vec{G} = \frac{\alpha}{k_o}\vec{r} + \beta\vec{v} + \gamma\frac{k_1}{k_o}\vec{v}_1 \text{ and } T = \frac{\alpha}{k_o}t + \beta + \gamma\frac{k_1}{k_o}\text{; where} \\ \alpha, \beta \text{ and } \gamma \text{ are Lorentz-invariants.} \end{cases}$

We recall that $T \underset{\text{Def}}{=} \vec{G}.\vec{v}$. Multiplying the first equation in (15) by $\vec{v}$, then subtracting it from the second equation, we obtain:

$$o = \alpha(t - \vec{r}.\vec{v})/k_o + \beta(1-v^2) + \gamma k_1(1 - \vec{v}.\vec{v}_1)/k_o.$$

Since $k_o \underset{\text{Def}}{=} (1-v^2)^{-1/2}$, the last equation, when multiplied by $k_o^2 = (1-v^2)^{-1}$, yields:

$$o = \alpha(t - \vec{r}.\vec{v}).(1-v^2)^{-1/2} + \beta + \gamma(1 - \vec{v}.\vec{v}_1)(1-v^2)^{-1/2}.(1-v_1^2)^{-1/2} \text{. i.e.}$$

(16) $\alpha A + \beta + \gamma C = 0$ (by 7)).

Hence, at most 2 of the 3 quantities: α,β,γ, are independent of each other. If we choose $\beta = 0$, then $\alpha A + \gamma C = 0$, i.e. $\gamma = -\alpha A/C$. Substituting in (15):

$$\textbf{(17)}\quad \vec{G} = \frac{\alpha}{k_o}\vec{r} - \alpha\frac{Ak_1}{Ck_o}\vec{v}_1 = \frac{\alpha}{k_o}\left[\vec{r} - k_1\frac{A}{C}\vec{v}_1\right]$$

In this case it remains for us to determine α, e.g. from the correspondence principle.

Every equation of the form (15) is Lorentz-covariant and will therefore satisfy the first condition imposed by Poincaré, namely the relativity principle. Let us now consider the second condition, i.e. the correspondence principle. We have to determine, on the one hand the time interval t by means of a covariant equation of the form $\varphi\,(t, \vec{r}, \vec{v}, \vec{v}_1) = 0$ and, on the other hand, the coefficients: α,β,γ in such a way that the new theory yields the Newtonian law as a limiting case. We have already seen that, in view of (16), it suffices to fix two of the three quantities: α,β,γ.

Let us begin by determining φ, i.e. the speed of propagation of the gravitational action. Since φ must be an invariant, we could for example take: $\varphi \equiv r^2 - t^2$ (cf. (8)). Equation (3) becomes $0 = r^2 - t^2$. Hence: $t = \pm r$. In view of (2):

$$\textbf{(18)}\quad t = -r.$$

This means that gravitational disturbances are propagated with unit speed, i.e. with the speed of light. In view of the analogy with electromagnetism, Poincaré considers this assumption very plausible. There is another reason for opting for $\varphi = r^2 - t^2$: among the 4 invariants $(r^2 - t^2)$, A, B, and C, only $\varphi = A$ or $\varphi = B$ qualify as alternative solutions; i.e. $\varphi = (t - \vec{r}.\vec{v}).\,(1 - v^2)^{-1/2}$ or $\varphi = (t - \vec{r}.\vec{v}_1).\,(1 - v_1^2)^{-1/2}$; but then $\varphi = 0$ would lead to one of the two equations: $t = \vec{r}.\vec{v}$ or $t = \vec{r}.\vec{v}_1$, which admit positive solutions for t. This would violate the causality principle.

If we take $\beta = 0$, then equation (18) enables us to express $\acute{G}$ in a form which is similar to that of the Lorentz-force $\vec{L}$. Taking $t = -r$, we have by (7): $A = (-r - \vec{r}.\vec{v})\,(1 - v^2)^{-1/2}$. Remembering that:

$$C \underset{\text{Def}}{=} (1 - \vec{v}.\vec{v}_1).\,(1 - v^2)^{-1/2}.\ (1 - v_1^2)^{-1/2},\quad k_o \underset{\text{Def}}{=} (1 - v^2)^{-1/2}$$
$$\text{and } k_1 = (1 - v_1^2)^{-1/2},$$

we obtain from (17):

$$\textbf{(19)}\quad \vec{G} = \frac{\alpha}{k_o}\left(\vec{r} - k_1\frac{A}{C}\vec{v}_1\right)$$

$$= \frac{\alpha\sqrt{1-v^2}}{(1-\vec{v}\,.\,\vec{v}_1)}[(1-\vec{v}\,.\,\vec{v}_1)\vec{r} + (r + \vec{r}\cdot\vec{v})\vec{v}_1]$$

$$= \frac{\alpha\sqrt{1-v^2}}{(1-\vec{v}\cdot\vec{v}_1)}[(\vec{r} + r\vec{v}_1) + \vec{v}\times(\vec{v}_1\times\vec{r})]$$

The vectors $(\vec{r} + r\vec{v}_1)$ and $(\vec{v}_1 \times \vec{r})$ can be assimilated to the electric and magnetic fields respectively.

Neglecting in the expression of $\vec{G}$ all 2nd order terms in v and v_1, we obtain:

(20) $\vec{G} \doteqdot \alpha(\vec{r} + r\vec{v}_1)$

since v^2, $\vec{v}\cdot\vec{v}_1$ and $\vec{v}\times(\vec{v}_1\times\vec{r})$ are all of the second order.

Now consider Newton's theory, according to which the attraction experienced by P at t_o is proportional to $(-\vec{r}_1/r_1^3) = -(x_1,y_1,z_1)/r_1^3$; where $\vec{r}_1 = (x_1,y_1,z_1)$ is the vector which connects P to the attracting body P_1^* *considered at the same instant* t_o.

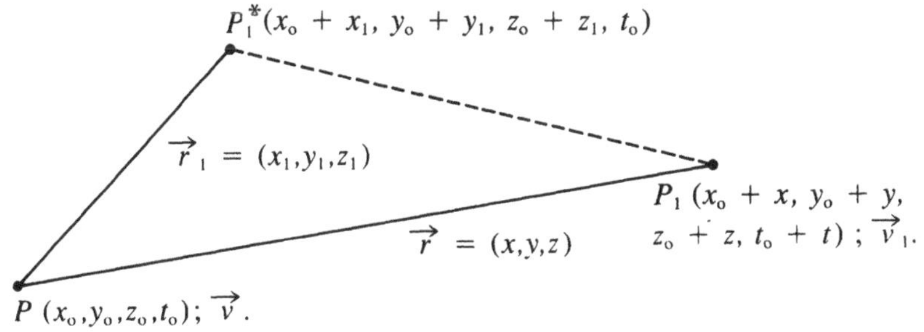

We recall that $\vec{v}_1$ is the velocity of the attracting body, that $t = -r$ and that t is therefore negative. Thus, in the time interval $-t = r$, the attracting body moves from P_1 to some new position P_1^*. Taking the change in the velocity between P_1 and P_1^* to be negligible, we can write:

$$\overrightarrow{P_1P_1^*} \doteqdot -t\vec{v}_1 = r\vec{v}_1.$$

Hence:

(21) $\vec{r}_1 = \vec{r} + \overrightarrow{P_1P_1^*} \doteqdot \vec{r} + r\vec{v}_1.$

Substituting in (20), we obtain:

(22) $\vec{G} \doteqdot \alpha\,[\vec{r} + r\vec{v}_1] \doteqdot \alpha\vec{r}_1.$

Remembering that, in accordance with the correspondence principle, $\vec{G}$ must be approximately equal to $-\vec{r}_1/r_1^3$ for small velocities, we conclude:

(23) $\alpha \doteqdot -1/r_1^3.$

Let us recall that α must be Lorentz-invariant. We shall therefore examine what becomes of the invariants A, B and C, when all second order quantities in v and v_1 are neglected. The invariant C presents no difficulties since:

(24) $C = \dfrac{1 - \vec{v}.\vec{v}_1}{\sqrt{(1 - v^2)(1 - v_1^2)}} \doteqdot 1.$ Also, by (10): $(A - B)^2 \doteqdot o.$

By (7) and (18) we have

(25) $A = (-r - \vec{r}.\vec{v})\,/\,\sqrt{1 - v^2}\,;$

$$B = (-r - \vec{r}.\vec{v}_1)\,/\,\sqrt{1 - v_1^2}.$$

Hence:

(26) $A \doteqdot (-r - \vec{r}.\vec{v})$ and $B \doteqdot (-r - \vec{r}.\vec{v}_1).$

We have thus to examine the two quantities $(r + \vec{r}.\vec{v})$ and $(r + \vec{r}.\vec{v}_1)$. By (21): $\vec{r}_1 \doteqdot \vec{r} + r\vec{v}_1$. Through multiplying both sides by $\vec{v}$, then neglecting $\vec{v}_1.\vec{v}$, we obtain:

(27) $\vec{r}_1.\vec{v} \doteqdot \vec{r}.\vec{v}.$

Similarly, by multiplying both sides by $\vec{v}_1$, then neglecting v_1^2, we have:

(28) $\vec{r}_1.\vec{v}_1 \doteqdot \vec{r}.\vec{v}_1$

From $\vec{r}_1 \doteqdot \vec{r} + r\vec{v}_1$, it follows that: $\vec{r} \doteqdot \vec{r}_1 - r\vec{v}_1$. Squaring both sides, then neglecting the term $r^2v_1^2$: $r^2 \doteqdot r_1^2 - 2r\vec{r}_1.\vec{v}_1$. Hence:

$$\begin{aligned} r_1^2 &\doteqdot r^2 + 2r\vec{r}_1.\vec{v}_1 \doteqdot r^2 + 2r\vec{r}.\vec{v}_1, \text{ (by (28))} \\ &\doteqdot (r^2 + 2r\vec{r}.\vec{v}_1 + (\vec{r}.\vec{v}_1)^2)\,, \\ &\text{(since } (\vec{r}.\vec{v}_1)^2 \text{ is of the 2nd order)} \end{aligned}$$

Thus:

$$r_1^2 \doteqdot r^2 + 2r\vec{r}_1.\vec{v}_1 + (\vec{r}.\vec{v}_1)^2 = (r + \vec{r}.\vec{v}_1)^2.$$

Therefore:

(29) $r_1 \doteqdot r + \vec{r}\,.\,\vec{v}_1.$

Hence also:

(30) $r \doteqdot r_1 - \vec{r}\,.\,\vec{v}_1 \doteqdot r_1 - \vec{r}_1.\,\vec{v}_1$,(by (28)).

Equations (26) and (29) entail:

$$B \doteqdot -(r + \vec{r}\,.\,\vec{v}_1) \doteqdot -r_1;$$

and, in view of (26), (27) and (30):

$$A \doteqdot -(r + \vec{r}\,.\,\vec{v}\,) \doteqdot -(r_1 - \vec{r}_1.\,\vec{v}_1 + \vec{r}_1.\,\vec{v}\,) =$$

$$-\,[r_1 + \vec{r}_1.\,(\vec{v} - \vec{v}_1)]$$

(31) Thus: $A \doteqdot -\,[r_1 + \vec{r}_1.\,(\vec{v} - \vec{v}_1)]$ and $B \doteqdot -r_1$.
Hence:

$$1/B^3 = -\,1/r_1^{\,3}$$

Since we require that $\alpha \doteqdot -\,1/r_1^{\,3}$, the most natural assumption is $\alpha = 1/B^3$. This choice however is not unique. Since $C \doteqdot 1$ and $(A - B)^2 \doteqdot o$ we can also take $\alpha = C/B^3$, or $\alpha = [C + (A - B)^2]/B^3$, etc. Note that, in view of (7) and of (25):

(32) $$\frac{1}{B^3} = -\,\frac{(1 - v_1^{\,2})^{3/2}}{(r + \vec{r}\,.\,\vec{v}_1)^3} \text{ and } C = \frac{1 - \vec{v}\,.\,\vec{v}_1}{\sqrt{(1 - v^2)\,(1 - v_1^{\,2})}}$$

(33) Taking $\alpha = 1/B^3$, we obtain, by (19) and (32):

$$\vec{G} = -\,\frac{(1 - v^2)^{1/2}.\,(1 - v_1^{\,2})^{3/2}}{(r + \vec{r}\,.\,\vec{v}_1)^3.\,(1 - \vec{v}\,.\,\vec{v}_1)}\,.$$

$$[(\vec{r} + r\vec{v}_1) + \vec{v} \times (\vec{v}_1 \times \vec{r}\,)]$$

(34) Taking $\alpha = C/B^3$, we have:

$$\vec{G} = -\,\frac{(1 - v_1^2)}{(r + \vec{r}\,.\,\vec{v}_1)^3}\,[(\vec{r} + r\vec{v}_1) + \vec{v} \times (\vec{v}_1 \times \vec{r}\,)]$$

In "On the Dynamics of the Electron" Poincaré does not dwell at any length on the problem of a possible variation of the gravitational mass with the speed. He is nonetheless aware of the importance of the equivalence principle according to which the inertial and gravitational masses of the same body are equal. As is well-known, this principle entails that all test particles placed in the same gravitational field experience, irrespective of their physical compositions, the same acceleration. Unlike Einstein however, Poincaré does not ascribe absolute validity to the equivalence principle; he only requires that it constitute a limiting case of some consequence of his new theory. This perhaps enabled him to keep special relativity as a suitable framework for his gravitational hypothesis. As for Einstein, his unqualified adherence to the equivalence principle eventually led him to abandon flat space and adopt Riemannian geometry as the basis of a new revolutionary theory of gravitation (see §8.2–§8.3).

— 6 —

HOW CRUCIAL WAS KAUFMANN'S EXPERIMENT?

§6.1 THE INTENDED ROLE OF KAUFMANN'S EXPERIMENT

The present chapter is entirely devoted to discussing one famous experiment. It was carried out by Kaufmann in 1905 and was intended to be crucial between two rival theories of the electron: a classical theory elaborated by Abraham and the new relativistic theory proposed by Lorentz and Einstein.

An immediate question comes to mind: why should the general reader, who may be interested in the *philosophy* of science but not in the technicalities of the history of science, invest effort in reading the present detailed case study of an ingenious but rather elaborate test between two rather complicated models of the electron? Would not a simpler example have been better for the philosophical problems at hand?

I shall give two different reasons why general methodology may demand detailed, sophisticated case studies (see introduction §2). My first reason is this. I subscribe to Lakatos' thesis that the history of science acts as an arbiter between different methodologies. It is of course true that methodologies are systems of appraisal consisting of normative propositions; and, as is well-known, there is no *direct* logical contact between normative propositions and statements of historical fact. However, Lakatos proposed that a methodology M′ be preferred to another methodology M if, other things being equal, M′ characterises as progressive more of what scientists themselves intuitively consider as good science than does M (or if M′ excludes more of what is intuitively regarded as pseudo-science than does M). Thus, one possible way of differentiating between M and M′ is to find a historical shift from a theory T to another theory T′, where T′ was intuitively judged to constitute progress over T and where M′, but not M, pronounces T′ superior to T. Such a shift would provide historical support for M′ as against M.

If M and M′ were widely different methodologies, then such a crucial episode should be easy to come by. In our present case however, M is falsificationism and M′ the methodology of scientific research programmes (MSRP). It is obvious that MSRP is a refinement of falsificationism or, equivalently, that falsificationism is a coarse-grained version of MSRP (see Worrall 1978). (This does not prevent MSRP and falsificationism from contradicting each other in the strict logical sense.) It is therefore to be expected that MSRP and falsificationism should give equivalent appraisals in most straightforward situations. It is even conceivable that MSRP and falsificationism should yield similar appraisals in *all* actual historical situations: had this been the case, then MSRP and falsificationism would be analogous to two empirical theories T and T′ such that T′ is a theoretical development of T, but T and T′ explain exactly the same known facts; i.e., T and T′ have so far been observationally equivalent. In such a case it might reasonably be claimed that the excess sophistication of T′ over T, or of MSRP over falsificationism, constitutes not progress but *unnecessary* complication. This is why it was important to try to find an actual historical example over which MSRP and falsificationism disagreed. However, since MSRP is a rather subtle refinement of falsificationism, it was almost inevitable that the best example I could find should be rather more complicated than the usual ones.

I now turn to my second reason for regarding a detailed case study of the present kind as relevant to general methodology. So far in this chapter I have spoken of methodologies purely as systems of appraisal which pass judgement on finished products, on hypotheses laid on the table (see §1.1). The question arises as to whether methodologies can play any role in scientific *discovery*. Many philosophers claim that methodologies do not play such a role and the construction of methodologies is an academic exercise irrelevant to scientific praxis. Following the Vienna Circle, Lakatos made a distinction between the context of 'discovery' and the context of 'justification'; he also inclined to the view that methodologies appraise but give no advice. Such a view however lands us with the following difficulty. It is agreed on all sides that science has (globally) progressed, at least since the 16th century. It is an adequacy requirement for any methodology that its general account of scientific progress be in accord with this historical progress. Suppose that MSRP fulfils this requirement and that Lakatos is right in denying that methodology plays an effective role in the actual development of science. How is it then that scientists have systematically achieved what they did *not* set out to do: i.e. achieved progress according to MSRP without being guided by the principles elaborated in MSRP? This suggests a mysterious 'List der Vernunft' whereby scientists are led 'sleepwalking' down the path adumbrated by methodology. My own view is that methodology plays a positive role in scientific progress. I do not of course maintain that scientists possess fully articulated methodologies; but that, in concrete situations, they apply intuitive methodological principles which enable them both to make crucial choices and to construct important hypotheses. And I further maintain that MSRP provides the best available rational reconstruction of the intuitive methodology which some of the best scientists have acted on. It is obvious that this claim calls for detailed case studies in which the genesis of some important theory is examined in order to decide whether and how a certain (intuitive) methodology was operative during the process of discovery. This is the second reason why, in my opinion, the general reader may find the present case study of some interest.

Let us now return to Kaufmann's experiment. The discussion of this experiment will illustrate two points of methodological interest: first, the point that deductive logic may play a creative role in the development of the empirical sciences. This claim is denied for example by Descartes and by Feyerabend (see Feyerabend 1972). (Descartes based his denial of the claim on the triviality of logic; Feyerabend based his denial of the claim on the subserviance of logic to the demands of the physical sciences, or rather to the demands of physicists: the scientist freely bends logical rules so as to make them conform to whatever scientific system he wishes to construct.) Against Descartes and Feyerabend I claim that detailed logical analyses—logic-chopping if you like—can further the progress of science in important ways, while logical oversights can seriously impede it.

My second methodological point will be that MSRP, at least in the case of this 'crucial' experiment, reflects rather accurately scientists' intuitive judgements about the significance of the experimental result for the theories in question: Planck subjected Kaufmann's experiment, together with the rival theories under test, to a detailed logical analysis. The conclusions he thereby reached completely upset the received view about the logical impact of Kaufmann's result. Moreover, it turns out that Planck's own appraisal conforms very well to MSRP criteria.

§6.2 The General Structure of Kaufmann's Experiment and its Reinterpretation by Planck as a Crucial Test between Competing Methodologies

In 1905 Kaufmann performed an experiment in which he observed the deflection of electrons in an electromagnetic field. The results he obtained were originally taken to decide crucially in favour of Abraham's classical theory (henceforth referred to as T_A) and against the Lorentz-Einstein theory (henceforth referred to as T_E).

Kaufmann's experiment actually consisted of a sequence of nine sub-experiments, whose respective outcomes may be represented by the observation reports of the form $a_1 \Lambda b_1$, $a_2 \Lambda b_2$, . . . , $a_9 \Lambda b_9$, where a_1, . . . , a_9 may be taken as initial conditions and b_1, . . . , b_9 as the corresponding *observed* outcomes. As to the predicted outcomes: those predicted by T_E will be denoted by b_{E1}, b_{E2}, . . . , b_{E9} and those by T_A by b_{A1}, b_{A2}, . . . , b_{A9}. So the system can be schematically represented as follows:

(1) $T_E \Rightarrow (a_i \Rightarrow b_{E_i})$

and

(2) $T_A \Rightarrow (a_i \Rightarrow b_{A_i})$

for all i = 1, 2, . . . , 9.

What were the relations between the outcomes predicted by T_E and T_A and the observed outcomes? It turns out that in all nine cases, the 'values' b_{A_i} predicted by Abraham are closer to the observed 'values' b_i than are the 'values' b_{E_i} predicted by Lorentz-Einstein. This can be pictorially represented as follows:

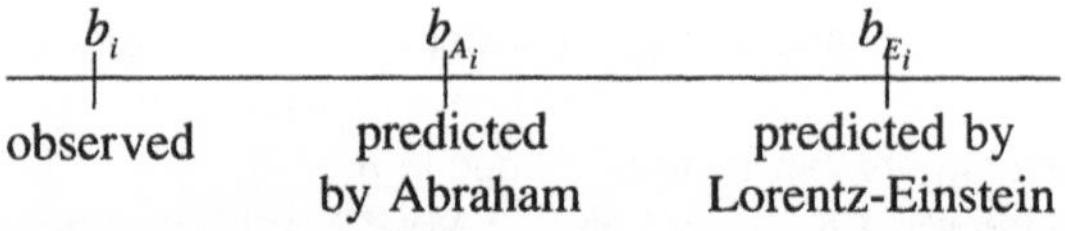

At first Planck (1906a) shared the general opinion that Kaufmann's results told against T_E and in favour of T_A; but later he challenged this opinion (1906b). He never challenged the truth of the observation reports $a_1 \Lambda b_1$, . . . , $a_9 \Lambda b_9$. What he did was to subject the theoretical structure of this whole experimental situation to strict logical analysis.

To begin with, Planck showed that, implicit in the derivation of the predictions b_{E_i} and b_{A_i} from T_E and T_A was an assumption K: K was assumed by both Abraham and Lorentz-Einstein and was moreover essential for the derivation. Thus the real situation is more accurately represented as follows:

(1′) $(T_E \Lambda K) \Rightarrow (a_i \Rightarrow b_{E_i})$

and

(2′) $(T_A \Lambda \mathrm{K}) \Rightarrow (a_i \Rightarrow b_{A_i})$ for all i = 1, . . . , 9.

Planck showed that K can be put in the form:

(3) $K \equiv (P \Lambda Q(w) \Lambda (w = w_0))$,

where w is a free parameter, while P and Q(w) are assumptions about the strengths of the magnetic and electric fields respectively. Note that w, which plays an important role in what follows, occurs in Q(w) but not in P.

Planck also established the following. Both T_E and T_A involve a common principle (which roughly states that the motion of the electron is governed by a Lagrangian function dependent on the velocity of the electron and that the latter remains smaller than the speed of light c). Let us call this principle S. Thus:

(4) $T_E \Rightarrow S$ and $T_A \Rightarrow S$.

Moreover, Planck proved that Kaufmann's observation reports logically contradict S Λ K, i.e.:

(5) $[((a_1 \wedge b_1) \wedge (a_2 \wedge b_2) \wedge \ldots \wedge (a_9 \wedge b_9)) \Rightarrow \neg (S \wedge K)]$

((5) was established by a *reductio ad absurdum*. S includes the principle that the velocity of the electron remains smaller than c. Planck proved that $[(a_1 \wedge b_1) \wedge \ldots \wedge (a_9 \wedge b_9) \wedge S \wedge K]$ implies that the electron reaches velocities greater than c.)

Thus an important interim result of Planck's analysis is that Kaufmann's results are neutral between T_E and T_A. Those results show that S Λ K is false, where S is entailed by both theories and K is not logically implied by either theory. Thus, if S were false, T_E and T_A would sink together, whereas if K were false, neither of them need sink with it. We may formally summarise this as follows:

(6) *S Λ K* is false (given Planck's acceptance of Kaufmann's observation results).

By (4) and (6) it follows that:

(7) $(T_E \wedge K)$ and $(T_A \wedge K)$ are both false.

Were there any reasons for imputing the falsity of S Λ K to one rather than the other of its conjuncts? There were. To Planck it seemed that *any* theory of the electron, whether classical or not, would have to contain S. By contrast there were physical reasons for suspecting K to be false. (These were roughly that, contrary to K, the electric field ought to be altered by the electrons ionising the gas between the plates of the condenser.) Thus Planck investigated the possibility of modifying K (see Planck 1907).

Planck proposed two different replacements for K, which we may call K_E and K_A. K_E and K_A are such that both $(T_E \wedge K_E)$ and $(T_A \wedge K_A)$ yield Kaufmann's results. Although Planck had independent reasons (already indicated) for doubting K, these did not uniquely determine a replacement for

K. To obtain this, he had to use Kaufmann's results. But these yielded alternative solutions, namely K_E and K_A, according to whether T_E or T_A was to be reconciled with those results. Let us now look more closely at what Planck did.

Planck's method in the case of the Lorentz-Einstein theory can be described schematically as follows. (As we shall see, Planck's method in the case of Abraham's theory is the same.) Remember that K is of the form (see equivalence (3) above) [P Λ Q(w) Λ (w = w_0)]. To simplify this expression I shall henceforth write R(w) for the conjunction P Λ Q(w). The deductive test-structure of the Lorentz-Einstein theory as articulated by Planck now reads:

(8) $(T_E \wedge R(w) \wedge (w = w_0)) \Rightarrow (a_i \Rightarrow b_{E_i})$ *for all* $i = 1, 2, \ldots, 9$.

Planck proceeded to regard Kaufmann's results not as tests of T_E but (in the first place at least) as means of determining the value of the free parameter w in the auxiliary hypothesis R(w) needed to deduce Kaufmann's results from T_E. In fact he found that, for each i $\epsilon\{1, 2, 3, \ldots, 9\}$, the equation $w = w_0$ can be altered so as to obtain $w = w_{Ei}$, where w_{Ei} is a numerical value uniquely determined by $[T_E \wedge R(w) \wedge a_i \wedge b_i]$. That is:

(9) $(T_E \wedge R(w) \wedge a_i \wedge b_i) \Rightarrow (w = w_{E_i})$.

It turns out that we also have:

(10) $(T_E \wedge R(w) \wedge (w = w_{E_i})) \Rightarrow (a_i \Rightarrow b_i)$.

In other words, each of Kaufmann's nine results could be explained on the basis of T_E by making a suitable assumption about the parameter w.

If we let

(11) $K_{Ei} \equiv (R(w) \wedge (w = w_{E_i}))$

then (10) becomes:

(12) $(T_E \wedge K_{E_i}) \Rightarrow (a_i \Rightarrow b_i)$,

which holds for all $i = 1, 2, \ldots, 9$.

Now the important point is that, since the K's are all statements about the electric and magnetic fields in a certain experimental situation, and since the same apparatus is used in all of Kaufmann's nine sub-experiments, we ought always to have the same auxiliary hypothesis. In other words, we should have:

(13) $K_{E1} \equiv K_{E2} \equiv \ldots \equiv K_{E9}$; i.e. $w_{E1} = w_{E_2} = \ldots = w_{E9}$

Planck showed that it does indeed turn out that the w_{Ei}'s are all nearly equal. If we allow ourselves the idealising assumption that the *near* equality is due to experimental error and that really the w_{Ei}'s are all equal, then Planck's procedure (as thus idealised) is as follows.

One singles out one of Kaufmann's experimental results, say $(a_m \wedge b_m)$, and determines the auxiliary hypothesis K_{Em} such that:

$$K_{Em} \equiv (R(w) \wedge (w = w_{Em})),$$

where

$$(T_E \wedge R(w) \wedge a_m \wedge b_m) \Rightarrow (w = w_{Em});$$

one then finds that:

(14) $(T_E \wedge K_{Em}) \Rightarrow (a_i \Rightarrow b_i)$ for all i $= 1, 2, \ldots, 9$.

Writing K_E for K_{Em}, then, if we accept this idealisation, it follows from MSRP together with my criterion of ad hocness that $(T_E \wedge K_E)$ provides a non ad hoc explanation of eight out of Kaufmann's nine experimental results. In fact Planck thought that all nine results would thus be explained. His view seems to have been that if $w_{E1} = w_{E2} = \ldots = w_{E9}$, then any one among the observation statements $a_1 \wedge b_1, \ldots, a_9 \wedge b_9$ can be used in (9) in order to determine w_E and hence K_E. None of the statements $a_1 \wedge b_1, \ldots, a_9 \wedge b_9$ plays a privileged role with respect to the remaining ones. Planck therefore intuitively concluded that, if one of these statements is explained in a non ad hoc way, then all of them are. However, this intuitive conclusion is incorrect for it fails to distinguish between a possible case in which K_E would be constructed totally independently of Kaufmann's results and the actual case, where at least one result is needed. In other words, we have two conflicting intuitions: first that, if *any one* of Kaufmann's results is satisfactorily explained, then *all of them* are; secondly, that Planck's achievement would have been greater had he not worked backwards from at least one experimental result to the determination of the auxiliary hypothesis K_E. I would argue that the conflict ought to be resolved in favour of the second intuition, which anyway does not involve much modification of the first (see §6.3).

So much for Planck's view of the explanation of Kaufmann's results by T_E. What was the situation with regard to T_A? Planck showed that the relation of T_A to Kaufmann's result is almost exactly parallel to that of T_E: a series of auxiliary assumptions K_{A_i} can be constructed in such a way that:

(15) $(T_A \wedge K_{A_i}) \Rightarrow (a_i \Rightarrow b_i)$.

The procedure is completely parallel to the one used in the Lorentz-Einstein case.

Planck again found that the $w_{A1}, \ldots, w_{A9}$, which determine $K_{A1}, \ldots, K_{A9}$, are nearly equal. Hence, under the same assumptions as above, T_A explains eight out of nine experimental results in a non ad hoc way.

However the parallel between T_A and T_E is not perfect. It turned out that the numbers $w_{E1}, \ldots, w_{E9}$ are more nearly equal to one another than are $w_{A1}, \ldots, w_{A9}$. Thus T_E has a definite (but inconclusive) advantage over T_A.

How do these historical considerations about T_E, T_A and Kaufmann's experiment support the philosophical claims outlined at the beginning of this section?

The claim that 'logic-chopping' may be important in physics is borne out in at least three ways. First, Planck derived, by means of a purely logical analysis, the negative result that Kaufmann's experiments were not crucial between Abraham and Lorentz-Einstein: what the experiments actually refuted was the conjunction of the common component S of both theories plus the non-trivial auxiliary hypothesis K.

Secondly, with the help of the logical implication (9), which has no physical content, Planck was able to modify K into K_E, so that $(T_E \wedge K_E)$ yields Kaufmann's results.[1] He was likewise able to modify K into K_A so that $(T_A \wedge K_A)$ also yields Kaufmann's results.

Lastly and very importantly, consider what effect Kaufmann's experiment had on the scientific development of someone who did not have the benefit of Planck's logical analysis, namely Lorentz. In his *Theory of Electrons* Lorentz claimed that, because of Kaufmann's experiment, he could not accept the full covariance of Maxwell's equations (Lorentz 1906, p. 212). Had Lorentz been aware of the correct logical situation as just presented, had he realised that Kaufmann's experiment is neutral between Abraham's and Einstein's theories, then he might have been converted to relativity more quickly—especially when we remember that Planck's logical analysis shows that the Lorentz-Einstein theory fits the facts more closely than does Abraham's classical hypothesis. That is, had Lorentz realised what logical connections link the experiment to the theories, then: first, he would have accepted Poincaré's corrections which made Lorentzian electrodynamics indistinguishable from Einstein's; secondly, there is good reason to suppose that, once he had accepted the covariance of Maxwell's equations, Lorentz would have gone the whole hog and joined the relativity camp. In an earlier chapter (§3.3) I argued that the heuristic power of a programme can be judged independently of its empirical success or failure. The covariance of Maxwell's equations, which implies that the electromagnetic

[1] Though this involved an idealisation, as indicated above.

ether is in principle undetectable, meant that the ether as it stood in 1905 had lost all heuristic power.

It is this complete loss of heuristic power which would have impelled Lorentz to join the relativity camp. The only (apparent) obstacle was Kaufmann's result which seemed decisively to go against the fully covariant T_E.

My conjecture that Lorentz was prepared to switch to relativity and that Kaufmann's result was a main obstacle to this, is further supported by the following facts. In the notes to his *Theory of Electrons* (Lorentz 1906, p. 339 n. 6) Lorentz cited Bucherer's 1908 experiment (see Bucherer 1909) as one decisive reason for his switch to relativity. But Bucherer's experiment is in fact a variant of Kaufmann's with this difference: it takes account of Planck's logical analysis. We have seen that the auxiliary hypotheses K (a conjunction of the form $[P \wedge Q(w) \wedge (w = w_0)]$ where Q(w) is an assumption about the intensity of the electrical field) contains the free parameter w. It is by suitably adjusting w to fit the experimental outcome that *both* T_E and T_A can be squared with Kaufmann's results. Every value of w reflects an assumption Q(w) about the strength of the electric field. One way of ruling out this possibility of post hoc adjustment (which saves both theories) is to devise an experiment whose outcome is independent of any specific assumption about the electric field. This is precisely what Bucherer did, and his results told unequivocally in favour of the Lorentz-Einstein theory and against Abraham's theory.

Lorentz's decision to switch to relativity would not have been directly motivated by Bucherer's result, since that result provides equal support for Lorentz's and for Einstein's programmes. Furthermore, in a paper on gravitation published in 1914 Lorentz indicated that he had long accepted the relativity principle as a heuristic tool. Thus Lorentz's acceptance of relativity antedates Einstein's explanation of Mercury's perihelion in 1915, i.e. it antedates the *empirical* supersession of his old programme by Einstein's. It seems plausible therefore that Lorentz was converted to relativity by the realisation that, while covariance was opening up new possibilities, the ether had become heuristically sterile (see §3.3). Thus, had Lorentz known right from the beginning that Kaufmann's experiment was not crucial, he would most probably have accepted the covariance of Maxwell's equations and joined the relativity programme at its inception in 1905.

In order to vindicate my second philosophical claim made at the beginning of this section, I shall examine how falsificationism compares with MSRP as regards the historical episode I have described. Let us see how falsificationism and MSRP might answer the question of whether or not Kaufmann's results support either of the competing theories.

First, had the original belief that Kaufmann's experiment decided in favour of Abraham and against Lorentz-Einstein been correct, then falsificationism and MSRP would have yielded parallel appraisals. According to falsificationism, Kaufmann's experiment would have confirmed Abraham and refuted Lorentz-Einstein. According to MSRP, Kaufmann's experiment would have constituted an anomaly for Lorentz-Einstein and a novel fact successfully predicted by Abraham.

Secondly, had Kaufmann's experimental results been predicted in advance on the basis of both theories, then again no difference between MSRP and falsificationism would have emerged. According to both methodologies both theories would have been confirmed by Kaufmann's results.

Thirdly, in the actual case, the two methodologies yield different appraisals. The question with which I shall conclude is: which appraisal fits in better with the intuitions of working scientists—or, more specifically in this case, with Planck's intuitions?

According to falsificationism, Planck's modifications of K (into, respectively, K_E and K_A) were ad hoc, perhaps even 'conventionalist stratagems': they saved both T_E and T_A from Kaufmann's results. Those results provide no corroboration for either T_E or T_A: they were not the results of tests on these two theories, but were already part of background knowledge. Nor were those results adequately *explained* by theories which had been adjusted to fit them.

This appraisal clashes with Planck's. Admittedly, Planck did not speak of confirmation or corroboration in this connection, but he did say that, if $w_{E1} = w_{E2} = \ldots = w_{E9}$, then "all the deflections actually observed by Mr Kaufmann *were completely explained*" by T_E (Planck 1907, p. 213; my italics and translation).

The MSRP appraisal comes out much closer to Planck's than does the falsificationist appraisal. According to MSRP, Planck's claim that T_E (or T_A) explains all of Kaufmann's *nine* results was a slight exaggeration (it only explains any eight of them); but it was only a slight exaggeration.

§6.3 A More Detailed Account of Kaufmann's Experiment

In this final section I propose to give a more detailed account of Kaufmann's experiment. I do not of course intend to give a *complete* description of the experiment; such a description would be irrelevant from the standpoints both of theoretical physics and of the philosophy of science.

The second section of this chapter contains, as we have seen, the uninterpreted symbols 'S', 'K', 'a_i', 'b_i', etc. One major purpose of this third section is, on the one hand, to state the hypotheses and the observation reports which 'S', 'K', 'a_i', 'b_i', . . stand for and, on the other, to establish those logical relations between 'S', 'K', 'a_i', 'b_i', . . on which the philosophical arguments presented in the second section hinge. There is also an important methodological by-product of this interpretative analysis: it explains why, in the end, Planck took the view that Kaufmann's results provide more support (or less disconfirmation) for the relativistic theory T_E than for its rival T_A.

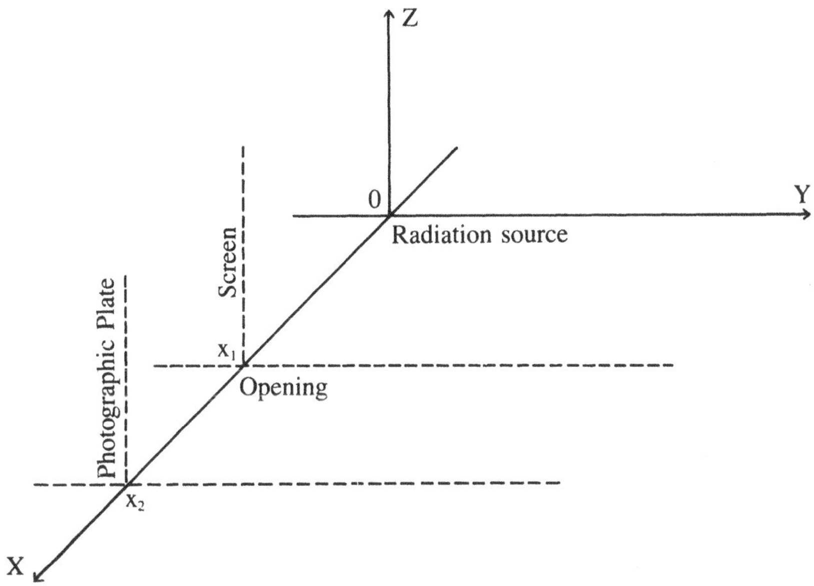

Figure 1.

Kaufmann's experiment consists in letting electrons move in an electromagnetic field and in then observing the deflection of these electrons by the field. The electric field $\vec{E}$ is generated by a condenser whose plates are parallel to the xz-plane of the chosen frame of reference. Thus $\vec{E}$ is of the form: $\vec{E} = (0, E, 0)$.

A magnet, whose poles are parallel to the plates of the condenser, creates a constant magnetic field $\vec{H}$ which, being parallel to $\vec{E}$, is of the form: $\vec{H} = (0, H, 0)$.

At the origin there is a grain of radium which acts as an electron source. The plane $x = x_1$ is occupied by a screen which has an opening at the point $(x_1, 0, 0)$ [x_1 stands for the numerical value 1.994]. At $x = x_2$ there is a photographic plate on which the electrons impinge [x_2 stands for the numerical value 3.963]. The plates of the condenser extend from $x = 0$ to $x = x_1$ and are parallel to the xz-plane. Thus the equations of the plates are of the form $y = \pm d$, where d is some numerical value. The poles of the magnet can be regarded as two infinite planes parallel to X0Z.

Some electrons leave the origin, pass through the opening at $(x_1, 0, 0)$ and then impinge on the plate at $x = x_2$. The outcome of Kaufmann's experiment can be described by saying that nine electrons hit the plate where the measured coordinates of the electrons are $(x_2, \bar{y}_1, \bar{z}_1)$, $(x_2, \bar{y}_2, \bar{z}_2)$, . . . , $(x_2, \bar{y}_9, \bar{z}_9)$. '$\bar{y}_1$,' '$\bar{z}_1$', '$\bar{y}_2$', '$\bar{z}_2$', . . . , $\bar{y}_9$, $\bar{z}_9$ represent certain numerical values; eg. $\bar{y}_3 = 0.0506$ and $\bar{z}_3 = 0.2423$. For each $i \in \{1, 2, \ldots, 9\}$, the observation reports a_i and b_i are as follows:

(16) $a_i \equiv$ (the z coordinate of the ith electron as it hits the plate is $\bar{z}_i$),

and

$b_i \equiv$ (the y-coordinate of the ith electron as it hits the plate is $\bar{y}_i$).

Let us now turn to the interpretation of 'K', 'S', 'T_E' and 'T_A'. By (3) above K is of the form $[P \wedge Q(w) \wedge (w = w_0)]$ where w is a free parameter, w_0 is a numerical value, and P and Q(w) are propositions about the strengths of the magnetic and electric fields respectively. We have seen that the modulus H of the magnetic field $\vec{H}$ is constant; the calculated value of H turns out to be 142.8. Thus:

(17) $P \equiv (H = 142.8 = \text{constant})$.

Turning to Q(w), let us first observe that, since the plates of the condenser do not extend beyond the screen, $E = 0$ for all $x \in [x_1, x_2]$, where $[x_1, x_2]$ denotes the closed interval from x_1 to x_2. Between the origin and the screen the electric field $\vec{E}$ is for the most constant and then linearly drops to zero in the neighbourhoods of $x = 0$ and $x = x_1$.

Let us write $E = 10^8 wE_1$, where w is a free parameter and E_1 is the function of x represented by the graph shown in figure 2 below.

We see that E_1 is normalised (i.e. $E_1 = 1$) in the region where E is homogenous. If we change the independent variable from x to $\xi = x - x_1/2$, E_1 becomes a function $E_1(\xi)$ of ξ such that:

$$(18)\begin{cases} E_1(\xi) = 0 \text{ for } \xi \in [x_1/2,\, x_2 - x_1/2], \text{ and} \\ E_1(\xi) = 1 \text{ for } \xi \in [0, \xi'], \text{ and} \\ E_1(\xi) = \alpha - \lambda\xi \text{ for } \xi \in [\xi', x_1/2], \text{ and} \\ E_1(-\xi) = E_1(\xi) \text{ for } \xi \in [0, x_1/2], \text{ where:} \end{cases}$$

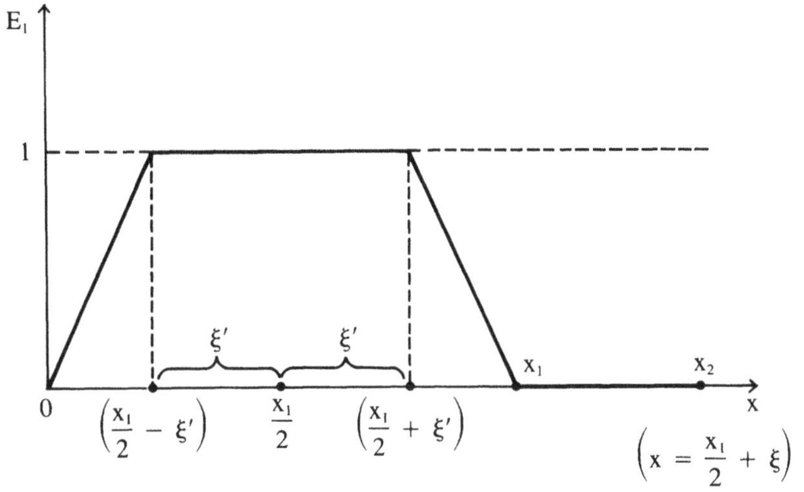

Figure 2.

(19) $\begin{cases} x_1 = 1.994,\ x_2 = 3.963,\ \xi' = 0.593, \\ \alpha = 2.468 \text{ and } \lambda = 2.475. \end{cases}$

Kaufmann took w to be equal to 2500/0.1242. Thus, if we write w_0 for the numerical value 2500/0.1242, the auxiliary hypothesis K can be expressed as follows:

(20) $K \equiv [P \wedge Q(w) \wedge (w = w_0)]$
$\equiv [(H = 142.8) \wedge (E = 10^8\, wE_1) \wedge (w = 2500/0.1242)]$,
where E_1 is the function of ξ determined by (18).

Let us now turn to the interpretation of T_E and T_A. Since T_E and T_A are high-level all-embracing theories, I shall not spell them out in detail. I shall rather specify those consequences of T_E and T_A that are relevant to Kaufmann's results. Let us start by considering the component S common to T_E and T_A which is such that $S \wedge K$ is falsified by Kaufmann's experiment. S states that the speed q of the electron remains smaller than c and that there exists a function L of q such that:

$$\frac{d}{dt}\left(\frac{\partial L}{\partial \dot{x}}, \frac{\partial L}{\partial \dot{y}}, \frac{\partial L}{\partial \dot{z}}\right) = \text{Lorentz force acting on the electron.}$$

By a suitable choice of units the Lorentz force can be expressed as: $e(\vec{E} + \vec{q} \times \vec{H})$, where e = (charge of the electron)/c, $\vec{q}$ = velocity of the electron = $(\dot{x}, \dot{y}, \dot{z})$, and q = speed of the electron = $|\vec{q}| = (\dot{x}^2 + \dot{y}^2 + \dot{z}^2)^{1/2}$.

Therefore:

$$\frac{d}{dt}\left(\frac{\partial L}{\partial \dot{x}}, \frac{\partial L}{\partial \dot{y}}, \frac{\partial L}{\partial \dot{z}}\right) = e(\vec{E} + \vec{q} \times \vec{H}),$$

$$= e((0, E, 0) + (\dot{x}, \dot{y}, \dot{z}) \times (0, H, 0))$$

$$= (-eH\dot{z}, eE, eH\dot{x}).$$

This vector equation splits into the following three scalar equations:

(21) $$\frac{d}{dt}\left(\frac{\partial L}{\partial \dot{x}}\right) = -eH\dot{z}$$

(22) $$\frac{d}{dt}\left(\frac{\partial L}{\partial \dot{y}}\right) = eE$$

(23) $$\frac{d}{dt}\left(\frac{\partial L}{\partial \dot{z}}\right) = eH\dot{x}$$

The component S can therefore be spelled out as follows:

(24) $S \equiv [(q \leq c) \land (21) \land (22) \land (23)]$

$$\equiv \left[(\dot{x}^2 + \dot{y}^2 + \dot{z}^2 \leq c^2) \land \left(\frac{d}{dt}\left(\frac{\partial L}{\partial \dot{x}}\right) = -eH\dot{z}\right) \land \left(\frac{d}{dt}\left(\frac{\partial L}{\partial \dot{y}}\right) = eE\right) \land \left(\frac{d}{dt}\left(\frac{\partial L}{\partial \dot{z}}\right) = eH\dot{x}\right)\right]$$

Note that L, as it occurs in S, is an unspecified function of q. L acts somewhat like a free parameter. T_E and T_A differ in that they ascribe different forms to the function L. According to T_E:

$$L = -m_0c^2\,((1 - q^2/c^2)^{1/2} - 1),$$

and according to T_A:

$$L = -\frac{3}{4}\, m_0c^2\left(\frac{(c^2 - q^2)}{2cq} \cdot \log\left(\frac{c + q}{c - q}\right) - 1\right)$$

where m_0 = mass of the electron at rest.

For the purpose of assessing the methodological significance of Kaufmann's results, 'T_E' and 'T_A' can be regarded as representing the following two propositions:

(25) $T_E \equiv [S \wedge (L = -m_0c^2((1 - q^2/c^2)^{1/2} - 1))]$ and

(26) $T_A \equiv \left[S \wedge \left(L = -\frac{3}{4} m_0c^2\left(\frac{(c^2 - q^2)}{2cq}\log\left(\frac{c + q}{c - q}\right) - 1\right)\right)\right]$

We are now ready to give an interpretation of b_{Ei} and b_{Ai}. Let $(x_2, \bar{y}, \bar{z})$ be the coordinates of an arbitrary electron as it hits the plate. By (16), a_i is the proposition that $\bar{z}_i$ is the z-coordinate of the ith electron; b_i is the proposition that $\bar{y}_i$ is the value of $\bar{y}$ when $\bar{z} = \bar{z}_i$. Our notation already suggests that b_{E_i} will be a proposition about the value of $\bar{y}$ which, according to T_E, ought to correspond to $\bar{z} = \bar{z}_i$. But how does T_E predict such a value of $\bar{y}$? Let me give a simplified answer which will be more fully clarified below (see $(65)_E$ below). The conjunction $T_E \wedge K$ determines an equation of the form $\bar{y} = f_E(\bar{z})$ where the right hand side is an expression involving $\bar{z}$. Substituting $\bar{z}_i$ for $\bar{z}$ and computing $f_E(\bar{z}_i)$, we obtain a numerical value $\bar{y}_{E_i}$ such that $\bar{y}_{E_i} = f_E(\bar{z}_i)$. The proposition b_{E_i} is defined by:

(27) $b_{E_i} \equiv$ (the y-coordinate of the ith electron as it hits the plate is $\bar{y}_{E_i}$).

Thus:

(28) $(T_E \wedge K \wedge a_i) \Rightarrow b_{Ei}$. That is: $(T_E \wedge \text{K}) \Rightarrow (a_i \Rightarrow b_{Ei})$.

Similarly, $T_A \wedge K$ determines an equation (see $(65)_A$ below) of the form $\bar{y} = f_A(\bar{z})$.

Substituting $\bar{z}_i$ for $\bar{z}$, we obtain $\bar{y}_{A_i} = f_A(\bar{z}_i)$, where $\bar{y}_{Ai}$ is some numerical value. Defining:

(29) $b_{A_i} \equiv$ (the y-coordinate of the ith electron as it hits the plate is $\bar{y}_{A_i}$),

we have:

(30) $(T_A \wedge \text{K}) \Rightarrow (a_i \Rightarrow b_{A_i})$.

It was found that $\bar{y}_i$ is different from both $\bar{y}_{A_i}$ and $\bar{y}_{E_i}$, but that $|\bar{y}_i - \bar{y}_{A_i}| < |\bar{y}_i - \bar{y}_{E_i}|$ for all $i = 1, 2, \ldots, 9$. This is why Kaufmann's experiment was taken to refute Lorentz-Einstein while confirming Abraham.

This brings us to the end of the interpretative part of this section. I shall now tackle the problem of establishing the logical relations on which my philosophical claims rest.

§6.3.1 Important Consequences of $S \wedge P \wedge Q(w)$

We consider a single electron which leaves the origin, passes through the opening at $(x_1,0,0)$ and then hits the plate at $(x_2,\bar{y},\bar{z})$. Let

(31) $p \underset{\text{Def}}{=} \dfrac{dL}{dq}$

Remembering that q = speed = $(\dot{x}^2 + \dot{y}^2 + \dot{z}^2)^{1/2}$, we have

$$\frac{\partial L}{\partial \dot{x}} = \frac{dL}{dq} \cdot \frac{\partial q}{\partial \dot{x}} = p \cdot \frac{1}{2} (\dot{x}^2 + \dot{y}^2 + \dot{z}^2)^{-1/2} \cdot 2\dot{x} = p \cdot \frac{\dot{x}}{q}.$$

Thus equation (21) becomes:

(32) $\dfrac{d}{dt}\left(p \cdot \dfrac{\dot{x}}{q}\right) = -eH\dot{z}.$

Similarly, equations (22) and (23) can be rewritten as:

(33) $\dfrac{d}{dt}\left(p \cdot \dfrac{\dot{y}}{q}\right) = eE$

and

(34) $\dfrac{d}{dt}\left(p \dfrac{\dot{z}}{q}\right) = eH\dot{x}$

respectively.

We recall that H is constant, so that (32) and (34) can be immediately integrated to yield:

(35) $p \cdot \dfrac{\dot{x}}{q} = -eHz + \mu$

and

(36) $p \cdot \dfrac{\dot{z}}{q} = eHx + \nu,$

where μ and ν are constant. Dividing (35) by (36):

$$\frac{\dot{x}}{\dot{z}} = \frac{-z + a}{x + b},$$

where a and b are also constant.

Hence: $\dot{x}(x + b) = \dot{z}(-z + a)$. Integrating with respect to t:

$(1/2)(x + b)^2 = -(1/2)(z - a)^2 + k$, i.e.

$(x + b)^2 + (z - a)^2 = 2k$. [$k$ = constant].

This is the equation of a circle Γ in the xz-plane. Since the electron passes through the points (0,0,0), $(x_1,0,0)$ and $(x_2,\bar{y},\bar{z})$, Γ must go through the three points (0,0), $(x_1,0)$ and $(x_2,\bar{z})$ in the xz-plane. Of course these three points uniquely determine Γ, i.e. Γ is uniquely determined by x_1, x_2 and $\bar{z}$. Let us give a parametric representation of Γ, taking as our parameter the angle φ which the tangent to Γ at a variable point P makes with the x-axis (see figure 3).

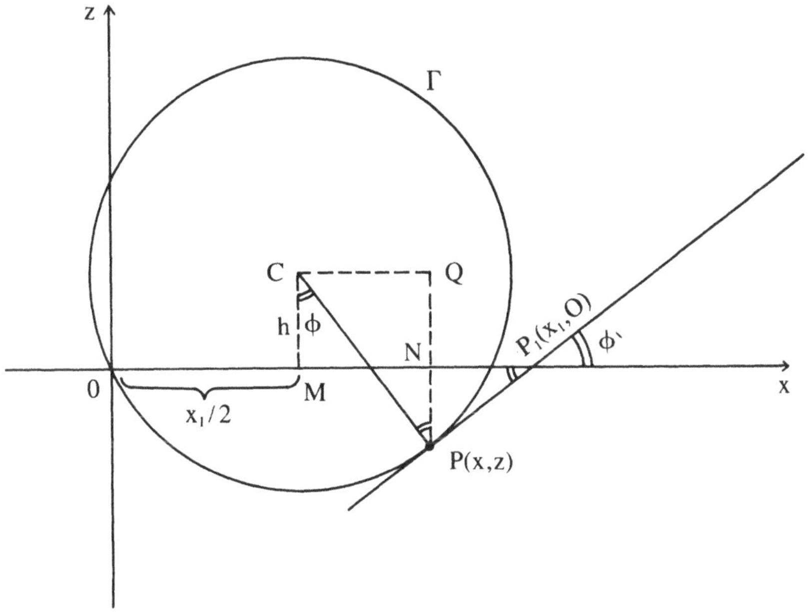

Figure 3.

Let r = radius of the circle Γ and h = MC, where M is the midpoint of OP_1 and C is the centre of Γ. The coordinates x and z of a variable point P of Γ can be expressed in terms of φ as follows:

$$\textbf{(37)}\quad \begin{cases} x = OM + MN = x_1/2 + r \sin \phi, \\ z = NP = -(PQ - NQ) = -(PQ - MC) = -r \cos \phi + h \end{cases}$$

We now propose to express r and h in terms of x_1, x_2 and $\bar{z}$. This is possible because x_1, x_2 and $\bar{z}$ uniquely determine Γ. Let ϕ_1 and ϕ_2 be the values assumed by φ at the points $P_1(x_1,0)$ and $P_2(x_2,\bar{z})$ respectively (see figure 4).

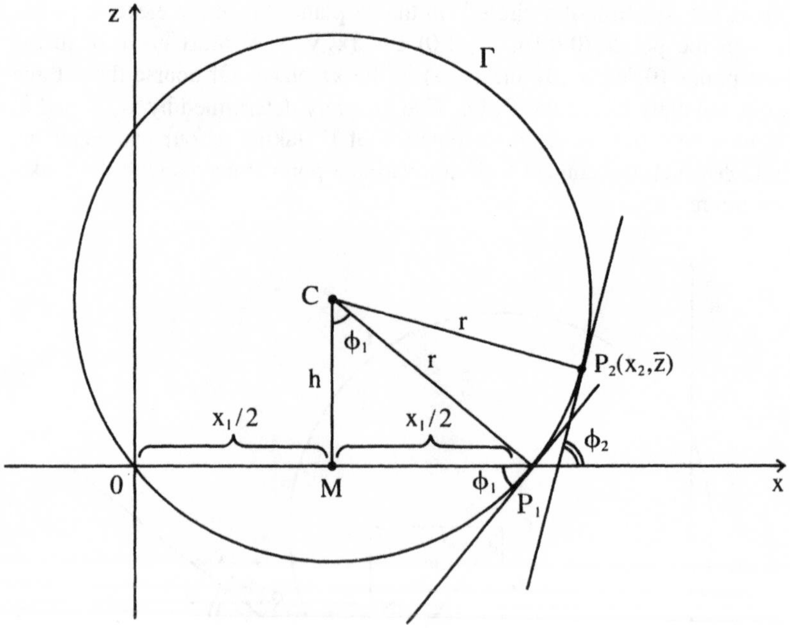

Figure 4.

In the right-angled triangle MCP_1: $\tan \phi_1 = (x_1/2)/h$. Hence $h = (x_1/2)/\tan \phi_1$, i.e. $h = (x_1/2) \cot \phi_1$.

Substituting in (37):

$$\textbf{(38)} \quad \begin{cases} x = r \sin \phi + x_1/2, \\ z = -r \cos \phi + (x_1/2) \cot \phi_1. \end{cases}$$

Similarly, in the triangle MCP_1: $(x_1/2)/r = \sin \phi_1$, i.e.

(39) $r = x_1/(2 \sin \phi_1)$.

It remains for us to express ϕ_1 in terms of x_1, x_2 and $\bar{z}$. We recall that ϕ_2 is the value of ϕ at $P(x_2,\bar{z})$.

Substituting in (38):

$x_2 = r \sin \phi_2 + x_1/2$ and

$\bar{z} = -r \cos \phi_2 + (x_1/2) \cot \phi_1$.

Substituting for r from (39) into $x_2 = r \sin \phi_2 + x_1/2$, we obtain:

(40) $\sin \phi_2 = (2x_2 - x_1) \sin \phi_1/x_1$.

Equation (40) expresses ϕ_2 in terms of ϕ_1. Let us now turn to

$$\bar{z} = -r \cos \phi_2 + (x_1/2) \cot \phi_1.$$

Thus:

$$r \cos \phi_2 = (x_1/2) \cot \phi_1 - \bar{z}.$$

Squaring:

$$r^2 \cos^2\phi_2 = \Big((x_1/2) \cos \phi_1 - \bar{z}\Big)^2; \text{ i.e.}$$

$$r^2(1 - \sin^2\phi_2) = \Big((x_1/2)\cot \phi_1 - \bar{z}\Big)^2.$$

By (39) and (40):

$$\frac{x_1^2}{4 \sin^2\phi_1}\left(1 - \frac{(2x_2 - x_1)^2 \sin^2\phi_1}{x_1^2}\right) = \left(\frac{x_1}{2} \cot\phi_1 - \bar{z}\right)^2$$

Expanding both sides of this last equation and noting that

$$1 + \cot^2\phi_1 = \frac{1}{\sin^2\phi_1},$$

we obtain:

$$\textbf{(41)} \quad \tan \phi_1 = \frac{x_1\bar{z}}{(x_2 - x_1)x_2 + \bar{z}^2}$$

We are now home and dry. (41) gives us ϕ_1 in terms of x_1, x_2 and $\bar{z}$. Substituting in (39) and (40), we obtain r and ϕ_2 in terms of x_1, x_2 and $\bar{z}$. Equations (38) express x and z in terms of ϕ, ϕ_1, x_1 and r, and hence, indirectly, in terms of ϕ, x_1, x_2 and $\bar{z}$.

We shall now determine ϕ as a function of t. Differenting equations (38):

$$\textbf{(42)} \quad \dot{x} = r \cos\phi \cdot \dot{\phi} \text{ and } \dot{z} = r \sin\phi \cdot \dot{\phi}.$$

Substituting from (42) and (38) into (35):

$$\frac{p}{q}(r \cos\phi \cdot \dot{\phi}) = eHr \cos \phi + k.$$

where k is some constant. Substituting $\pi/2$ for ϕ in this last equation: $0 = 0 + k$. Thus $k = 0$. Therefore

$$\frac{p}{q}(r \cos\phi \cdot \dot{\phi}) = eHr \cos\phi + k = eHr \cos\phi.$$

Dividing by r cos ϕ:

(43) $\frac{p}{q}\dot{\phi} = eH$; that is $\dot{\phi} = eHq/p$.

We recall that $q^2 = (\dot{x}^2 + \dot{y}^2 + \dot{z}^2) = (r^2\cos^2\phi \cdot \dot{\phi}^2 + \dot{y}^2 + r^2\sin^2\phi \cdot \dot{\phi}^2)$ (by (42)) $= (r^2\dot{\phi}^2 + \dot{y}^2)$. Since $\dot{y}^2$ is negligibly small in comparison with $(\dot{x}^2 + \dot{z}^2)$,

(44) $q^2 = r^2\dot{\phi}^2$; that is, $q = r\dot{\phi}$ (approximately).

Substituting in (43):

(45) $p = eHr = \dfrac{eHx_1}{2 \sin \phi_1}$ [by (39)].

p is therefore constant. We recall that p $\underset{\text{Def}}{=}$ dL/dq, where L is a function of q. Solving p = dL/dq for q in terms of p, we conclude that q is a function of p. Like p, q is therefore constant. Since both p and q are constant, it follows from (43) that

(46) $\dot{\phi} =$ constant. Integrating:

(47) $\phi = at + b$, where a and b are both constant.

Let us now consider equation (33), i.e.

$$\frac{d}{dt}\left(\frac{p}{q}\dot{y}\right) = eE$$

Since both p and q are constant, this equation can be written as:

(48) $eE = \dfrac{p}{q}\dfrac{d\dot{y}}{dt} = \dfrac{p}{q}\dfrac{d^2y}{dt^2}$

But:

$$\frac{d}{dt} = \frac{d\phi}{dt} \cdot \frac{d}{d\phi} = \dot{\phi} \cdot \frac{d}{d\phi}$$

Noting that $\dot{\phi}$ is constant:

$$\frac{d^2}{dt^2} = \dot{\phi}\frac{d}{d\phi}\left(\dot{\phi}\frac{d}{d\phi}\right) = \dot{\phi}^2\frac{d^2}{d\phi^2}$$

Substituting in (48)

(49) $\frac{p}{q}\dot{\phi}^2\frac{d^2y}{d\phi^2} = eE$

By (43) and (45)

(50) $\frac{d^2y}{d\phi^2} = \frac{r}{qH}E$

It remains for us to integrate (50) with respect to ϕ. We recall that $E = 10^8 \cdot wE_1(\xi)$ where $\xi = x - x_1/2$. (See figure 2 and equations (18).)

Substituting in (50):

(51) $\frac{d^2y}{d\phi^2} = 10^8\frac{rw}{qH}E_1(\xi)$

We have to express ξ in terms of ϕ. By (38):

(52) $\xi = x - x_1/2 = r\sin\phi + x_1/2 - x_1/2 = r\sin\phi.$

Hence $r\sin(-\phi) = -r\sin\phi = -\xi$; so $-\xi$ corresponds to $-\phi$.

(53) Put: $E_1(\xi) = \overline{E}_1(\phi)$.

Since $E_1(-\xi) = -E_1(\xi)$, we have: $\overline{E}_1(-\phi) = E_1(-\xi) = -E_1(\xi) = -\overline{E}_1(\phi)$. In order to integrate (51), we have to express $\overline{E}_1(\phi)$ explicitly in terms of ϕ. Let ϕ' be the value of ϕ which corresponds to $\xi' = 0.593$ (See (19)). By (52):

(54) $\xi' = r\sin\phi'$ i.e. $0.593 = r\sin\phi'$.

Since r is a known function of x_1, x_2 and $\bar{z}$, this last equation determines ϕ' in terms of x_1, x_2 and $\bar{z}$. By (18):

(55)
$$\begin{cases} \overline{E}_1(\phi) = E_1(\xi) = 1 \text{ for } \xi\epsilon[0, \xi'], \\ \text{i.e. for } \phi\epsilon[0,\phi']; \text{ and} \\ \overline{E}_1(\phi) = E_1(\xi) = \alpha - \lambda\xi = \alpha - \lambda r\sin\phi \\ \text{for } \xi\epsilon[\xi', x_1/2], \text{ i.e. for } \phi\epsilon[\phi',\phi_1]; \text{ and} \\ \overline{E}_1(-\phi) = E_1(-\xi) = -E_1(\xi) = -\overline{E}_1(\phi) \\ \text{for } \xi\epsilon[0, x_1/2], \text{ i.e. for } \phi\epsilon[0,\phi_1]. \end{cases}$$

Integrating (51) from 0 to ϕ_1:

$$\left(\frac{dy}{d\phi}\right)_{\phi_1} = 10^8\frac{rw}{qH}\int_0^{\phi_1}\overline{E}_1(\phi)\cdot d\phi$$

$$= 10^8 \frac{rw}{qH} \left(\int_0^{\phi'} \overline{E}_1(\phi) \cdot d\phi + \int_{\phi'}^{\phi_1} \overline{E}_1(\phi) \cdot d\phi \right)$$

$$= 10^8 \frac{rw}{qH} \left(\int_0^{\phi'} 1 \cdot d\phi + \int_{\phi'}^{\phi_1} (\alpha - \lambda \cdot r \cdot \sin \phi) d\phi \right)$$

Hence:

$$\textbf{(56)} \left(\frac{dy}{d\phi}\right)_{\phi_1} = 10^8 \frac{rw}{qH} (\phi' + \alpha(\phi_1 - \phi') + \lambda r (\cos\phi_1 - \cos\phi'))$$

By the first equation in (18), $E_1(\xi) = 0$ for $\xi \epsilon\ [x_1/2,\ x_2 - x_1/2]$; i.e. $\overline{E}_1(\phi) = 0$ for $\phi \epsilon [\phi_1, \phi_2]$. By (51):

$$\frac{d^2y}{d\phi^2} = 0 \text{ for all } \phi \epsilon [\phi_1, \phi_2].$$

Integrating between ϕ and ϕ_1:

$$\frac{dy}{d\phi} - \left(\frac{dy}{d\phi}\right)_{\phi_1} = 0; \text{ i.e.}$$

$$\frac{dy}{d\phi} = \left(\frac{dy}{d\phi}\right)_{\phi_1} = \text{constant, for } \phi \epsilon [\phi_1, \phi_2].$$

Integrating between ϕ_1 and ϕ_2:

$$\textbf{(57)}\ (y)_{\phi_2} - (y)_{\phi_1} = (\phi_2 - \phi_1) \cdot \left(\frac{dy}{d\phi}\right)_{\phi_1}$$

But $(y)_{\phi_1} = 0$ because the electron passes through the point $(x_1,0,0)$ and $(y)_{\phi_2} = \overline{y}$ because the electron hits the plate at $(x_2, \overline{y}, \overline{z})$. Substituting in (57):

$$\textbf{(58)}\ \overline{y} = (\phi_2 - \phi_1) \left(\frac{dy}{d\phi}\right)_{\phi_1}.$$

By (56) and (58):

$$\textbf{(59)}\ \overline{y} = 10^8 \frac{rw}{qH} (\phi_2 - \phi_1) [\phi' + \alpha(\phi_1 - \phi') + \lambda r (\cos \phi_1 - \cos \phi')].$$

Note that ϕ_1, ϕ_2, ϕ' and r are known functions of x_1, x_2 and $\overline{z}$, and that x_1, x_2, α and λ are numerically given quantities. Thus (59) can be written as:

(60) $\bar{y} = F(w,q,\bar{z})$, where F is a known function of the three variables w, q and $\bar{z}$.

Equation (60) plays a central role in Planck's assessment of Kaufmann's results. This is because (60) enables us, given $\bar{z}$, to calculate each of the three quantities $\bar{y}$, w, q in terms of the remaining two. Note that *the function F does not depend on the form of L,* i.e. on the particular functional dependence of L on q. Thus F is the same function, whether we adopt T_E or T_A.

§6.3.2 Experimental Refutation of S Λ K

We recall that $K \equiv [P \wedge Q(w) \wedge (w = w_o)]$, where w_0 stands for the numerical value 2500/0.1242 (see (20)). By (60):

(61) $\bar{y} = F(w_0, q, \bar{z}) = F(2500/0.1242, q, \bar{z})$.

Consider the observation report $a_i \wedge b_i$ which tells us that $\bar{y} = \bar{y}_i$ when $\bar{z} = \bar{z}_i$ (see (16)). Substituting $\bar{z}_i$ for $\bar{z}$ and $\bar{y}_i$ for $\bar{y}$ in (61), we obtain:

(62) $\bar{y}_i = F(w_0, q_i, \bar{z}_i)$,

where q_i denotes the speed of the ith electron as it hits the photographic plate. (62) enables us to calculate q_i, since $\bar{y}_i$, w_0 and $\bar{z}_i$ are all numerically given. Planck found that, for $\bar{y}_i = 0.0247$ and $\bar{z}_i = 0.1354$, $q_i/c = 1.034$ (approximately) where c is the velocity of light. Hence $q_i > c$, which contradicts an implication of S, namely that the speed of an electron remains smaller than c. Thus S Λ K is experimentally refuted by Kaufmann's results. Since S is entailed by both T_E and by T_A, it follows that $T_E \wedge K$ and $T_A \wedge K$ are also refuted.

§6.3.3 Important Consequences of $T_E \wedge P \wedge Q(w)$ and $T_A \wedge P \wedge Q(w)$

We recall that $T_E \Rightarrow S$ and $T_A \Rightarrow S$ because S is a component common to T_E and T_A (see (25) and (26)). Hence:

$$(T_E \wedge P \wedge Q(w)) \Rightarrow (S \wedge P \wedge Q(w))$$

and

$$(T_A \wedge P \wedge Q(w)) \Rightarrow (S \wedge P \wedge Q(w)).$$

We can therefore make use of all the results established in §6.3.1.

By (25), $L = -m_o c^2 ((1 - q^2/c^2)^{1/2} - 1)$. Noting that $p \overset{=}{\underset{\text{Def}}{=}} dL/dq$, it follows that T_E implies:

$$p = \frac{d}{dq}(-m_0c^2((1 - q^2/c^2)^{1/2} - 1)) = m_0q/\sqrt{1 - q^2/c^2}.$$

In conjunction with (45), this last equation yields:

$$\left(\frac{e}{m_o}\right)\frac{Hx_1}{2\ sin\phi_1} = \frac{q}{\sqrt{1 - q^2/c^2}}$$

q will thus depend on ϕ_1, i.e. on $\bar{z}$ [by (41)].

This enables us to calculate the (constant) speed of the electron. Let us denote this calculated value of q by $q_E(\bar{z})$ where the subscript E indicates the dependence of the calculation on the theory T_E. Thus $T_E \wedge P \wedge Q(w)$ implies:

(63_E) $$\left(\frac{e}{m_o}\right)\cdot\frac{Hx_1}{2\ sin\phi_1} = \frac{q_E}{\sqrt{1 - 9^2_E/c^2}}$$

Substituting $q_E(\bar{z})$ for q in (60):

(64_E) $\bar{y} = F(w, q_E(\bar{z}), \bar{z})$.

Similarly, $T_A \wedge P \wedge Q(w)$ implies that:

$$L = -\frac{3}{4}m_oc^2\left(\left(\frac{c^2 - q^2}{2qc}\right)\log\left(\frac{c + q}{c - q}\right) - 1\right);$$

hence:

$$p = \frac{dL}{dq} = \frac{3}{4}\frac{m_oc^2}{q}\left(\left(\frac{c^2 + q^2}{2qc}\right)\log\left(\frac{c + q}{c - q}\right) - 1\right)$$

In conjunction with (45), this last equation yields:

(63_A) $$\left(\frac{e}{m_o}\right)\frac{Hx_1}{2\ sin\phi_1} = \frac{3c^2}{4q_A}\left(\left(\frac{1 + q_A{}^2/c^2}{2q_A/c}\right)\log\left(\frac{c + q_A}{c - q_A}\right) - 1\right)$$

where: $q_A = q_A(\bar{z})$ = the speed of the electron according to the theory $T_A \wedge P \wedge Q(w)$. Substituting $q_A(\bar{z})$ for q in (60):

(64_A) $\bar{y} = F(w, q_A(\bar{z}), \bar{z})$.

Note that q_E and q_A are functions of $\bar{z}$ and that $q_E \neq q_A$. Thus T_E and T_A predict different values for the speed of the electron.

§6.3.4 'Confirmation' of $T_A \wedge K$ and Refutation of $T_E \wedge K$

Let us remember that $K \equiv [P \wedge Q(w) \wedge (w = w_0)]$ where w_0 =

2500/0.1242. Substituting w_0 for w in $(64)_E$, we see that $T_E \Lambda K$ implies: $\bar{y} = F(w_0, q_E(\bar{z}), \bar{z})$. Since w_0 denotes a known numerical value, the equation can be written as:

(65_E) $\bar{y} = f_E(\bar{z})$, where $f_E(\bar{z}) = F(w_0, q_E(\bar{z}), \bar{z})$.

For each value of $\bar{z}$, $(65)_E$ determines the value of $\bar{y}$ which, according to $T_E \Lambda K$, corresponds to $\bar{z}$. Substituting the observed value $\bar{z}_i$ for $\bar{z}$, let us calculate the value $\bar{y}_{E_i}$ predicted by $T_E \Lambda K$ for $\bar{y}$. Thus:

(66_E) $\bar{y}_{E_i} = f_E(\bar{z}_i)(= F(w_0, q_E(\bar{z}_i), \bar{z}_i))$.

Similarly, $T_A \Lambda K$ entails:

(65_A) $\bar{y} = f_A(\bar{z})$ where $f_A(\bar{z}) = F(w_0, q_A(\bar{z}), \bar{z})$;

and

(66_A) $\bar{y}_{A_i} = f_A(\bar{z}_i)(= F(w_{0,} q_A(\bar{z}_i), \bar{z}_i))$.

As we have already remarked, it turns out that:

$|\bar{y}_i - \bar{y}_{A_i}| < |\bar{y}_i - \bar{y}_{E_i}|$ for all $i = 1, \ldots, 9$ (see, above, comments following relation (30))

This inequality led people to the erroneous conclusion that Kaufmann's experiment confirmed Abraham while refuting Lorentz-Einstein.

§6.3.5 Adjustment of Parameters in $T_E \Lambda P \Lambda Q(w)$ and in $T_A \Lambda P \Lambda Q(w)$

We recall that $T_E \Lambda P \Lambda Q(w)$ implies:

(64_E) $\bar{y} = F(w, q_E(\bar{z}), \bar{z})$.

By (16), $(a_i \Lambda b_i)$ tells us that $\bar{y} = \bar{y}_i$ when $\bar{z} = \bar{z}_i$. If we adjoin $(a_i \Lambda b_i)$ to $(T_E \Lambda P \Lambda Q(w))$, we can substitute $\bar{y}_i$ for $\bar{y}$ and $\bar{z}_i$ for $\bar{z}$ in $(64)_E$. Thus:

(67_E) $\bar{y}_i = F(w, q_E(\bar{z}_i), \bar{z}_i)$.

Solving (67) for w, we obtain:

(68_E) $w = w_{E_i}$, say.

Hence:

(69_E) $(T_E \Lambda P \Lambda Q(w) \Lambda a_i \Lambda b_i) \Rightarrow (w = w_{E_i})$.

Note that w_{E_i} is a solution of $(67)_E$. Thus, substituting w_{E_i} for w in $(67)_E$ we obtain the identity:

(70_E) $\bar{y}_i = F(w_{E_i}, q_E(\bar{z}_i), \bar{z}_i)$.

From $(70)_E$, it follows that $\bar{y} = \bar{y}_i$ when: $\bar{z} = \bar{z}_i$, $q = q_E(\bar{z}_i)$ and $w = w_{E_i}$. But $T_E \wedge P \wedge Q(w)$ entails $q = q_E$. (see §6.3.3). Hence:

(71_E)
$$\begin{cases} (T_E \wedge P \wedge Q(w) \wedge (w = w_{E_i}) \wedge a_i) \Rightarrow b_i, \text{ i.e.} \\ (T_E \wedge P \wedge Q(w) \wedge (w = w_{E_i})) \Rightarrow (a_i \Rightarrow b_i). \\ \text{(For the meanings of } a_i \text{ and } b_i \text{ see (16) above)} \end{cases}$$

Similarly it can be shown that:

(69_A) $(T_A \wedge P \wedge Q(w) \wedge a_i \wedge b_i) \Rightarrow (w = w_{A_i})$;

and

(71_A) $(T_A \wedge P \wedge Q(w) \wedge (w = w_{A_i})) \Rightarrow (a_i \Rightarrow b_i)$,

where w_{A_i} is a solution of

(67_A) $\bar{y}_i = F(w, q_A(\bar{z}_i), \bar{z}_i)$.

That is:

(70_A) $\bar{y}_i = F(w_{A_i}, q_A(\bar{z}_i), \bar{z}_i)$

is an identity.

Planck found that $w_{E_1}, \ldots, w_{E_9}$,[2] are more nearly equal among themselves than are $w_{A_1}, \ldots, w_{A_9}$.

This tips the balance slightly in favour of T_E.

---

[2]It turns out that $w_{E_1} > w_{E_2} > \ldots > w_{E_9}$ and $w_{A_1} > w_{A_2} > \ldots > w_{A_9}$; where eg., $w_{E_1} = 17970$; $w_{E_9} = 17590$; $w_{A_1} = 18840$; $w_{A_9} = 18040$.

— 7 —

FURTHER DEVELOPMENTS OF THE RELATIVITY PROGRAMME

In this chapter we take up, once more, the problem of how Einstein's heuristics was used in order to obtain new hypotheses through modifying existing laws in a prescribed way. We shall see that these further developments of the relativity programme issued in Einstein's derivation of his celebrated equation $E = mc^2$.

Let us recall that the heuristics of STR consists of two components: a condition of Lorentz-covariance to be satisfied by every physical law (requirement (i)) and, wherever possible, the requirement that a relativistic theory yield some classical law as a limiting case (requirement (ii)).

§7.1 Planck's Modification of Newton's Second Law of Motion

As shown in chapter three, STR leaves Maxwell's field *equations* unchanged. But one modification brought about by Einstein's approach is that all electromagnetic laws, instead of holding only in the ether frame, assume the same form in every inertial system. Thus, Maxwell's equations present us with the first example of a relativistic theory. However, this example is somewhat uninteresting, in that it exhibits a method, not of constructing a new hypothesis, but of entrenching the existing Maxwellian system. With hindsight, we can see that such had to be the case. Einstein had to start from some existing theory, investigate its invariance properties and then alter the rest of physics from this fixed vantage point. In this section we shall deal with Planck's application of requirements (i) and (ii) to the equation of motion, which actually led to the discovery of a new law.

In 1906 Planck started from the same problem which Einstein had tackled in 1905, namely the dynamics of the slowly accelerated electron. By using Einsteinian methods, Planck corrected a mistake made by Einstein in his

EMB. Ironically, this error, which is interpretative rather than logical in character, is similar to the mistake made by Lorentz in determining the transformation of the charge density. We shall see that, like Einstein with regard to Lorentz, Planck arrived at exactly the same equations as did Einstein; but the latter had made the faux pas of calling 'force' a three-vector which has no physical significance in the 'stationary' frame.

Let us now give Planck's derivation of the new equation of motion in greater detail. Newton's law

$$\vec{f} = \frac{d}{dt}(m_0\vec{v}) = m_0\vec{a} = \text{mass} \times \text{acceleration},$$

confronts us with a special difficulty. For two centuries, a controversy had raged as to whether $\vec{f} = m_0\vec{a}$ is a synthetic proposition or a mere definition of force. If Newton's law of motion is really analytic, then nothing can prevent us from adopting it as the definition of force in relativistic mechanics. In such a case, the problem of modifying the law does not arise at all; or rather, the problem is reduced to the pragmatic one of finding the most convenient definition of force in the new physics.

Planck was aware of this problematic character of Newton's law but it is not clear where he stood on this issue. He certainly envisaged defining force by means of $\vec{f} = m_0\vec{a}$. Thus it looks as though he did after all regard Newton's law as a mere definition. Nevertheless, in the case of relativistic mechanics he rejects such a definition on two non-logical grounds.

His first claim is that, since the acceleration $\vec{a}$ transforms according to a complicated law having no obvious physical meaning, then the same would apply to $\vec{f}$, were $\vec{f}$ to be *defined* as $m_0\vec{a}$. This argument is weak, for in the sequel Planck does not tell us why his own law, namely

$$\vec{f} = \frac{d}{dt}\left(\frac{m_0\vec{v}}{\sqrt{1 - v^2/c^2}}\right),$$

fares any better, in this respect, than

$$\vec{f} = m_0\frac{d\vec{v}}{dt}.$$

It is of course true that the four-force $m_0 dV/d\tau$ transforms like the coordinate differentials and is in this sense 'simple'. In 1906 however Planck seemed to be interested only in the three-force $\vec{f}$.

Secondly Planck rejects the equation $\vec{f} = m_0\vec{a}$ because it destroys the relation between a potential function and the force derived from it. In other words: suppose that $\vec{f} = m_0\vec{a}$ and that, in some inertial frame K, $\vec{f} = -\nabla V$ for some scalar V.

$$\text{(i.e. } f_x = -\frac{\partial V}{\delta x}, f_y = \frac{\partial V}{\delta y}, f_z = -\frac{\partial V}{\delta z}).$$

In another frame K′ we generally have $\vec{f}\,' \neq -\nabla' V$. *In the form in which it is presented by Planck,* this second objection applies equally well to his own law,

$$\vec{f} = \frac{d}{dt}\left(\frac{m_0 \vec{v}}{\sqrt{1 - v^2/c^2}}\right)$$

Once again, this is due to the fact that he considers 3-forces and not 4-vectors (see appendix two).

Fortunately, the two reasons adduced by Planck for rejecting $\vec{f} = m_0 \vec{a}$ are superfluous, as is shown by the method he actually used in modify ing Newton's law. This method clearly demonstrates that $\vec{f} = m_0 \vec{a}$, far from being analytic, is in fact false. In order to see this, we need first to consider a situation in which $\vec{f} = m_0 \vec{a}$ is synthetic; i.e. in which the left-hand side of the equation, the force, is determined independently of the acceleration. Secondly, we must be able to transform both sides of $\vec{f} = m_0 \vec{a}$ and thus show that the equation is not Lorentz-covariant.

At the turn of the century the only well-known fields were those created by the gravitational and electromagnetic actions. Since gravitation had not yet been forced into the relativistic framework, there was no choice left for Einstein but to resort, once again, to electrodynamics (unless of course he wanted to embark on a new theory of gravitation; which he eventually did, but much later). Fortunately, it turned out that electrodynamics was adequate for the purpose of obtaining a new law of motion. We have seen that, given the charge and velocity distributions, Maxwell's equations determine $\vec{E}$ and $\vec{H}$. A test-electron of charge e, mass m_0 and velocity $\vec{v}$, when placed in an electromagnetic field, is subject to the so-called Lorentz-force $e(\vec{E} + (\vec{v}/c) \times \vec{H})$. Hence, if we accept Newton's law of motion, we must have:

$$m_0 \vec{a} = m_0 \frac{d^2 \vec{r}}{dt^2} = e\left(\vec{E} + \frac{\vec{v}}{c} \times \vec{H}\right)$$

These relations are synthetic because $\vec{E}$ and $\vec{H}$ are determined by Maxwell's equations, hence independently of the electron's motion. The transformation rules for $\vec{E}$, $\vec{H}$, x, y, z and t are all known; so it is easy to verify that $m_0 \vec{a} = e(\vec{E} + (\vec{v}/c) \times \vec{H})$ is not Lorentz-covariant. It is in fact unnecessary to carry out this verification, for we shall show that the Lorentz-covariance condition entails a law incompatible with $\vec{f} = m_0 \vec{a}$.

As explained in chapter three, the Lorentz transformation collapses into the Galilean one when $u/c \to 0$, hence when $c \to \infty$. Let $\mathbb{E}\left(\frac{\vec{v}}{c}\right) = \mathbf{0}$ denote the relativistic equation of motion which is to replace $\vec{f} - m_0\vec{a} = \mathbf{0}$. ($(\vec{v}/c)$ exhibits the likely dependence of $\mathbb{E}$ on the vector $\vec{v}/c$). The condition that $\vec{f} - m_0\vec{a} = \mathbf{0}$ be a limiting case of the new equation will be clearly met if

$$\mathbb{E}\left(\frac{\vec{v}}{c}\right) \to (\vec{f} - m_0\,\vec{a}\,), \text{ as } \frac{v}{c} \to 0.$$

We naturally assume that $\mathbb{E}$ is a continuous function of $\vec{v}/c$. Thus

$$\mathbb{E}\left(\frac{\vec{v}}{c}\right) \to \quad \mathbb{E}(0), \text{ as } \frac{v}{c} \to 0.$$

This, together with the requirement that $\mathbb{E}\left(\frac{\vec{v}}{c}\right) \to (\vec{f} - m_0\vec{a})$ yields $\mathbb{E}(0) = \vec{f} - m_0\vec{a}$. Thus: $\mathbf{0} = \mathbb{E}(0) = \vec{f} - m_0\vec{a}$ for $v/c = 0$, hence for $v = 0$. This means that Newton's law must strictly hold for every mass-point instantaneously at rest; moreover, by the relativity principle, this result applies to every inertial frame.

Both Einstein and Planck start from the following special case. They consider a charged point mass, eg. an electron, which moves with velocity $\vec{v}$ in an electromagnetic field. Einstein's ingenious device consists in choosing, at the considered time t, an inertial system K′ moving with the same velocity $\vec{v}$ as the electron. In K′ the electron is instantaneously at rest; hence its motion must initially ensue according to Newton's law. This enables Einstein to write down the equations connecting the force with the acceleration. Using the Lorentz-transformation, Planck turned the special equations obtained by Einstein for the system K′ into a general relation holding in an arbitrary inertial frame. In this way, Planck arrived at a general Lorentz-covariant formulation of a new law of motion. Note that both the relativity principle and the proposition that Newton's law is true for $\vec{v} = \mathbf{0}$ play an absolutely crucial role in this argument; in other words, both requirements (i) and (ii) are (tacitly) used.

I have mentioned that Planck corrected the 'definition' of force proposed by Einstein in the EMB. However, this aspect of Planck's contribution can be understood only through a detailed examination of the problem of the moving electron, to which we now turn.

Consider an arbitrary inertial frame K_0 in which an electron of mass m_0 and charge e moves with velocity $\vec{v}$ under the effect of electromagnetic forces. We suppose that the contribution of the electron to the field is negligible. At time t choose two inertial frames K and K′, both of whose

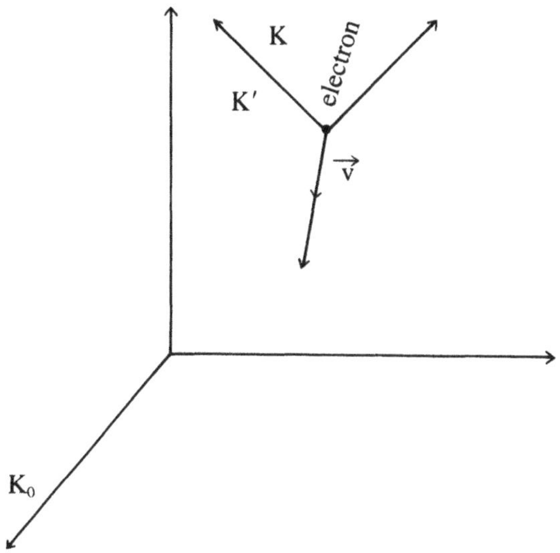

origins coincide with the position of the electron; the three space axes of K' coincide with those of K at time t, the x-axis being chosen to lie along the direction of the velocity $\vec{v}$. K is supposed to be at rest, whereas K' moves with velocity $\vec{v}$ relatively to K_0. All these relations hold only at the chosen time t. (As t varies, K remains fixed in K_0, K' moves away from K with uniform velocity $\vec{v}$ and the electron will generally alter its position and its speed with respect to both K and K'.)

Let $\vec{E} = (E_x, E_y, E_z)$, $\vec{H} = (H_x, H_y, H_z)$ and $\vec{E}' = (E'_x, E'_y, E'_z)$, $\vec{H}' = (H'_x, H'_y, H'_z)$ be the fields in K and K' respectively. The expressions of $\vec{E}'$ and $\vec{H}'$ in terms of $\vec{E}$ and $\vec{H}$ are: $E'_x = E_x$; $E'_y = \gamma(E_y - (v/c)H_z)$; $E'_z = \gamma(E_z + (v/c)H_y)$; $H'_x = H_x$; $H'_y = \gamma(H_y + (v/c)E_z)$; $H'_z = \gamma(H_z - (v/c)E_y)$; where $\gamma \underset{\text{Def}}{=} (1 - v^2/c^2)^{-1/2}$.

In order to exploit Newton's second law we need to determine the transformation equations of the acceleration $\vec{a}$. We know that the velocity $\vec{w} = (w_x, w_y, w_z)$ transforms according to (see appendix one):

$$\textbf{(1)}\quad w'_x = \frac{w_x - v}{1 - v \cdot w_x/c^2};\ w'_y = \frac{w_y}{\gamma(1 - v \cdot w_x/c^2)};$$

$$w'_z = \frac{w_z}{\gamma(1 - v \cdot w_x/c^2)}.$$

By definition, $a_x = dw_x/dt$, $a'_x = dw'_x \cdot dt'$, etc. Differentiating (1) then dividing by $dt' = \gamma(dt - (v/c^2)dx) = \gamma dt(1 - v \cdot w_x/c^2)$, we obtain (see appendix one):

$$(2)\begin{cases} a'_x = \dfrac{a_x}{\gamma^3(1 - v \cdot w_x/c^2)^3} \\ a'_y = \dfrac{1}{\gamma^2(1 - v \cdot w_x/c^2)^2}\left[a_y + \dfrac{v}{c^2}\dfrac{a_x \cdot w_y}{(1 - v \cdot w_x/c^2)}\right] \\ a'_z = \dfrac{1}{\gamma^2(1 - v \cdot w_x/c^2)^2}\left[a_z + \dfrac{v}{c^2}\dfrac{a_x \cdot w_z}{(1 - v \cdot w_x/c^2)}\right] \end{cases}$$

Substituting $(-v)$ for v yields the inverse transformation:

$$(3)\begin{cases} a_x = \dfrac{a'_x}{\gamma^3(1 + v \cdot w'_x/c^2)^3} \\ a_y = \dfrac{1}{\gamma^2(1 + v \cdot w'_x/c^2)^2}\left[a'_y - \dfrac{v}{c^2}\dfrac{a'_x \cdot w'_y}{(1 + v \cdot w'_x/c^2)}\right] \\ a_z = \dfrac{1}{\gamma^2(1 + v \cdot w'_x/c^2)^2}\left[a'_z - \dfrac{v}{c^2}\dfrac{a'_x \cdot w'_y}{(1 + v \cdot w'_x/c^2)}\right] \end{cases}$$

For $\vec{w}' = \mathbf{0}$, i.e. for $0 = w'_x = w'_y = w'_z$, we have the special case:

$$(4)\begin{cases} a_x = \dfrac{a'_x}{\gamma^3};\ a_y = \dfrac{a'_y}{\gamma^2};\ a_z = \dfrac{a'_z}{\gamma^2};\ \text{i.e.} \\ (a'_x, a'_y, a'_z) = (\gamma^3 a_x, \gamma^2 a_y, \gamma^2 a_z). \end{cases}$$

In K′ the velocity of the electron vanishes; so we can apply the equation 'mass × acceleration = force'; i.e. $m_0 \vec{a}\,' = e\vec{E}\,'$. (Note: since the electron is at rest in K′, it is acted upon only by the electric field.) This vector equation is equivalent to the following:

$$(5)\ m_0 a'_x = eE'_x;\ m_0 a'_y = eE'_y \text{ and } m_0 a'_z = eE'_z.$$

In deriving these relations we made use of the assumption that, whenever the velocity vanishes, Newton's law holds and the force per unit charge is equal to the electric field. Transforming back to the frame K, i.e. expressing the primed quantities in terms of the unprimed ones, we obtain:

$$(6)\begin{cases} m_0\gamma^3 a_x = eE_x;\ m_0\gamma^2 a_y = e\gamma(E_y - (v/c)H_z);\ \text{and} \\ m_0\gamma^2 a_z = e\gamma(E_z + (v/c)H_y). \end{cases}$$

Both Planck and Einstein derived these equations, which however marked the parting of their ways. Einstein took $e\vec{E}' = (eE'_x, eE'_y, eE'_z) = (eE_x, e\gamma(E_y - (v/c)H_z), e\gamma(E_z + (v/c)H_y))$ to be the force in K. He also followed Newton in writing force = mass × acceleration. In view of the above equations, the mass had to become a directed magnitude which splits into a longitudinal mass $m_0\gamma^3 = m_0(1 - v^2/c^2)^{-3/2}$ and a transverse mass $m_0\gamma^2 = m_0(1 - v^2/c^2)^{-1}$.

It is true that $e\vec{E}'$ is the force acting on the electron in K′. Einstein however mistakenly interpreted $e\vec{E}'$ as representing the force *both* in K′ and in K. But K has a hybrid status; it is fixed with respect to K_0, but the direction of its x-axis is determined by the velocity $\vec{v}$ of the electron at the considered time t. (Note that $\vec{v}$ also depends on K_0, since $\vec{v}$ is the velocity of the electron *with respect to* K_0). If the three-vector $e\vec{E}' = (eE'_x, eE'_y, eE'_z)$ is taken to be the force in K, then, since K is fixed in K_0, $e\vec{E}'$ will also be the force in K_0. There is however no reason to suppose that $e\vec{E}'$ has any physical significance in K_0. In other words: we know, on the one hand, that K_0 is a genuinely arbitrary inertial system and, on the other, that K′ is the rest frame of the electron. K occupies an uncertain intermediate position between K_0 and K′; yet Einstein takes the 3-force (eE'_x, eE'_y, eE'_z) out of K′, refers the three components of this vector to the axes of K and calls the result of these arbitrary operations the three-force in K (and hence also in K_0). The correct course of action is to go over, *first* from K′ to K, which has already been done by passing from (5) to (6); *then* to shift from K to K_0. In the end, we have to go right back to the really arbitrary inertial frame K_0; only in this way can we hope to obtain relations holding in *any* inertial system. This is precisely what Planck did through projecting equations (6) along the directions of the axes of K_0. He then looked at the result, recognised the Lorentz-force on one side of the equation and called the other side the net force acting on the electron. I shall not reproduce Planck's own argument in which he considers each vector component separately. Since K is fixed relatively to K_0, the time variable is the same in these two frames; hence all three-vector equations are invariant as we pass from K to K_0. We shall therefore have achieved our aim if we succeed in putting (6) in three-vector form. Dividing the second and third equations by γ, we obtain: $m_0\gamma^3 a_x = eE_x$; $m_0\gamma a_y = e(E_y - (v/c)H_z)$ and $m_0\gamma a_z = e(E_z + (v/c)H_y)$. These three relations can be compactly written as:

$$(7)\quad m_0(\gamma^3 a_x, \gamma a_y, \gamma a_z) = e(\vec{E} + (\vec{v}/c) \times \vec{H}),$$

since $\vec{v} = (v, o, o)$ in K.

On the right hand side we recognise the Lorentz-force which, in classical electrodynamics, acts on a point-charge e moving with velocity $\vec{v}$. As for the left hand side, note that $a_x = dv_x/dt$; $a_y = dv_y/dt$; $a_z = dv_z/dt$, where $\vec{v} = (v_x, v_y, v_z)$, $v_x = v$ and $v_y = v_z = 0$ at the considered time t. It is easily verified that in the coordinate system K:

$$\text{(8)}\quad \frac{d}{dt}\left(\frac{m_0 \vec{v}}{\sqrt{1 - v^2/c^2}}\right) = \frac{m_0 \vec{a}}{\sqrt{1 - v^2/c^2}} + \frac{(\vec{v} \cdot \vec{a})}{c^2} \frac{m_0 \vec{v}}{(\sqrt{1 - v^2/c^2})^3} = m_0(\gamma^3 a_x, \gamma a_y, \gamma a_z)$$

By (7) and (8) we finally obtain:

$$\text{(9)}\quad \frac{d}{dt}\left(\frac{m_0 \vec{v}}{\sqrt{1 - v^2/c^2}}\right) = e\left(\vec{E} + \frac{\vec{v}}{c} \times \vec{H}\right) \underset{\text{Def}}{=} \text{Lorentz-force.}$$

This vector equation holds in the frame K. We have already remarked that, since K is rigidly fastened to K_0, the time variable is the same in both frames; hence equation (9) is also true in K_0. The following equation therefore connects the velocity with the force in an *arbitrary* inertial frame K_0:

$$\text{Lorentz-force} = \frac{d}{dt}\left(\frac{m_0 \vec{v}}{\sqrt{1 - v^2/c^2}}\right).$$

It is now easy, though tedious, to verify that this equation is Lorentz-covariant. This is why Planck used this relation, which replaces the classical equation: Lorentz-force $= \frac{d}{dt}(m_0 \vec{v})$, as the basis of his new law of motion:

$$\text{(10)}\quad \vec{f} = \frac{d}{dt}\left(\frac{m_0 \vec{v}}{\sqrt{1 - v^2/c^2}}\right)$$

We can now determine the new expression k(v) for the kinetic energy (see appendix one) of a body which, starting from rest at time t_0, attains the speed v_1 at the time t_1:

$$k(v_1) = \int_{t_0}^{t_1} \vec{f} \cdot \vec{v}\, dt = \int_{t_0}^{t_1} \frac{d}{dt}\left(\frac{m_0 \vec{v}}{\sqrt{1 - v^2/c^2}}\right) \cdot \vec{v} \cdot dt$$

Integrating by parts, we have:

$$k(v_1) = \left[\frac{m_0 \vec{v} \cdot \vec{v}}{\sqrt{1 - v^2/c^2}}\right]_{t_0}^{t_1} - \int_{t_0}^{t_1} \frac{m_0 \vec{v} \cdot (d\vec{v}/dt)}{\sqrt{1 - v^2/c^2}} \cdot dt.$$

Consequently:

$$\textbf{(11)}\ k(v_1) = \frac{m_0 v_1^2}{\sqrt{1 - v_1^2/c^2}} + \frac{m_0 c^2}{2}\left[\frac{(1 - v^2/c^2)^{1/2}}{1/2}\right]_{t_0}^{t_1} =$$

$$\frac{m_0 c^2}{\sqrt{1 - v_1^2/c^2}} - m_0 c^2$$

Substituting v for v_1 and writing m(v) for $m_0 \cdot (1 - v^2/c^2)^{-1/2}$, we obtain

(12) $k(v) = m(v) \cdot c^2 - m_0 c^2$. (Note that $m(o) = m_0$.)

In the EMB Einstein had derived equation (12), despite the fact that his concept of force differed from Planck's. This is because the Einsteinian force $\vec{f}_E$ and the Planckian force $\vec{f}$ have the same component along $\vec{v}$. Remember that, in the frame K;

$\vec{v} = (v, 0, 0)$, $\vec{f}_E = m_0(\gamma^3 a_1, \gamma^2 a_2, \gamma^2 a_3)$ and

$\vec{f} = m_0(\gamma^3 a_1, \gamma a_2, \gamma a_3)$,

hence: $\vec{f}_E \cdot \vec{v} = m_0 \cdot \gamma^3 \cdot a_1 \cdot v = \vec{f} \cdot \vec{v}$, and so:

$$(\text{kinetic energy})_{\text{Einstein}} = \int \vec{f}_E \cdot \vec{v}\, dt = \int \vec{f} \cdot \vec{v}\, dt = (\text{kinetic energy})_{\text{Planck}}$$

Since only variations in energy levels are important in classical physics and since m_0 has been treated as a constant, we are tempted to put: $E = \text{energy} = k(v) + m_0c^2 = m(v) \cdot c^2$. It looks as if we have already arrived at Einstein's formula for the interchangeability of mass and energy; but this is an illusion created by hindsight. For $v = 0$ this last equation yields $E = m(o) \cdot c^2 = m_0c^2$; in other words, we have, *by fiat,* laid down that the value of the energy is m_0c^2 when $v = o$; but we cannot, from a mere convention, conclude that the rest mass m_0 represents the equivalent amount of energy m_0c^2. This result, which is in fact valid, was independently deduced by Einstein in his 1905 paper: "Does the Inertia of a Body Depend on its Energy Content?". In this work Einstein showed that even the rest-mass m_0 of a material body changes by the amount $\Delta m_0 = \Delta E/c^2$ whenever the body loses an amount of energy equal to ΔE; and it is of course true that the expression $E = k(v) + m_0c^2 = m(v) \cdot c^2$ *suggests* that the rest-mass m_0 represents energy. Similarly, Planck's law:

$$\vec{f} = \frac{d}{dt}\left(\frac{m_0 \vec{v}}{\sqrt{1 - v^2/c^2}}\right)$$

suggests that the vector

$$\frac{m_0 \vec{v}}{\sqrt{1 - v^2/c^2}}$$

should be regarded as the momentum of the particle; but this is again a weak argument by analogy. The equation

$$\vec{p} = \text{momentum} = \frac{m_0 \vec{v}}{\sqrt{1 - v^2/c^2}}$$

had to be established on different grounds. (In the next example we shall deal with the close logical connection between the two notions of relativistic energy and relativistic momentum; we shall also show that the conservation of momentum implies $\Delta E = c^2 \cdot \Delta m$.) For the time being, let us note that the expression $k(v)$ of the kinetic energy yields the classical expression as a limiting case:

$$k(v) = m_0 c^2 (1 - v^2/c^2)^{-1/2} - m_0 c^2 =$$
$$m_0 c^2 [1 + {}^1/_2 v^2/c^2 - \ldots] - m_0 c^2 \doteqdot {}^1/_2 m_0 v^2,$$

if we neglect terms of higher order in v/c.

To sum up: Planck's modification of Newton's second law provides a brilliant illustration of the way in which the heuristic of the new programme can be applied. We have seen how Planck attacked a problem already tackled by Einstein but drew different conclusions from the same equations. Einstein had shown how to make use of Newton's law as a limiting case in the rest-frame of a charged particle; by the principle of relativity which was also formulated in the EMB, this rest-frame is as 'good' as any other inertial system. It simply remained for Planck to exploit the relativity principle in order to construct a law of motion which assumes the same form in all inertial frames.

In the light of MSRP we see why worked examples, like the one in which Einstein studied the motion of the electron, play an important role in the progress of science; this was also pointed out by Thomas Kuhn (1962). Such examples give a demonstration of how the heuristic can be applied in order to arrive at new laws. We can now affirmatively answer the question whether, in 1905, brilliant physicists had any indication that the new approach would prove fruitful. In the EMB they could see how Einstein had used requirements (i) and (ii) in order to give both the transformation equations for the field and a new transformation law for the charge density; how he had determined a new equation of motion for the electron; finally, how he had found a new expression for the kinetic energy. So, apart from proposing a new kinematics, an important function of the EMB was to exhibit a method by which one can scrutinize classical laws either in order to ascertain their Lorentz-covariance, as in the case of Maxwell's equations, or in order to modify them so as to obtain new theories from which the old ones follow as limiting cases. In other words, the EMB forcefully exhibits the effectiveness of the new heuristic.

§7.2 Lewis's and Tolman's Modification of the Conservation Laws of Momentum and Energy: The Problem of Essentialism

In 1908–1909 Lewis and Tolman proposed new laws for the conservation of momentum and energy. In constructing these relativistic laws they explicitly used the methods which I have labelled requirements (i) and (ii). That is, they altered the classical equations so as to make them accord with the relativity principle while yielding the old conservation laws as limiting cases. In his "Revision of the Fundamental Laws of Matter and Energy" G.N. Lewis wrote:

> The recent experiments which indicate a change in the mass of an electron with the speed, together with the phenomenon of radio-activity, have in some minds created some doubts as to the exact validity of some of the most general laws of nature. In the following pages I shall attempt to show that we may construct a simple system of mechanics which is consistent with all known experimental facts, and which rests upon the assumption of the truth of the three great conservation laws, namely the law of conservation of energy, the law of conservation of mass, and the law of conservation of momentum. To these we may add, if we will, the law of conservation of electricity. (Lewis 1908, p. 705)

In 1909 G.N. Lewis and R.C. Tolman stated explicitly:

> The following development will be based solely upon the conservation laws and the two postulates of the Principle of Relativity . . . This idea, that the velocity of light will seem the same to two different observers, even though one may be moving towards and the other away from the source of light, constitutes the really remarkable feature of the principle of Relativity, and forces us to the strange conclusions which we are about to deduce. (Lewis and Tolman 1909, p. 510)

Let us start by formulating the classical conservation laws of momentum and mass. Consider a system of particles $P_1, P_2, \ldots, P_n$, on which no external forces act. Let $m_1, \ldots, m_n$ be the (rest) masses of $P_1, \ldots, P_n$ and $\vec{v}_1, \ldots, \vec{v}_n$, their respective velocities. By Newton's second law,

$$\frac{d}{dt}(m_i \vec{v}_i)$$

is the force acting on the ith particle; hence the total force is

$$\sum_i \frac{d}{dt}(m_i \vec{v}_i) = \frac{d}{dt}(\sum_i m_i \vec{v}_i).$$

By Newton's third law, forces occur in opposite pairs, which must therefore cancel each other out. Thus:

$$\frac{d}{dt}(\sum_i m_i \vec{v}_i) = 0,$$

from which it follows that $\sum_i m_i \vec{v}_i$ = constant. This is the law of conservation of linear momentum. In particular, if the n point-masses collide in such a way that no external forces act on them during the collision and if $\vec{w}_1, \ldots, \vec{w}_n$ are their velocities after the collision, then: $\sum_i m_i \vec{v}_i = \sum_i m_i \vec{w}_i$. As for the law of the conservation of mass, i.e. the equation $\sum_i m_i$ = constant, it was taken so much for granted that few physicists bothered to write it down explicitly.

It can be easily verified that the conservation law of Newtonian momentum is not Lorentz-covariant. Once again, it is unnecessary to establish this result since, as will presently be shown, the relativity principle entails a conservation law incompatible with the classical one.

We now face the problem of so redefining momentum that the sum of momenta, if conserved in one Lorentzian frame, will be automatically conserved in all others. Starting from the classical conservation law $\sum_i m_i v_i$ = constant, we have to modify it so as to obtain a Lorentz-covariant equation which yields the old law as limiting case.

Before embarking on an historical account of how Lewis and Tolman actually proceeded, it is interesting, from an expository point of view, to present a method by which the new conservation laws *might* have been obtained. (This is in effect the way in which the concept of momentum is usually introduced in textbooks on relativity physics.) Such an unhistorical approach serves to clarify the strict logical connections which exist—or fail to exist—between the conservation laws and the relativity principle. These logical relations tend to be blurred by lengthy historical details, which reveal the hesitations and fumblings of the scientists making the discoveries.

The discovery of the new conservation laws could have taken place as follows. Newton prophetically stated that "the change of motion is proportional to the motive force impressed and is made in the direction of the right line in which that force is impressed." By his Definition II, "the quantity of motion is the measure of the same, arising from the velocity and quantity of matter conjointly." In other words, Newton expressed his second law in the form 'force equals rate of change of momentum' rather than in the form 'force equals mass × acceleration.' In view of Planck's equation

$$\vec{f} = \frac{d}{dt}\left(\frac{m_0 \vec{v}}{\sqrt{1 - v^2/c^2}}\right),$$

it follows that in relativistic mechanics the three-momentum must be represented by the vector

$$\frac{m_0 \vec{v}}{\sqrt{1 - v^2/c^2}}.$$

There arises the question whether the conservation law for this newly defined momentum is Lorentz-covariant.

Let us put:

$$m(v) = m_0(1 - v^2/c^2)^{-1/2}, \vec{p} = m(v) \cdot \vec{v}$$

$$= m_0 \vec{v}(1 - v^2/c^2)^{-1/2} \text{ and } E = m(v)c^2 = \frac{m_0 c^2}{\sqrt{1 - v^2/c^2}}.$$

Let us agree to call E the total energy-content of a body having rest-mass m_0 and velocity $\vec{v}$ *without*, for the time being, attaching any precise physical meaning to the word 'energy'. Since we know how to transform $\vec{v}$ and since m_0 is an invariant, we can easily verify that, if *both* $\vec{p}$ and E are conserved at one point of some inertial frame K, then *both* are also conserved at the corresponding point of every other inertial frame K′. Putting it more briefly, if momentum *and* energy are conserved in K, then they are both automatically conserved in K′.

All the steps in the above argument are valid except for the first one, which constitutes a blatant non-sequitur. It is bad essentialist philosophy to suppose that force is, *by its very nature*, rate of change of momentum and conclude that, since

$$f = \frac{d}{dt}\left(\frac{m_0 \vec{v}}{\sqrt{1 - v^2/c^2}}\right),$$

the momentum must equal

$$\frac{m_0 \vec{v}}{\sqrt{1 - v^2/c^2}}.$$

According to *metaphysical essentialism*, the world consists of basic entities which possess irreducible properties called essences. Essentialist metaphysics is tenable in that it is both internally consistent and empirically irrefutable. Methodological essentialism is a stronger position: it asserts that science should aim at capturing the ultimate qualities which scientists can effectively recognise as constituting essences. Essential definitions having once been established, all other scientific statements logically follow from such definitions.

It was of course possible for physicists to have regarded the equation: force $= \frac{d}{dt}$ (momentum) as an essential definition of force. In *this* particular case they would have been lucky, for this essentialist approach, in conjunction with the relation

$$\vec{f} = \frac{d}{dt}\left(\frac{m_0 \vec{v}}{\sqrt{1 - v^2/c^2}}\right),$$

leads to the 'correct' conservation laws; i.e., to laws which, for the time being, physicists take to be true. But essentialism is an unreliable ally, which can just as easily be made to 'refute' relativity. Kant was not an essentialist philosopher in that he thought science capable neither of discovering the properties of noumena nor even of talking meaningfully about them. However, he used an essentialist type of argument in order to establish that mass *must* be conserved. According to Kant, the category of substance-and-accident is such that substance can neither be created nor annihilated. The Kantian argument is roughly as follows: if a thing possesses changeable qualities (accidents), then the thing itself must remain unaltered; otherwise there exists no common substratum in which all these qualities inhere. Hence, an essential feature of substance is that its total quantum must be conserved. In the *Metaphysical Foundations of Natural Science* (section III theorem 2) he identified substance with matter; from this it follows that mass, which was defined as quantity of matter, must be conserved. This result contradicts relativistic mechanics; for if, as seen above, $\sum_i m_i/\sqrt{1 - v_i^2/c^2}$ is to be conserved, Then $\sum_i m_i$ cannot obviously also remain constant. Thus, Kantian philosophy provides an a priori refutation of special relativity. Nothing shows more clearly the reactionary character of essentialist *methodology*. There is in fact no harm in believing in the metaphysical existence of essences and even in their knowability, at least in principle; *provided* one also accepts the impossibility of deciding whether any given property constitutes an essence. It can be argued that the belief in man's ability to capture essences acts as an incentive to scientific research. Einstein himself was a metaphysical essentialist. At one time, he held the universe to be the realisation of what is mathematically simplest and was thus led to use variational principles which proved fruitful in general relativity. Thus, all the scientist has to guard against is regarding any existing concepts or propositions as ultimate.

Let us now resume our historical account and examine how Lewis and Tolman set about modifying the classical conservation laws.

Lewis proposed that the new momentum should be of the form $m(v) \cdot \vec{v}$, where m(v), instead of remaining constant throughout the motion of the point-mass, is now a function of the speed. An immediate question comes to mind: why this dependence on the speed? Let us immediately note that the inertial mass cannot depend on the *direction* of the vector $\vec{v}$; for such a dependence would violate the isotropy of space. A body moving in a given direction would possess less inertia than the same body moving with the same speed in a different direction. Hence inertia is at most dependent on the magnitude of the velocity, i.e., on the speed. But we may again ask: why this particular dependence on the speed? One short answer might be that Lewis made an inspired and inexplicable guess; certain passages in his paper however strongly indicate that he was inspired by Thompson's, Heaviside's and Lorentz's works on electrodynamics. Lewis wrote:

> That a charged particle must possess mass in virtue of its charge, and that this mass must vary with the velocity of the particle, was shown to be a consequence of the electromagnetic theory by J J Thompson and by Heaviside, and numerous attempts have been made to find the exact expression for the change of mass with the velocity. But before this can be done some assumption is necessary as to the shape of the particle and the distribution of its charge. The three theories of the simple negative particle or electron which are now most discussed are due to Abraham, Bucherer and Lorentz. . . . The second assumes an electron which is spherical when at rest but which in motion contracts in the direction of its translation and expands laterally so as to keep a constant volume. The third assumes an electron similar to the second, which contracts in the direction of translation but which does not change its other dimensions. On the basis of these theories and from known electromagnetic principles, three equations have been obtained for the value of m/m_0 as a function of β[1], namely:
> (a) $m/m_0 = (3/4\beta^2)\,[((1 + \beta^2)/2\beta) \log ((1 + \beta)/(1 - \beta)) - 1]$;
> (b) $m/m_0 = 1/(1 - \beta^2)^{1/3}$;
> (c) $m/m_0 = 1/(1 - \beta^2)^{1/2}$.
> The extraordinary significance of the similarity of the first two of these equations and the identity of the third with equation (15), which we have derived from strikingly different principles, needs no emphasis. (Lewis 1908, p. 713)

In chapter five we briefly described the way in which Lorentz achieved this remarkable result (see appendix three). He took the immobile electron to be a uniformly charged spherical shell. By virtue of the Lorentz-Fitzgerald

[1] $\beta = v/c$, where v is the speed of the particle and c is the velocity of light (not to be confused with the symbol γ which stands for $(1 - v^2/c^2)^{-1/2}$.)

contraction, the electron, when set in motion, acquires an ellipsoidal shape; it also creates an electromagnetic field which reacts on its source. In the rest frame of the electron there exists a purely electrostatic field, which can be easily calculated then referred back to the stationary system by means of formulae found by Lorentz in 1904. In the stationary system there exist both an electric and a magnetic field which, together, exert on the electron a force $\vec{I}$ equal to:

$$-\mu_0\frac{d}{dt}\left(\frac{\vec{v}}{\sqrt{1-v^2/c^2}}\right), \text{ where:}$$

$$\mu_0 = \frac{e^2}{6\pi c^2 R},$$

e is the total charge and R the radius of the electron. Lorentz projected the vector $\vec{I}$ first along the velocity $\vec{v}$, then normally to it as follows. Let us write $\vec{v} = v\vec{u}$, where v is the speed and $\vec{u}$ the unit vector along the tangent to the path. Hence: $\vec{u}.\vec{u} = u^2 = 1$. Differentiating:

$$\vec{u}.\frac{d\vec{u}}{dt} = 0; \text{ i.e. } \frac{d\vec{u}}{dt} \text{ is perpendicular to } \vec{u}.$$

We have:

$$\vec{a} = \text{acceleration} \underset{\text{Def}}{=} \frac{d\vec{v}}{dt} = \frac{d}{dt}(v\vec{u}) = \frac{dv}{dt}\vec{u} + v\frac{d\vec{u}}{dt}.$$

Let

$$\vec{j}\,' = \frac{dv}{dt}\vec{u}, \text{ and } \vec{j}\,'' = v\frac{d\vec{u}}{dt}.$$

$\vec{j}\,'$ is parallel to the direction of motion and $\vec{j}\,''$ is perpendicular to it. Thus: $\vec{a} = \vec{j}\,' + \vec{j}\,''$.

Remember that

$$I = -\mu_0\frac{d}{dt}\left(\frac{\vec{v}}{\sqrt{1-v^2/c^2}}\right) = -\mu_0\frac{d}{dt}\left(\frac{v\vec{u}}{\sqrt{1-v^2/c^2}}\right)$$

Therefore:

$$\textbf{(13)}\quad \vec{I} = -\mu_0\left[\frac{d}{dv}\left(\frac{v}{\sqrt{1-v^2/c^2}}\right)\frac{dv}{dt}\vec{u} + \frac{v}{\sqrt{1-v^2/c^2}}\frac{d\vec{u}}{dt}\right]$$

$$= [-m'\vec{j}\,' - m''\vec{j}\,'']$$

where:

$$\textbf{(14)}\quad m' = \mu_0 \frac{d}{dv}\left(\frac{v}{\sqrt{1 - v^2/c^2}}\right)$$

$$= \frac{\mu_0}{(\sqrt{1 - v^2/c^2})^3} \text{ and: } m'' = \frac{\mu_0}{\sqrt{1 - v^2/c^2}}$$

Let $\vec{f}$ be the sum of *all* the forces acting on the electron *except* for the force due to the self-field. If m_0 denotes the material mass of the electron, then, by Newton's second law, the following equation holds:

$$\textbf{(15)}\quad \begin{cases} \vec{f} - m'\vec{j}\,' - m''\vec{j}\,'' = m_0\vec{a} = m_0(\vec{j}\,' + \vec{j}\,''). \text{ i.e.} \\ \vec{f} = (m_0 + m')\vec{j}\,' + (m_0 + m'')\vec{j}\,''. \end{cases}$$

Both $(m_0 + m')$ and $(m_0 + m'')$ measure the particle's capacity for resisting acceleration. $(m_0 + m')$ and $(m_0 + m'')$ are, respectively, the electron's inertias along the tangent to the path and perpendicularly to it; i.e. $(m_0 + m')$ and $(m_0 + m'')$ are the longitudinal and transverse masses respectively. Whereas m_0 remains constant, both m' and m'' depend on the speed of the particle. Lorentz envisaged the possibility of m_0 vanishing altogether, which would mean that the electron is nothing but a distribution of charge in the ether. This conjecture was experimentally confirmed by Kaufmann. Going back to equation (15), and putting $m_0 = 0$, we thus obtain:

$$\textbf{(16)}\quad \vec{f} = m'\vec{j}\,' + m''\vec{j}\,'' = \frac{d}{dt}\left(\frac{\mu_0\vec{v}}{\sqrt{1 - v^2/c^2}}\right)$$

There is an amazing similarity between this equation and Planck's law of motion which was arrived at through totally different considerations.[2]

We recall that Planck was interested, not in the self-field of the electron, but in the total ponderomotive force, i.e., in the Lorentz-force. At this stage, it can be said that the classical and relativistic programmes ran parallel to each other but made use of different heuristics. In order to obtain relation (16), Lorentz resorts to Newton's law in which he lets the material mass vanish. That (16) should resemble Planck's law of force is paradoxical: whereas Planck's equation was designed to supplant Newton's second law, (16) was actually *deduced from* Newtonian dynamics taken together with

[2]Note that Lorentz's transverse mass $m'' = \mu_0(1 - v^2/c^2)^{-1/2}$ is very different from Einstein's, namely from $m_0(1 - v^2/c^2)^{-1}$. This is hardly surprising since equation (16) is similar to Planck's and not to Einstein's law of motion.

electromagnetic theory. This illustrates a point already made, namely that the classical results which possess relativistic equivalents were obtained by way of Maxwell's equations. Having once and for all decided to keep Newtonian mechanics, Lorentz shifted the explanatory burden on to electrodynamics and on to hypotheses structurally similar to Maxwell's theory. His rods shrink and his clocks are retarded because of the electro-magnetic interactions between their constituent parts and also because of molecular forces which transform like the electromagnetic ones. One can therefore say that Lorentzian 'relativity' is roughly coextensive with electrodynamics. Apart from relying on the structure of one particular theory, Lorentz also commits himself to a specific hypothesis about the constitution of matter: the electron is nothing but a charge distribution in the ether. (This hypothesis immediately poses the problem of the electron's stability: the repulsive forces between the component parts of the charge should cause the particle to expand indefinitely.)

A distinguishing feature of the relativity programme is that it turns the covariance postulate into a universal meta-principle subsuming every particular physical hypothesis. Every theory has to conform to this principle and electrodynamics is nothing more than one particular physical hypothesis. We have seen that Einstein developed his kinematics independently of Maxwell's equations. Following Einstein, Lewis tried to construct a new mechanics which was logically independent of electrodynamics and of any specific atomic hypothesis. This new mechanics was meant to pervade the whole of physics, not to be subservient to any particular theory. Thus inertia was treated, in purely dynamic terms, as the mobile's resistance to acceleration, no matter what the origin of this resistance might be. Lewis wrote:

> The mass of a negative particle is usually spoken of as electromagnetic mass, but if we are to hold to our definitions we must recognise only one kind of mass. In general we have defined the mass of a moving body as the quotient of the time during which it will be brought to rest by unit force divided by the initial velocity. *It matters not what the supposed origin of this mass may be*. (Lewis 1908, p. 712)

Thus, despite being inspired by Lorentz's results concerning the dependence of inertia on speed, Lewis adopted an agnostic attitude towards the problem of the constitution of matter. We shall presently see that his first attempt at constructing a new dynamics was a partial failure: although the new mechanics presupposed no specific atomic theory, it remained dependent on electrodynamics.

For the time being let us confine ourselves to rectilinear motion. Lewis 'defines' momentum, mass, force and kinetic energy as follows: mass is the ratio of momentum to velocity and force is the rate of change of momentum. In other words:

(17) $p = mv$ and $f = \frac{dp}{dt}$.

The infinitesimal increase in energy equals the force multiplied by the infinitesimal displacement; i.e., dE = f.dx = f.v.dt. We have already seen that Lewis was inspired by Lorentz's work; so he let m depend on the velocity, or rather on the speed. Hence: dp = d(mv) = dm.v + m.dv. It follows that:

$$f = \frac{dp}{dt} = \frac{dm}{dt}.v + m.\frac{dv}{dt};$$

consequently:

(18) $dE = f.v.dt = dt = v^2.dm + mvdv.$

These equations do not take us very far since they apply equally well to Newtonian mechanics where, in addition to the above relations, it is laid down that m should remain constant, i.e. dm = 0. This is why Lewis turns to the electromagnetic theory of radiation from which he borrows the following results.

> When a black body is placed in a beam of light it is subject to a pressure or force which tends to move it in the direction in which the light is moving. If dE/dt denotes the time rate at which the body receives energy, f the force and V the velocity of light, we have in rational units the formula $f = \frac{1}{V}.\frac{dE}{dt}$.
>
> Postulating the validity of the fundamental conservation laws mentioned above, we shall need in the following development only this one cardinal assumption, that a beam of radiation possesses not only momentum and energy, but also mass, travelling with the velocity of light, and that a body absorbing radiation is acquiring this mass as it also acquires the momentum and the energy of the radiation. Therefore a body which absorbs radiant energy increases in mass. The amount of this increase is readily found. If in general we write momentum as the product of mass and velocity, then the momentum of any part of a beam of radiation having the mass m will be given by the equation: M = mV. (Lewis 1908, p. 706)

Thus we have the two additional equations:

(19)
$$\begin{cases} f = \frac{1}{c}\frac{dE}{dt} \text{ and } p = mc. \text{ Hence } dp = c.dm; \text{ so:} \\ f = \frac{dp}{dt} = c\frac{dm}{dt}, \text{ and } \frac{1}{c}\frac{dE}{dt} = f = c\frac{dm}{dt}. \end{cases}$$

i.e.

(20) $dE = c^2.dm.$

Comparing (18) and (19) we obtain: $c^2dm = v^2dm + mvdv$. Separating the variables:

$$\frac{dm}{m} = \frac{vdv}{c^2 - v^2}$$

Integrating both sides:

$$\log m = -\frac{1}{2}\log(c^2 - v^2) + \log k;$$

i.e.

$$m = \frac{k}{\sqrt{c^2 - v^2}}, \text{ where } k \text{ is a constant.}$$

Let $m = m_0$ for $v = 0$. Hence $m_0 = k/\sqrt{c^2} = k/c$, i.e., $k = m_0c$. Finally:

$$\textbf{(21)}\ m = \frac{k}{\sqrt{c^2 - v^2}} = \frac{m_0 c}{\sqrt{c^2 - v^2}} = \frac{m_0}{\sqrt{1 - v^2/c^2}}$$

Note once again that, in the above argument, the mass m was assumed to depend exclusively on v. For $v = 0$, m takes the fixed value m_0, no matter what values are given to other variables, eg. to acceleration.

Since $m = m_0.\ (1 - v^2/c^2)^{-1/2}$ and $p = mv$, it follows that

$$p = \frac{m_0 v}{\sqrt{1 - v^2/c^2}}.$$

Strictly speaking, this result applies only to the case of rectilinear motion. However, by assuming that the momentum has the same direction as the velocity and by assimilating motion along an infinitesimal arc to rectilinear motion, we obtain:

$$\textbf{(22)}\ \vec{p} = \frac{m_0 \vec{v}}{\sqrt{1 - v^2/c^2}}$$

This proof is obviously unsatisfactory: although it does not tie mechanics to any particular atomic theory, it makes dynamics highly dependent on Maxwell's equations. In 1909 Lewis and Tolman took the final step towards severing the remaining links between mechanics and electrodynamics. The following quotation shows that the two authors were perfectly aware of pursuing one definite aim:

> The consequences which one of us has obtained from a simple assumption as to the mass of a beam of light, and the fundamental conservation laws of mass, energy and momentum, Einstein has derived from the principle of Relativity and the electromagnetic theory. We propose in this paper to show that these consequences may also be obtained merely from the conservation laws and the principle of Relativity, *without any reference to electromagnetics*. (Lewis and Tolman, 1909, p. 512; my italics)

This passage reveals the method used by the authors in order to arrive at the new laws. We shall presently show that they effectively applied the two heuristic devices which we called requirements (i) and (ii).

We have seen why the mass m was taken to be a function m(v) of the speed. The new conservation law of momentum thus assumes the form: '$\sum_i m(v_i)\vec{v}_i$ = constant', which has to yield the classical law '$\sum_i m_i \vec{v}_i$ = constant' as a limiting case. Consequently, we require m(v) to tend to the Newtonian mass m_0 as v/c tends to zero, i.e.

$m(v) \rightarrow m_0$ as $(v/c) \rightarrow 0$. In particular

$m(v) \rightarrow m_0$ as $c \rightarrow \infty$. If $m(v)$ is to be a continuous function of v,

$m(v) \rightarrow m(0)$ as $v \rightarrow 0$; hence $m(v) \rightarrow m0$ as $v \rightarrow 0$.

It follows that:

(23) $m(0) = m_0$ = Newtonian mass (or rest-mass).

By the relativity principle the law '$\sum_i m(v_i)\vec{v}_i$ = constant' must be satisfied in all inertial frames.

Consider two 'symmetric' observers A and B. The velocity $\vec{v}$ of A with respect to B is equal and opposite to the velocity of B relatively to A. Suppose A and B have identical particles, each possessing a rest-mass equal to m_0. A and B impart to their respective particles equal and opposite velocities, $\vec{u}$ and $-\vec{u}$, perpendicularly to the direction of their relative motion. $\vec{u}$ and $-\vec{u}$ are as measured in the frames of A and B respectively. The two particles meet in such a way that the velocity of each particle, as seen by the corresponding observer, is exactly reversed. In other words: in A's frame K_A the velocity of A's particle jumps from $+\vec{u}$ to $-\vec{u}$ during the collision; an exactly symmetrical process takes place in K_B. The momenta of A's particle before and after the impact are $m(u)\vec{u}$ and $-m(u) \cdot \vec{u}$

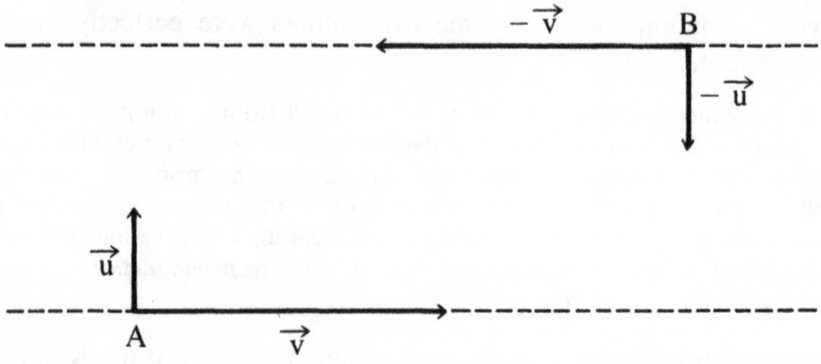

respectively. In virtue of the exact symmetry between the situations of A and of B, A knows that, prior to the collision, B judges the velocity of B's particle to be $-\vec{u}$ perpendicularly to $\vec{v}$. By virtue of the Lorentz transformation, A knows that B's clocks are retarded but that B's rods along the normal to $\vec{v}$ are unaffected by the motion of B. Hence, in K_A, B's particle must have velocity $(\sqrt{1 - v^2/c^2})u$ perpendicularly to the motion of the two observers (remember that $(1 - v^2/c^2)^{-1/2}$ is the time-dilatation factor). The total speed of B's particle in K_A is $(v^2 + (1 - v^2/c^2)u^2)^{1/2}$ because the two components of the velocity are $(-v)$ and $(-(1 - v^2/c^2)^{1/2} \cdot u)$. Hence, normally to the direction of relative motion, the momenta of the particle before and after the impact are:

$$-m((v^2 + (1 - v^2/c^2)u^2)^{1/2}) \cdot \sqrt{1 - v^2/c^2} \cdot u \text{ and}$$

$$+m((v^2 + (1 - v^2/c^2)u^2)^{1/2}) \cdot \sqrt{1 - v^2/c^2} \cdot u, \text{ respectively.}$$

A assumes that the law of the conservation of momentum holds and concludes that:

$$[m(u) \cdot u - m((v^2 + (1 - v^2/c^2)u^2)^{1/2}) \cdot \sqrt{1 - v^2/c^2} \cdot u] =$$
$$[-m(u) \cdot u + m((v^2 + (1 - v^2/c^2)u^2)^{1/2}) \cdot \sqrt{1 - v^2/c^2} \cdot u].$$

Hence:

$$(24) \quad \frac{m(u)}{\sqrt{1 - v^2/c^2}} = m((v^2 + (1 - v^2/c^2)u^2)^{1/2}).$$

Letting u tend to zero, we have:

$$m(u) \rightarrow m(o) = m_0 \text{ and } (v^2 + (1 - v^2/c^2)u^2)^{1/2} \rightarrow v.$$

Hence:

$$(25) \quad \frac{m_0}{\sqrt{1 - v^2/c^2}} = m(v).$$

By definition of momentum:

$$\vec{p} = m(v)\vec{v} = \frac{m_0 \vec{v}}{\sqrt{1 - v^2/c^2}}$$

The above argument is somewhat informal and it also seems to presuppose the presence of the conscious observers A and B. It is however clear that the 'observers' are introduced merely for expositive purposes. All the same, let us give a more perspicuous proof which, though similar to the one proposed by Lewis and Tolman, is more rigorous in that it relies on no assumption of symmetry between inertial 'observers'.

Consider the usual inertial frames K and K′ (i.e. K′ moves with velocity (v,0,0) with respect to K). Let two particles, which have the same rest-mass m_0, collide in such a way that the velocities before and after the collision are as follows:

	Before	After
1st particle	$(v,u,0)$	$(v,-u,0)$
2nd particle	$(-v,-u,0)$	$(-v,u,0)$

These are the velocities in K, where momentum is thus clearly conserved. In the frame K′ the velocities are as follows (see appendix one):

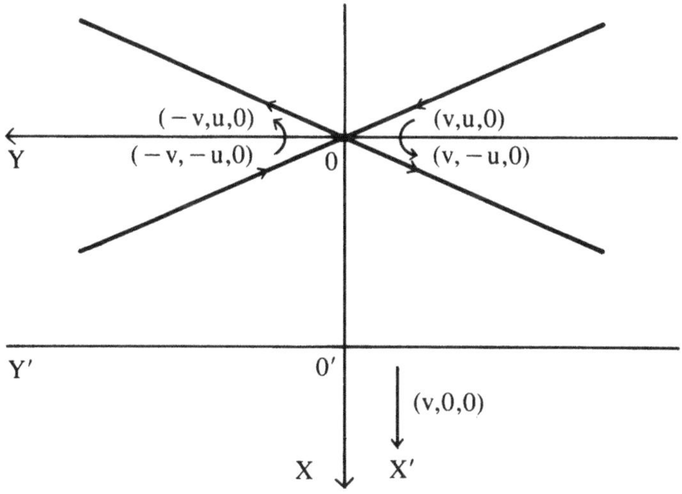

Before	After
$(0,\gamma u,0)$	$(0,-\gamma u,0)$

$$\left[\frac{-2v}{\left(1+\frac{v^2}{c^2}\right)}, \frac{-u}{\gamma\left(1+\frac{v^2}{c^2}\right)}, 0\right]\left[\frac{-2v}{\left(1+\frac{v^2}{c^2}\right)}, \frac{u}{\gamma\left(1+\frac{v^2}{c^2}\right)}, 0\right]$$

where: $\gamma \underset{\text{Def}}{=} (1 - v^2/c^2)^{-1/2}$.

The proposition that the second component of the momentum must be conserved in K′ is expressed by the equation:

$$m(\gamma u) \cdot \gamma u + m(w) \cdot \left[\frac{-u}{\gamma(1 + v^2/c^2)}\right] = -m(\gamma u) \cdot \gamma u + m(w) \cdot \left[\frac{u}{\gamma(1 + v^2/c^2)}\right]$$

where:

$$\begin{aligned} w &= \text{magnitude of } (-2v/(1 + v^2/c^2), -u/\gamma(1 + v^2/c^2), 0) \\ &= (4v^2 + u^2/\gamma^2)^{1/2} (1 + v^2/c^2)^{-1}. \end{aligned}$$

Remember that $\gamma = (1 - v^2/c^2)^{-1/2}$. It follows that:

$$\textbf{(26)}\quad \gamma \cdot m(\gamma u) = \frac{m(w)}{\gamma(1 + v^2/c^2)}.$$

Let u tend to zero while v is kept fixed. Thus

$$u \rightarrow 0;\ w \rightarrow \frac{2v}{(1 + v^2/c^2)};\ \text{and } m(\gamma u) \rightarrow m(0) = m_0.$$

By equation (25) we have:

$$\textbf{(27)}\quad m\left(\frac{2v}{1 + v^2/c^2}\right) = m_0 \cdot \gamma^2 \cdot (1 + v^2/c^2) = m_0 \cdot \left(\frac{1 + v^2/c^2}{1 - v^2/c^2}\right).$$

Let us put: $z = 2v/(1 + v^2/c^2)$; i.e. $v^2 - (2c^2/z)v + c^2 = 0$. Solving this quadratic in v then substituting back into (27) we obtain:

$$\textbf{(28)}\quad m(z) = \frac{m_0}{\sqrt{1 - z^2/c^2}}.$$

To sum up: in determining that the momentum should be defined as

$$\frac{m_0 \vec{v}}{\sqrt{1 - v^2/c^2}},$$

both the relativity principle and the requirement that the classical law '$\sum_i m_i \vec{v}_i$ = constant' be yielded as a limiting case are invoked; in other words, both requirements (i) and (ii) are effectively used. Lewis and Tolman took a nominalist view of their new definition of momentum. Starting from a result obtained by Lorentz, Lewis decided that the classical equation '$\sum_i m_i \vec{v}_i$ = constant' should be replaced by '$\sum_i m(v_i)\vec{v}_i$ = constant', where $m(v_i)$ is some function of the speed. The condition: $m(v_i) \rightarrow m_i$ as $(v_i/c) \rightarrow o$ ensures that the old law is a limiting case of the new one. The vector quantity $m(v_i)\vec{v}_i$ was called 'the relativistic momentum of the ith particle', so that the new conservation law could be expressed as: 'the sum of the momenta is conserved'. Requirement (i), i.e. the relativity principle, was brought to bear on the problem of determining the exact form of the function $m(v_i)$; it was found that the inertial mass $m(v_i)$ must be equal to $m_i \cdot (1 - v_i^2/c^2)^{-1/2}$.

The fact that the conservation law, on the one hand, and Newton's equation 'force = rate of change of momentum' taken together with Planck's law $\vec{f} = d(m_0\vec{v}/\sqrt{1 - v^2/c^2})/dt$, on the other hand, lead to the same definition of momentum may seem more than mere coincidence. Such convergence of different methods towards the same result gives the scientist the feeling that he must have stumbled upon some basic law, or upon some fundamental entity in nature. There is in fact little harm in assuming that Lewis's definition *may* have captured the 'essence' of momentum; in this sense, methodological nominalism is compatible with metaphysical essentialism. The only proviso is that the scientist should not allow himself to be misled, by what could after all be a mere coincidence, into dogmatically adopting some definition which may well later have to be replaced by a better, i.e. by a more convenient one. Thus fallibilism reconciles methodological nominalism with metaphysical essentialism (see Popper 1945, vol. 2 ch. 11).

Let us conclude this section by establishing a fundamental theorem of STR; namely the proposition that the two conservation laws of momentum and of energy imply each other. We recall that by 'energy' we mean the quantity

$$E = \frac{m_0 c^2}{\sqrt{1 - v^2/c^2}},$$

whose variation was shown to be equal to the work done by the net force. (We do not presuppose that every material body of rest-mass m_0 possesses a rest energy equal to m_0c^2. This proposition will be proved in the next section.)

Put $\vec{r} = (x,y,z)$, so that $(x,y,z,t) = (\vec{r},t)$. Since the Lorentz transformation: $x' = \gamma(x - vt)$, $y' = y$, $z' = z$, $t' = \gamma\left(t - \frac{v}{c^2}x\right)$, is both homogenous and linear, (dx, dy, dz, dt) obeys the same transformation rule as $(\vec{r},t)$. Call any such 4-tuple of numbers a four-vector.

We know that the rest-mass m_0 and the proper time $d\tau = dt\sqrt{1 - w^2/c^2}$, where:

$$\vec{w} = \left(\frac{dx}{dt}, \frac{dy}{dt}, \frac{dz}{dt}\right) = \text{velocity},$$

are invariants. Multiplying (dx, dy, dz, dt) by

$$\frac{m_0}{d\tau} = \frac{m_0}{\sqrt{1 - w^2 \cdot c^2 \cdot dt}},$$

we conclude that the 4-tuple:

$$\left[\frac{m_0\vec{w}}{\sqrt{1 - w^2/c^2}}, \frac{m_0}{\sqrt{1 - w^2/c^2}}\right]$$

transforms like $(\vec{r}, t)$, i.e. constitutes a four-vector. But:

$$\left[\frac{m_0\vec{w}}{\sqrt{1 - w^2/c^2}}, \frac{m_0}{\sqrt{1 - w^2/c^2}}\right] = \left(\vec{p}, \frac{E}{c^2}\right), \text{ where}$$

$$\vec{p} = (p_x, p_y, p_z) = \left[\frac{m_0 w_x}{\sqrt{1 - w^2/c^2}}, \frac{m_0 w_y}{\sqrt{1 - w^2/c^2}}, \frac{m_0 w_z}{\sqrt{1 - w^2/c^2}}\right]$$

$$\text{and } E = \frac{m_0 c^2}{\sqrt{1 - w^2/c^2}}.$$

Since $(\vec{p}, E/c^2)$ transforms like $(\vec{r},t)$, i.e. in accordance with the Lorentz transformation which is both linear and homogenous, the same applies to the increments $(\Delta\vec{p}, (1/c^2)\Delta E)$; i.e. $(\Delta\vec{p}, (1/c^2)\Delta E) = (\Delta p_x, \Delta p_y, \Delta p_z, (1/c^2)\Delta E)$ is also a four-vector.

Consider a system of particles $Q_1, Q_2, \ldots, Q_n$ which collide at some world point (x,y,z,t). Let $\vec{p}_i$ and E_i be, respectively, the momentum and energy of Q_i. $\Delta\vec{p}_i$ and ΔE_i will thus respectively denote the changes in $\vec{p}_i$ and E_i caused by the collision. Thus $(\Delta\vec{p}_i, (1/c^2)\Delta E_i)$ is a four-vector; the same holds for the sums $\left(\sum_i \Delta\vec{p}_i, \frac{1}{c^2}\sum_i \Delta E_i\right)$, once again by virtue of the linearity and homogeneity of the Lorentz transformation.

Now consider another inertial system K′ moving with constant velocity

$\vec{v}$ along the x-axis of K. By the Lorentz transformation:

$$\textbf{(29)}\quad \sum_i \Delta p'_{ix} = \gamma\Big(\sum_i \Delta p_{ix} - \frac{v}{c^2}\sum_i \Delta E_i\Big), \text{ and:}$$

$$\frac{1}{c^2}\sum_i \Delta E'_i = \gamma\Big(\frac{1}{c^2}\sum_i \Delta E_i - \frac{v}{c^2}\sum_i \Delta p_{ix}\Big).$$

That is:

$$\textbf{(30)}\quad \sum_i \Delta E'_i = \gamma\big(\sum_i \Delta E_i - \sum_i \Delta p_{ix}\big);$$

where

$$\vec{p}_i = (p_{ix}, p_{iy}, p_{iz}),$$
$$\vec{p}\,'_i = (p'_{ix}, p'_{iy}, p'_{iz})$$
$$= \text{momentum in K}' \text{ corresponding to } \vec{p}_i,$$

and E'_i = energy in K′ corresponding to E_i.

Suppose that relativistic momentum is conserved. This means that $\sum_i \Delta \vec{p}_i = \mathbf{0}$ in all inertial frames. Hence:

$$\sum_i \Delta \vec{p}_i = \sum_i \Delta \vec{p}\,'_i = 0$$

$$\text{i.e. } \sum_i \Delta p_{ix} = \sum_i \Delta p'_{ix} = \sum_i \Delta p_{iy} = \sum_i \Delta p'_{iy} = \sum_i \Delta p_{iz} = \sum_i \Delta p'_{iz} = 0.$$

By equations (29) and (30) above we have:

$$\sum_i \Delta E_i = \sum_i \Delta E'_i = 0;$$

i.e. energy is conserved.

Conversely, assume that energy is conserved. By the relativity principle, this must hold in all inertial frames. In particular $\sum_i \Delta E_i = \sum_i \Delta E'_i = 0$. It follows from equations (29) and (30) that:

$$\sum_i \Delta p_{ix} = 0.$$

By choosing to let K′ move along the y-axis, then along the z-axis, we similarly deduce that

$$\sum_i \Delta p_{iy} = 0 \text{ and } \sum_i \Delta p_{iz} = 0.$$

Hence,

$$\sum_i \Delta p_{ix} = \sum_i \Delta p_{iy} = \sum_i \Delta p_{iz} = 0.$$

Thus, momentum is conserved in K. Since K is arbitrary, it follows that momentum is conserved in every inertial frame. We have therefore proved that:

(31) (Relativity Principle) $\Rightarrow$ [(Conservation of Energy)
$\Leftrightarrow$ (Conservation of Momentum)].

Note however that, in the above argument, we did not establish the conservation of momentum, or of energy, *over time*. We considered not processes over finite time-intervals, but punctual events like the instantaneous collision of several particles at a single space-time point. Since relativity closely connects time with space, it should come as no surprise that the conservation of the 3-momentum, i.e. of the spatial component of the 4-momentum, should be equivalent to the conservation of the energy, which constitutes the time-component of the *same* 4-vector.

§7.3 EQUIVALENCE OF MASS AND ENERGY: EINSTEIN'S EQUATION: E = mc^2

We have already shown that the kinetic energy of a body of rest-mass m_0 is given by the equation:

$$k(v) = \int \vec{f} \cdot d\vec{r} = \int \vec{f} \cdot \vec{v}\, dt = \left[\frac{m_0\, c^2}{\sqrt{1 - v^2/c^2}} - m_0\, c^2\right]$$

We have agreed to denote by the word 'energy' the quantity $E = m(v) \cdot c^2 = m_0c^2/\sqrt{1 - v^2/c^2} = k(v) + m_0c^2$. However, the mere convention of adding m_0c^2 to k(v) does not imply that the kinetic energy and the product of the rest-mass by the square of c should denote entities continuous with each other. The kinetic energy arises from the state of motion of the particle and vanishes when the particle comes to rest. According to Newtonian dynamics, m_0 measures the amount of matter, i.e., of substance, contained in the particle. Compared with the 'stuff' of which the mobile is made, the kinetic energy appears as a mere 'construct', which may of course prove useful in certain calculations. But, so far, we have no indication of any possible continuity between k(v) on the one hand and m_0c^2 on the other.

We are however in possession of enough physical *theory* to be able to show that kinetic energy can be transformed into rest-mass.

Lewis and Tolman proposed '$\sum_i m_i \vec{v}_i/\sqrt{1 - v_i^2/c^2}$ = constant' as the new law for the conservation of momentum. We have already proved that this law, together with the relativity principle, entails the conservation of the 'energy' $\sum_i E_i$, where $E_i \underset{\text{Def}}{=} m_i c^2/\sqrt{1 - v_i^2/c^2}$. Consider two identical particles moving with velocities $\vec{v}$ and $-\vec{v}$ along the x-axis. Let the particles collide at the origin. Suppose that after the (inelastic) collision the particles remain stuck together. The 'energy' before collision is $2m_0c^2/\sqrt{1 - v^2/c^2}$, where m_0 is the rest-mass. After the collision, the velocity vanishes; so the 'energy' becomes equal to $2M_0c^2$, where M_0 is the rest-mass of each particle after the impact. We leave open the question whether $m_0 = M_0$ or $m_0 \neq M_0$. By the conservation law of 'energy': $2M_0c^2 = 2m_0c^2/\sqrt{1 - v^2/c^2}$. Hence $M_0 = m_0/\sqrt{1 - v^2/c^2}$. Thus the rest-mass of each particle has increased. Let us calculate the amount of this increase. We have:

$$\begin{aligned} M_0 - m_0 &= [m_0 \cdot (1 - v^2/c^2)^{-1/2} - m_0] \\ &= (1/c^2)[m_0 c^2 \cdot (1 - v^2/c^2)^{-1/2} - m_0 c^2] \underset{\text{Def}}{=} (1/c^2)k(v) \end{aligned}$$

Remember that k(v) is the kinetic energy of the particle before the collision and that k(v) vanishes with v. After the collision, each particle remains at rest, so the kinetic energy completely disappears; but, by the last equation, this energy reemerges as an increase $k(v)/c^2$ in the rest-mass.

The only flaw in the above argument is its unhistorical character. It makes use of Lewis's conservation law which was put forward in 1908-1909; but, already in 1905, Einstein had come to the conclusion that the inertia of a body depends on its energy-content. His first derivation of the equation '$E = mc^2$' is heavily dependent on electromagnetism, as will presently be shown. Consider an inertial system K in which the electromagnetic field is given by:

$$\begin{aligned} E_1 &= X \cdot \sin\psi \qquad & H_1 &= L \cdot \sin\psi \\ E_2 &= Y \cdot \sin\psi \qquad & H_2 &= M \cdot \sin\psi \\ E_3 &= Z \cdot \sin\psi \qquad & H_3 &= N \cdot \sin\psi \end{aligned}$$

where: X, Y, Z, L, M and N are constant, $\psi = w(t - (\ell x + my + nz)/c)$ and $\ell^2 + m^2 + n^2 = 1$. We can regard (ℓ, m, n) as the direction cosines of an electromagnetic wave in the system K. We now propose to find the direction cosines of the same ray in a frame K′ moving with uniform velocity $\vec{v}$ along the x-axis of K. In K′ we have, by the transformation equations for the electromagnetic field:

$$E'_1 = X \cdot \sin\psi \qquad H'_1 = L \cdot \sin\psi$$
$$E'_2 = \gamma(Y - vN/c) \cdot \sin\psi \qquad H'_2 = \gamma(M + vZ/c) \cdot \sin\psi$$
$$E'_3 = \gamma(Z + vM/c) \cdot \sin\psi \qquad H'_3 = \gamma(N - vY/c) \cdot \sin\psi$$

where $\gamma = (1 - v^2/c^2)^{-1/2}$ and

$$\psi = w\left(t - \frac{\ell x + my + nz}{c}\right) = w \cdot \left(\gamma\left(t' + \frac{v}{c^2}x'\right) - \frac{\ell\gamma(x' + vt') + my' + nz'}{c}\right)$$

$$= w\gamma(1 - \ell v/c)\left[t' - \frac{1}{c}\left(\left(\frac{\ell - v/c}{1 - \ell v/c}\right)x' + \frac{my'}{\gamma(1 - \ell v/c)} + \frac{nz'}{\gamma(1 - \ell v/c)}\right)\right]$$

Hence, $\psi = w'(t' - (\ell'x' + m'y' + n'z')/c)$, where $w' = w\gamma(1 - \ell v/c)$; $\ell' = (\ell - v/c)/(1 - \ell v/c)$; $m' = m/\gamma(1 - \ell v/c)$ and $n' = n/\gamma(1 - \ell v/c)$. Note that

$$\ell = (\ell,m,n) \cdot (1,0,0) = (\ell,m,n) \cdot \frac{(v,o,o)}{v} = \cos\phi,$$

where ϕ is the angle between the direction of the wave and the velocity $\vec{v} = (v,o,o)$ of the moving observer K′. Since

$$w' = w\gamma(1 - \ell v/c) = w\frac{(1 - \ell v/c)}{\sqrt{1 - v^2/c^2}} = \frac{w(1 - v \cdot \cos\phi/c)}{\sqrt{1 - v^2/c^2}}$$

and since the frequencies ν and ν' are proportional to w and w′ respectively, we conclude that:

$$\nu' = \nu\frac{(1 - v \cdot \cos\phi/c)}{\sqrt{1 - v^2/c^2}}.$$

It can be shown that the ratio of the energies in K and K′ is the same as that of the frequencies. This is hardly surprising. Thus:

(32) $$E' = E\frac{(1 - v.\cos\varphi/c)}{\sqrt{1 - v^2/c^2}}$$

Consider a body B which remains at rest in K while radiating energy in two opposite directions inclined at an angle φ to the x-axis. Let B_0 and B_1 be the total energies of the body before and after it has emitted the electromagnetic wave; and let Q be the energy lost through radiation. Thus equal amounts Q/2 of energy are sent out in the two directions φ and $(\pi + \varphi)$ to the x-axis. By the law of conservation of energy in the frame K:

$$(33)\ B_0 = B_1 + Q = B_1 + Q/2 + Q/2$$

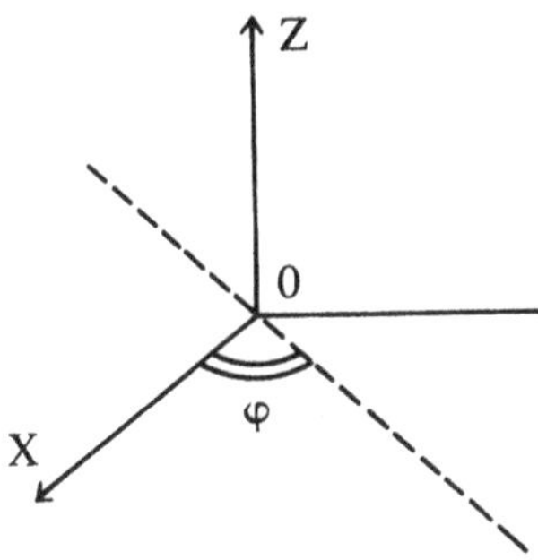

In K′ we have, by the transformation equation for the energy:

$$(34)\ Q' = \left[\frac{Q}{2}\frac{(1 - v.\cos\varphi/c)}{\sqrt{1 - v^2/c^2}} + \frac{Q}{2}\frac{(1 - v.\cos(\pi + \varphi)/c)}{\sqrt{1 - v^2/c^2}}\right] = \frac{Q}{\sqrt{1 - v^2/c^2}}$$

By the relativity principle, the energy must be conserved in K′ as well as in K. Hence:

$$(35)\ B'_0 = B'_1 + Q' = B'_1 + Q/\sqrt{1 - v^2/c^2}.$$

By (33) and (35) we obtain:

$$(36)\ (B'_0 - B_0) = (B'_1 - B_1) + Q[(1/\sqrt{1 - v^2/c^2}) - 1].$$

Einstein interpreted $(B'_0 - B_0)$ as follows: B_0 is the energy of the body in K before the radiation is emitted. B'_0 is the energy of the same body as seen from the moving frame K′. In K′ the body moves with velocity $-v$, while in K it remains at rest. Hence $(B'_0 - B_0)$ represents the kinetic energy of the body to within an additive constant; i.e.

$$(37)\ B'_0 - B_0 = [(M_0c^2/\sqrt{1 - v^2/c^2}) - M_0c^2] + h;$$

where h is some constant. Similarly $(B'_1 - B_1)$ is the kinetic energy of the body after the radiation is emitted. Therefore:

(38) $B'_1 - B_1 = [(m_0 c^2/\sqrt{1 - v^2/c^2}) - m_0 c^2] + h.$

At first sight, it may appear that this crucial step in Einstein's argument is carried out far too quickly, not to say too simplistically. Many authors have in fact doubted its validity, regarding it as viciously circular. However, as was shown by John Stachel and Robert Torretti (1982), Einstein's reasoning is both subtle and completely watertight. Suppose that, in its rest-frame K, the body B has an internal state characterised by certain parameters denoted by S. By the relativity principle, it must be possible for B to have the same internal state S when moving with any uniform velocity $\vec{u}$ in K; otherwise, we should in principle be able to detect the absolute motion of B and hence refute relativity. If E denotes the total energy of B, then E is a function both of the speed u and of the internal state S of B. (We have already seen that, by virtue of the isotropy of space, E cannot depend on the direction of $\vec{u}$.) Thus we can write: E = E(u,S).

This result holds in every inertial frame in which the speed and the internal state of the body are expressed by u and S respectively.

Now put $\psi(u,S) = -E(0,S) + E(u,S)$. We have $\psi(0,S) = 0$ for all S. In other words, $\psi(u,S)$ is the energy which accrues to B by virtue of its speed u alone. Thus $\psi(u,S)$ is the kinetic energy of B so that, as proved above:

$$\psi(u,S) = -E(o,S) + E(u,S) = \frac{m_0(S) \cdot c^2}{\sqrt{1 - u^2/c^2}} - m_0(S) \cdot c^2$$

(we write $m_0(S)$ to indicate the possible dependence of m_0 on S.)

We therefore have (in K and K′):

$$B_0 = E(0,S_0),\ B_1 = E(0,S_1),$$

$$B'_0 = E(v,S_0),\ B'_1 = E(v,S_1);$$

where S_0 and S_1 denote, respectively, the internal state of B immediately before and immediately after the emission of radiation. Thus:

$$B'_0 - B_0 = E(v,S_0) - E(0,S_0)$$

$$= \frac{m_0(S_0) \cdot c^2}{\sqrt{1 - v^2/c^2}} - m_0(S_0) \cdot c^2$$

and $B'_1 - B_1 = E(v,S_1) - E(0,S_1)$

$$= \frac{m_0(S_1) \cdot c^2}{\sqrt{1 - v^2/c^2}} - m_0(S_1) \cdot c^2$$

Putting $m_0(S_0) = M_0$, $m_0(S_1) = m_0$ and $h = o$, we retrieve equations (37) and (38). Substituting from (37) and (38) into (36):

$$\textbf{(39)}\ c^2(M_0 - m_0)\,[(1/\sqrt{1 - v^2/c^2}) - 1] = Q\,[(1/\sqrt{1 - v^2/c^2}) - 1]\,.$$

Hence:

$$\textbf{(40)}\ M_0 - m_0 = Q/c^2\,;\ \text{i.e. } \Delta m_0 = Q/c^2\,,$$

where Q is the energy lost through radiation. Note that Q is measured in the rest-frame of the body.

Let us once again point out the crucial role which the relativity principle plays in Einstein's argument: he looks at the same process from two distinct inertial frames K and K′; he then insists that energy must be conserved in K′ as well as in K. However, equation (40) remains too dependent on electromagnetic theory to be considered an integral part of relativistic mechanics. In conformity with his programme, Einstein tried to arrive at the same result from purely mechanical considerations. Radiation energy, whose study belongs to electromagnetics, was replaced by a purely dynamical quantity, namely the kinetic energy. By considering the case of a body consisting of a number of agitated corpuscles (Einstein 1907), Einstein showed that the inertia of the body depends on the kinetic energy of its constitutent parts: the more agitated the corpuscles the greater the inertia. Exactly as in the previous case, his method consists in 'looking' at the body from two different inertial frames, K and K′. In K the body B is at rest; hence in K′ it moves with the velocity of the whole frame K relatively to K′. An immediate difficulty arises: what do we mean by B being at rest in K when B is a collection of moving particles? We have somehow to look upon B as being globally at rest in K. In classical mechanics we can say that a system of particles is stationary if its center of gravity does not move; i.e. if $\sum_i m_i \vec{v}_i = 0$ where m_i is the rest-mass and $\vec{v}_i$ the velocity of the ith particle. Note that $\sum_i m_i \vec{v}_i$ = total classical momentum.

Einstein took a body to be at rest in a given inertial frame if its total *relativistic* momentum vanishes in that frame. Hence K must be such that $\sum_i m_i \vec{v}_i/\sqrt{1 - v_i^2/c^2} = \mathbf{0}$, i.e.

$$\textbf{(41)}\ \sum_i m_i \vec{v}_{ij}/\sqrt{1 - v_i^2/c^2} = \mathbf{0},\ \text{for } j = 1, 2, 3;$$

where $\vec{v}_i = (v_{i1}, v_{i2}, v_{i3})$. Therefore:

$$\textbf{(42)}\ (\sum_i m_i v_{i1}/\sqrt{1 - v_i^2/c^2},\ \sum_i m_i v_{i2}/\sqrt{1 - v_i^2/c^2},\ \sum_i m_i v_{i3}/\sqrt{1 - v_i^2/c^2},\ \sum_i m_i/\sqrt{1 - v_i^2/c^2}) = (0, 0, 0, \sum_i m_i/\sqrt{1 - v_i^2/c^2}).$$

We know that the left hand side of (42) transforms like the coordinates (x,y,z,t). Let K′ be a frame of reference moving with velocity −u along the x-axis of K. In K′ we have, by the Lorentz transformation:

$$(43)\quad \begin{cases} \sum_i m_i v'_{i1}/\sqrt{1 - v_i'^2/c^2} = \gamma(0 + u\sum_i m_i/\sqrt{1 - v_i^2/c^2}) = \\ \qquad\qquad \gamma u \sum_i m_i/\sqrt{1 - v_i^2/c^2} \\ \sum_i m_i v'_{i2}/\sqrt{1 - v_i'^2/c^2} = 0 = \sum_i m_i v'_{i3}/\sqrt{1 - v_i'^2/c^2} \\ \sum_i m_i/\sqrt{1 - v_i'^2/c^2} = \gamma[(\sum_i m_i/\sqrt{1 - v_i^2/c^2}) + (u/c^2)\cdot 0] = \\ \qquad\qquad \gamma \sum_i m_i/\sqrt{1 - v_i^2/c^2};\ \text{where } \gamma \underset{\text{Def}}{=} 1/\sqrt{1 - u^2/c^2} \end{cases}$$

In the case of a single particle the momentum is equal to $m_0\vec{v}/\sqrt{1 - v^2/c^2}$, where $m_0/\sqrt{1 - v^2/c^2}$ is the inertial mass. In K′ the velocity of the body B, taken as a whole, is identical with the velocity $\vec{u}$ of K with respect to K′. By equations (43), the momentum of B is $\gamma(\sum_i m_i/\sqrt{1 - v_i^2/c^2})\vec{u}$ (remember that $\vec{u}$ = (u,0,0) in K′). Hence we can identify the inertial mass of B with the coefficient of $\vec{u}$; i.e. with μ, where:

$$(44)\quad \mu = \gamma\sum_i \frac{m_i}{\sqrt{1 - v_i^2/c^2}} = \frac{1}{\sqrt{1 - u^2/c^2}} \cdot \sum_i \frac{m_i}{\sqrt{1 - v_i^2/c^2}} = \frac{\mu_0}{\sqrt{1 - u^2/c^2}}$$

where $\mu_0 = \sum m_i/\sqrt{1 - v_i^2/c^2}$. Since $\vec{u}$ is the velocity of B in K′, it follows from (44) that μ_0 is the rest-mass of B. Note that, if k_B denotes the internal kinetic energy of B in K, i.e. if k_B denotes the sum of the kinetic energies of all the particles composing B, then:

$$(45)\quad k_B = \sum_i \left[\frac{m_i c^2}{\sqrt{1 - v_i^2/c^2}} - m_i c^2\right] = \mu_0 c^2 - (\sum_i m_i c^2).$$

Taking the rest-masses of the separate corpuscles to be constant, we obtain from the last equation:

$$(46)\quad \Delta k_B = (\Delta\mu_0).\ c^2;\ \text{i.e. } \Delta\mu_0 = \Delta k_B/c^2.$$

Hence any increase Δk_B of the inertial kinetic energy of a body issues in an increase $\Delta k_B/c^2$ of its rest-mass.

Einstein's method in both of the above examples can be summed up as follows. We study the behaviour of the same body B in two different inertial frames K and K′. In K, B is at rest and, in K′, it moves with the velocity $\vec{u}$ of the frame K with respect to K′. Let $\bar{B}$ and $\bar{B}'$ be the total energies of B in K and K′ respectively. In view of the relativity principle which entails the physical equivalence of K and K′, we accept that $\bar{B}$ would also be the total energy of B in K′, had B been at rest in K′. Hence, whatever the origin of the energies $\bar{B}$ and $\bar{B}'$, the difference $(\bar{B}' - \bar{B})$ vanishes for $\vec{u} = \mathbf{0}$ because the increment $(\bar{B}' - \bar{B})$ accrues to the body B solely by virtue of its motion.

Einstein then considers cases in which the body loses some energy. Let us call B_0 and B_1 the total energies just before and just after the loss has taken place in K. B'_0 and B'_1 are similarly defined in K′. The conservation law which, by the relativity principle, applies to both K and K′, enables us to derive expressions both for $(B_0 - B_1)$ and $(B'_0 - B'_1)$. Subtracting, we obtain (trivially):

$$(B'_0 - B'_1) - (B_0 - B_1) = (B'_0 - B_0) - (B'_1 - B_1).$$

As shown above, $(B'_0 - B_0)$ and $(B'_1 - B_1)$ represent the kinetic energies respectively before and after the loss of total energy takes place in K′. Since the velocity of the body in K′ remains unchanged and since the kinetic energy is still given by $[m_0c^2/\sqrt{1 - v^2/c^2}) - m_0c^2]$, the rest-mass must have altered. The variation $[(B'_0 - B_0) - (B'_1 - B_1)]$ in the kinetic energy thus enables us to calculate the corresponding variation in m_0. We obtain: $\Delta m_0 = Q/c^2$, where Q is the lost energy, as measured in the rest-frame of the body.

It would be difficult to overestimate the importance of Einstein's law about the equivalence of mass and energy. Its technological significance needs no emphasis and its devastating consequences are still with us. In view of the magnitude of the factor c^2, the formula $Q = c^2.\Delta m_0$ entails that ordinary ponderable matter can disintegrate and thereby release an enormous amount of energy. This is certainly one novel prediction implied by the conclusions reached by Einstein in 1905–1907. However it is doubtful that such a prediction was actually testable at the time of Einstein's discovery. Whatever the case may be, the significance of $E = m(v) \cdot c^2$ transcends the mere novelty of some empirical prediction. The law about the equivalence of mass and energy brought about a revolution in philosophical outlook.

According to the accepted mechanistic view, ordinary 'ponderable' matter, i.e., the extended hard 'stuff' with which we are all familiar, constitutes the most fundamental layer of physical reality. Electric charge was added to this layer and one had considered adding a further constituent, the ether, which possessed properties similar to those of ponderable matter: the ether was extended and it was subject to pressure and stress. Matter moved in space and, out of its motion, charge, position and mass, i.e., out of its mode of existence, both the potential and the kinetic energies were constructed. It is true that in the nineteenth century energy came to be regarded as a quantity whose total amount remained constant. However, energy existed on an ontological level which was both distinct from and less fundamental than that of matter. Energy derived its 'reality' from the existence of matter, charge and motion. In any event, there was no question of any continuity *between* the two levels: matter and energy were both conserved, but each separately from the other. Einstein showed that the two levels could be regarded as identical, that 'energy' and 'mass' *could* be treated as two names for the same basic entity. The stuff which appears to the senses as hard extended substance and the quantity of energy which characterises a process are in fact one and the same thing. Each of these two quantities can disappear and thereby give rise to an equivalent amount of the other. We imagine that we can make an absolute ontological distinction between mass and energy merely because our senses perceive them in different ways. But relativity teaches us otherwise. Thus the acceptance of relativity theory involves a great loss of intelligibility. The usual notions of space, time, simultaneity and substance have to be replaced by counter-intuitive concepts embedded in sophisticated mathematical theories. This massive loss of intelligibility has pushed many philosophers and scientists into adopting a positivist view of the new physics: our theories say nothing about such metaphysical entities as space, time or substance; they simply connect, or bring order into, various observational results; the systems of mathematical physics are artifacts created in order to codify the results of experiments, which either have been or else will eventually be carried out. Thus Eddington wrote:

> Every item of physical knowledge must therefore be an assertion of what has been or what would be the result of carrying out a specified observational procedure.

And

> Mathematical formulation is very economic of hypothesis. Subject to a certain reservation, it enables us to state a conclusion which does not go beyond the ascertained facts; it is no more than a systematised statement of what is observed. The reservation is that, whereas the ascertained facts justify the mathematical formula within a limited degree of approximation, under limited conditions and in a limited number of instances, the mathematical formula omits reference to these limitations. . . . In one sense Einstein's (or Newton's) law of gravitation is not a hypothesis; it is a summary statement of what we have observed within certain limits of approximation. (Eddington 1939, pp. 10, 43)

This is an amazing claim since the only 'fact' on which Einstein based his GTR was the equality of inertial and gravitational masses. Even today, the empirical basis of the GTR is rather slight and there is no question of 'unpacking' it into a list of observation statements. Moreover, as clearly indicated in the EMB, relativity theory as a whole is not about observations but about events. Actual *and* potential events, not observed matter-in-motion, constitute the domain of both the special and the general theories. Relativistic mechanics asserts that something is conserved; only this 'something' is no longer the familiar hard substance but a new entity which can be interchangeably called matter or energy. With relativity an important shift in the direction of greater mathematical abstraction occurred. In classical physics we had clear, intuitive ideas of space, simultaneity and quantity of matter; and the mathematics was intended to capture these intuitive notions. In relativistic physics the mathematics comes first (though only in the heuristic sense). It is through the mathematics of Lorentz-covariance that one comes to accept the interchangeability of matter and energy *despite* one's intuitive conviction to the contrary. It is because conservation laws have to be Lorentz-covariant that kinetic energy can be converted into rest-mass; and Lorentz-covariance is a mathematical property with no obvious intuitive appeal. Of course, the difference between classical and relativistic physics is one of degree and not of kind. The law of inertia, which at first sight seems absurd, is part of Newtonian dynamics. Hence, compared with Aristotelian physics, Newtonian dynamics already involves a violation of common sense. Relativity carries this counter-intuitive process of abstraction a lot further. (The situation was to become much worse with the advent of quantum mechanics.) Apart from admitting the law of inertia, the relativist has to accept that simultaneity is frame-dependent, that the time order of certain events can be reversed and that solid matter can disappear into thin air; his only guides in this labyrinth are the form of his mathematical equations and the outcome of a few empirical tests; he can no longer trust his 'intuition'. In a foreward to Newton's 'Opticks' Einstein wrote, with a touch of nostalgia:

> Fortunate Newton, happy childhood of science! . . . Nature to him [Newton] was an open book, whose letters he could read without effort. The conceptions which he used to reduce the material of experience to order seemed to flow spontaneously from experience itself . . .

Let us however repeat that this loss of *psychological* intelligibility in no way entails a breakdown of realism. On the contrary: the less intelligible an empirically successful theory is, the more one ought to feel that, through it, one gains access to a reality independent of the human mind, hence also of its prejudices.

— 8 —

GENERAL RELATIVITY AND EARLY UNIFIED FIELD THEORIES

§8.1 The Status and Heuristic Role of the Covariance Principle

In this final chapter, the experimental confirmations of general relativity theory, namely: precession of Mercury's perihelion, bending of light rays and red shift are taken to be well-known. The problem to be tackled is whether this empirical success can be claimed by a single research programme comprising both STR and GTR. These two theories *appear* to be very different scientific hypotheses. On the one hand, according to some physicists, the empirical status of GTR is doubtful. Whittaker, on the other, distinguishes between the relativity principle, which he attributes largely to Lorentz and Poincaré, and GTR which he regards merely as a new theory of gravitation. In fact, prima facie, there should be no direct connection between the relativity principle on the one hand and any particular physical phenomenon like gravitation on the other. In his EMB Einstein first develops a new kinematics without referring to any specific theory; he then goes on to formulate mechanics and electrodynamics in a Lorentz-covariant way. The kinematical section of his paper remains logically, if not heuristically, independent of the subsequent sections. One might have expected the same to happen in GTR; namely, the formulation of some new kinematics followed by a gravitational theory set against this already articulated kinematical background. Nothing of the sort happens: gravitation enters the discussion right from the start.

This is the first reason for posing the continuity problem; the second one is that the heuristic of the programme experienced a major shift during the period which led to GTR. The methods applied by Einstein, Planck, Tolman and Lewis before 1910 differ widely from those used after 1910 by

Einstein, Weyl, Kaluza, Eddington and Schrödinger, say. Whereas, in the EMB, Einstein takes great pains to give an operational definition of length and duration, in the "Foundations of General Relativity" (1916) the coordinates have lost any direct physical meaning. We are simply told that events are described by a set of four real numbers. This is a far cry from the Einstein who, according to Bridgman, replaced absolute time and space by clock-time and rod-space. Whereas, in 1909, Tolman and Lewis base their new definition of momentum on considerations of how different observers measure distance and time, in *The Mathematical Theory of Relativity* [1923] Eddington sets out to geometrise physics a priori. From being experimental scientists mainly concerned with actual physical measurements, the relativists had become a priorist geometers. The whole tenor of their approach seemed to have undergone a fundamental change. This change was of course reflected at the experimental level. While special relativity is a highly confirmed empirical theory, many scientists like Trautmann (Trautmann, Pirani & Bondi 1965, p.7) wonder whether general relativity belongs to physics or to pure mathematics. Whatever the case may be, there exists an obvious disparity between the complicated mathematical machinery of GTR and the paucity of its experimental results. The question therefore arises as to whether the development leading to GTR occurred within the global heuristic of the same programme or was a radical change of tactics, a completely new departure.

In other words: Is there a relativistic programme as such or has *one word* come to denote a *heterogeneous* body of hypotheses simply because they were put forward by the same man?

One facile answer to the above question is that the relativity principle provides a continuous thread running through all the stages of the programme (i.e., the principle that the frames belonging to some class K are all physically equivalent). GTR simply extends this principle from the set of inertial frames to that of all possible (accelerated as well as unaccelerated) systems. The programme would thus be identified by a hard core, by one assumption shared by all scientists working in it. Although the relativity principle, as it stands, is both vague and metaphysical, we have seen that it played an important regulative role in the special relativistic modification of the laws of mechanics. However, in 1917 Kretschmann pointed out that *all* physical laws can be covariantly rewritten and Einstein accepted this criticism as valid. Thus it looks as if the relativity programme had lost its most distinctive feature, since any hypothesis whatever could now be regarded as a relativistic theory. This first attempt to answer our question seems to end in disaster. In his reply to Kretschmann, Einstein claimed that the covariance principle should be treated as a heuristic device and not as an ordinary statement. Of all the covariantly formulated laws, we pick out the simplest one as the most suitable. The important point to note is that a theory which at first sight looks simple may, on covariant reformulation, turn out to be very complicated. Einstein wrote:

> Although it is true that one can put every empirical law in a generally covariant form, yet principle a) [the principle of relativity] possesses a great heuristic power which has already brilliantly proved its mettle in the case of gravitation and which is based on the following. Of two theoretical systems, both of which are in agreement with experience, that one is to be preferred which, from the point of view of the absolute differential calculus, is the simpler (einfachere) and more transparent (durchsichtigere) one. Let one express Newtonian gravitational mechanics in the form of generally covariant equations (four-dimensionally) and one will surely be convinced that principle a) excludes this theory from the practical, if not from the theoretical, point of view. (Einstein 1918, p. 242; my translation)

It is obvious from the above passage that Einstein meant 'simplicity' and 'clarity' in a pragmatic sense. Although words can be arbitrarily redefined, 'simplicity' is to my mind a misleading term which ought to be replaced by something like 'compactness' or 'organic unity'. Even when covariantly formulated, Newton's theory cannot be said to be more complicated than Einstein's[1]; its main disadvantage is that it contains two heterogenous parts: the (degenerate) metric and the gravitational field. The latter is described by an affine connection not determined by the metric alone, so that we have to go beyond the metrical aspect of geometry in order to capture the field. In Einstein's theory the metric tensor and the gravitational potential are one and the same thing; the affinity Γ^i_{jk} and the Christoffel quantities $\left\{ {i \atop jk} \right\}$ are identical (in John Stachel's words, there is 'minimal coupling' between the field and the geometry). This is what is meant by saying that Einstein tried to geometrise physics or, equivalently, to make geometry empirical.

However, even if it were accepted that the covariance principle played a major heuristic role in the genesis of relativity theory, Kretschmann's objection would not thereby have been fully answered. For we may still ask: how can a tautological (empty) proposition yield a fruitful heuristic device? In chapter one it was maintained that heuristic reasoning is in most cases *deductive* metareasoning (§1.3); and it is well-known that, in a deductive argument, the addition of logical truths to the premises does not increase the latter's content. Thus, in order to meet Kretschmann's criticism, we have to demonstrate that the covariance principle is not a logical truth. Once again, the distinction between object-level and metalevel proves crucial.

[1]For this point I am indebted to Professor J Stachel.

Let ℂ be any physical statement. The covariance principle R[ℂ] is the proposition that ℂ assumes the same form under all allowable (one-one, bi-continuous, C^∞) transformations. Kretschmann's objection is that all physical theories can be written in a generally covariant form. In other words: given any (putative) physical law ℂ, there always exists a generally covariant physical law ℂ* such that ℂ⇔ ℂ*. Note: to say that ℂ* is generally covariant is to say that R[ℂ*] holds good.

If we look upon 'R[ℂ]' as an object-level statement in which 'ℂ' occurs, then from ℂ⇔ℂ* and R[ℂ*] there follows R[ℂ]. Since ℂ is an arbitrary physical law, R[ℂ] expresses a vacuous constraint. However, even if we accept Kretschmann's thesis that, for any ℂ, there exists a generally covariant ℂ* equivalent to ℂ, the above argument goes through only if R[ℂ] is regarded as an object-level sentence. But we know that the relativity principle is not vacuous, since we can exhibit propositions like $\Sigma m_0 \vec{v} = \mathbf{0}$ which are not Lorentz-covariant, let alone generally covariant. Let me therefore propose the following (tentative) solution to the problem posed by the status of the covariance principle.

The very formulation of this principle indicates that it is a proposition *about* 'ℂ'. In other words: the covariance principle belongs to the metalanguage and should be written R('ℂ'). Instead of being an object-level statement 'R[ℂ]' in which 'ℂ' occurs, the covariance principle is *about* 'ℂ'. Thus, although ⊢ℂ⇔ℂ* and ⊢R('ℂ*'), we cannot directly infer R('ℂ'); for we may well have ⊢'ℂ' ≠ 'ℂ*', hence possibly ⊢ ¬R(ℂ'). The upshot of all these remarks is that, although 'ℂ' can be rewritten in the generally covariant form 'ℂ*', 'ℂ' *itself* is not generally covariant.

Now let R stand for the following proposition: for all 'ℂ', if 'ℂ' is a law of nature, then R('ℂ'). And let R′ denote the (weaker) statement: for all 'ℂ', if 'ℂ' is a law of nature, then 'ℂ' is Lorentz-covariant. Trivially: ⊢R ⇒ R′. R′ can be considered as belonging to the hard core of the whole relativity programme. GTR strengthened R′ into R, so that covariance does after all provide an important link between STR and the later stages of the programme.

§8.2 The Abandonment of Special Relativity

In his 1916 paper Einstein explains his reasons for abandoning the STR and for recognising the close connection between gravitation and general relativity. His argument is as follows. Both classical and special relativistic physics give a privileged status to an infinite set of inertial frames *without* relating these frames to the distribution of matter in the universe. Such a distinction between inertial and noninertial frames is illegitimate from a

Machian standpoint (see §4.2). Einstein therefore proposed to put all frames on equal footing, so that the following situation arises. Let K be any frame with respect to which a particle, when left to itself, either remains at rest or else goes on moving uniformly along a straight line. Consider another frame K′ which is accelerating relatively to K. As viewed from K′, the particle P experiences an acceleration $\vec{a}$ which is due to the motion of K′ and must therefore be independent of the internal composition of P. Since we have agreed to look upon K′ as a legitimate frame of reference, we must regard P as acted upon by a real physical field $\vec{I}$ exerting a force $m\vec{a}$ proportional to the inertial mass m of P. In classical physics we are acquainted with a field, namely the gravitational field $\vec{G}$, to which bodies respond proportionately to their masses. Refusing, as usual, to believe in a mere coincidence, Einstein assumed that $\vec{I}$ is of the same nature as $\vec{G}$; hence any general theory of relativity would be closely bound up with gravitational phenomena.

Because of Einstein's references to Mach, the myth arose that a *positivist-idealist* philosophy, namely Machism, was the starting of a new revolutionary *physical* theory (see §4.5). However, as just shown, in order for the co-variance principle to be non-vacuous, it has to be regarded as a metastatement arising from the metaphysical thesis that there exist no privileged frames of reference. Needless to say, such general metaphysical principles, which do not refer to observables, are pronounced meaningless by Machist philosophy. We have also seen, in chapter one (§1.1.5), that metaphysical propositions can be turned into heuristic prescriptions governing the construction of physical theories. In this particular case, we have the prescription that all laws should assume the same form in all allowable frames. Even when interpreted in this prescriptive sense, the covariance principle has to be supplemented by a 'compactness' requirement in order to become effective: it is for example assumed that the field equations should involve at most the metric $[g_{ij}]$ and the first and second derivatives of the g_{ij}'s (see §8.4 below). The nearest analogue to compactness in Machist philosophy is the economy-of-thought principle, according to which scientific theories are convenient devices for subsuming an infinite number of facts under a limited number of laws. Thus, 'economy of thought' is more akin to 'simplicity' than to 'unity' or 'compactness'; and we have already seen that a theory may be more unified and yet less simple than another. There is of course no reason to doubt that Einstein was influenced by Mach who, in his 'conservation of energy', came very near to formulating a covariance principle (see §4.5). This influence has however nothing to do with Machism as such but only with Mach; moreover, it is not, as will presently be shown, the only reason why Einstein abandoned special relativity.

After deriving the Lorentz transformation, Einstein, then Planck, modified Newton's second law so as to make it Lorentz-covariant. It was to be expected that Einstein would try a similar procedure with the gravitational equations, notwithstanding his Machian qualms about the absolute nature of the inertial frames. In fact, this is precisely what he intermittently tried to do until 1908 when, in view of the equivalence principle, he decided that the undertaking was hopeless.

In the case of STR, we have seen that Einstein's starting point was a 'fact' discovered by Faraday: the fact that, in electromagnetic induction, the 'observed' result depends on the relative velocity of conductor and magnet, not on their absolute motion in the ether (see §3.2). In the case of GTR, the result from which Einstein started goes back even further; in fact, it goes back to the early 17th, or even late 16th, century. Galileo discovered that all bodies near the surface of the earth fall freely with the same acceleration. A two-fold constancy is involved here: the same body B has a constant acceleration throughout its time of fall; and all bodies, irrespective of their mass and composition, experience the same acceleration at the same point of their trajectory. The first type of constancy was rejected by Newton, but the second can be deduced from Newtonian theory as follows.

A body P has three different masses:
(1) An inertial mass m which measures the body's capacity for resisting the action of applied force;
(2) An active gravitational mass m′ through which the body contributes to the gravitational field; and
(3) A passive gravitational mass m″ by which it responds to an existing gravitational field.

Given two mass points P_1 and P_2, the force acting on P_2 due to the presence of P_1 is

$$\frac{Gm'_1m''_2}{r^2}\vec{u}.$$

Similarly the force on P_1 is

$$\frac{Gm''_1m'_2}{r^2}(-\vec{u}).$$

By Newton's third law:

$$\frac{Gm'_1m''_2}{r^2}\vec{u} = -\frac{Gm''_1m'_2}{r^2}(-\vec{u})$$

Hence $m'_1m''_2 = m''_1m'_2$ and so $m'_1/m''_1 = m'_2/m''_2$.

We can choose our units so as to make $m'_1 = m''_1$ and $m'_2 = m''_2$. The two gravitational masses are therefore equal. Moreover, by the second law of motion:

$$\frac{m'_1 m'_2 \vec{u}}{r^2} = m_2 \vec{a}_2,$$

where $\vec{a}_2$ is the acceleration of P_2. At this point, Newton blandly assumes that the gravitational and inertial masses are equal, so that $m_2 = m'_2$ and $m_1 = m'_1$; hence, in view of

$$\frac{m_1 m_2}{r^2} \vec{u} = m_2 \vec{a}_2, \frac{m_1 \vec{u}}{r^2} = \vec{a}_2.$$

It follows that, whatever its material composition, P_2 experiences the same acceleration $\vec{a}_2$ (provided m_2 is small enough not to affect the motion of P_1 in any appreciable way).

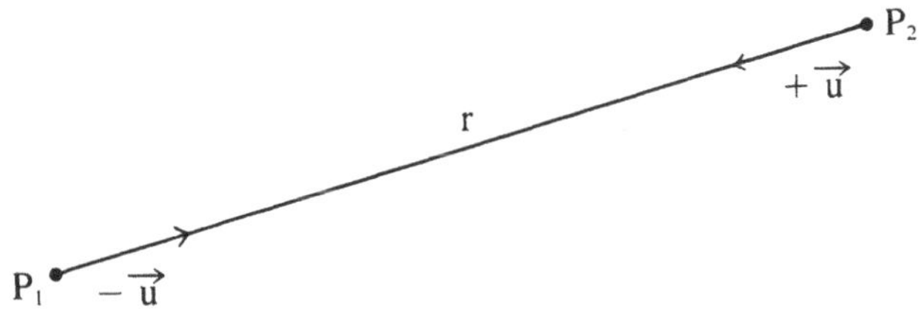

Figure 1.

The gravitational field is therefore a kinematical—more precisely, an acceleration—field. It is characterised by the direction and magnitude at each point of the acceleration which every test body experiences, irrespective of its composition and velocity. Hence, at each point of space, we can neutralise the field by choosing a frame whose acceleration is equal to the value of the field at that point; in other words, by letting the frame fall freely with the field.

We have here a 'fact', namely the equality of gravitational and inertial masses, which is unsatisfactorily accounted for by the classical theory. We shall see how GTR remedies this defect but, before pursuing this line of thought, let us examine what conditions a special relativistic theory would have to meet in order to account for gravitational phenomena. In a reply to Max Abraham, Einstein wrote:

> One of the most important results of relativity theory is the knowledge that every form of energy E possesses inertia E/c^2 proportional to E. Since, as far as our

experience goes, inertial mass is at the same time gravitational mass, we cannot but attribute to every form of energy E a gravitational mass equal to E/c^2. From this it immediately follows that the gravitational force acting on a body is greater when the body is in motion than when the body is at rest.

If the gravitational field were to be accounted for in present day Relativity Theory, this could happen in only one of the following ways. One can regard the gravitational field either as a 4-vector or as an antisymmetric tensor of the second order . . . One thereby obtains results which contradict the above-mentioned consequences concerning the gravitational mass of energy. It therefore looks as if the gravitational vector cannot be consistently fitted into the Relativistic scheme as it stands at the present moment. (Einstein 1912, pp. 1059-1064; my translation)

What follows is a technical illustration of the point made by Einstein in the above passage. Let $\vec{f}$ be the gravitational 3-field in some frame of reference. Consider a body P of rest mass m_o moving under the effect of the gravitational force. Since

$$\frac{m_o}{\sqrt{1 - v^2/c^2}}$$

is the inertial mass of the body, it is also, according to Einstein, the gravitational mass. In other words,

$$\frac{m_o}{\sqrt{1 - v^2/c^2}}$$

is the measure of P's response to the field. By Planck's equation of motion we have:

$$\frac{d}{dt}\left[\frac{m_o \vec{v}}{\sqrt{1 - v^2/c^2}}\right] = \frac{m_o}{\sqrt{1 - v^2/c^2}} \vec{f}.$$

Through expanding the left-hand side of this equation then dividing by $m_o/\sqrt{1 - v^2/c^2}$, we obtain:

$$\frac{1}{m_o} \cdot \frac{dm_o}{dt} \vec{v} + \sqrt{1 - v^2/c^2}\, \frac{d}{dt}\left[\frac{\vec{v}}{\sqrt{1 - v^2/c^2}}\right] = \vec{f}. \text{ Hence:}$$

$$\frac{1}{m_o} \cdot \frac{dm_o}{dt} \vec{v} + \vec{a} + \frac{\vec{v} \cdot \vec{a}}{(c^2 - v^2)} \vec{v} = \vec{f},$$

where $\vec{a}$ is the acceleration of P. If the body is heated and so gains energy during its motion, then, in view of $E = mc^2$, m_o varies with time. Hence $(dm_o/dt) \neq 0$; so $\vec{a}$ depends on the internal composition of P, which contradicts the principle of equivalence. Even if the rest mass remains constant, $\vec{f}$ still depends on $\vec{v}$ and has thus ceased to be a pure acceleration field.

The situation is no better if the field is described by an antisymmetric tensor $H_{\mu\nu}$. In the electromagnetic case the 4-force acting on a charged particle is given by $ke.F^{\mu}{}_{\sigma}V^{\sigma}$, where k is a constant, e is the charge and $F^{\mu\sigma}$ the electromagnetic tensor. Hence $d/ds(m_oV^{\mu}) = keF^{\mu}{}_{\sigma}.V^{\sigma}$. In the case of gravitation the equation would be as follows:

$$\frac{d}{ds}(m_oV^{\sigma}) = km_oH^{\sigma}{}_{\mu}V^{\mu}; \text{ i.e. } \frac{1}{m_o}\cdot\frac{dm_o}{ds}\cdot V^{\sigma} + \frac{dV^{\sigma}}{ds} = kH^{\sigma}{}_{\mu}V^{\mu}$$

Multiplying by V_{σ} and noting that $V_{\sigma}V^{\sigma} = 1$, hence that $V_{\sigma}.\frac{dV^{\sigma}}{ds} = 0$, we obtain $\frac{1}{m_o}\cdot\frac{dm_o}{ds} = 0$. Thus $\frac{dm_o}{ds} = 0$, which means that m_o *must* remain constant. This contradicts special relativity.[2]

To sum up: Einstein had the following two reasons for giving up special relativity as a suitable framework for the whole of physics: first philosophical dissatisfaction with having given a privileged status to the set of inertial frames; secondly, the technical difficulty, arising from $E = mc^2$, of accommodating gravitational theory within special relativity. The second reason appears to have been the more decisive one.

The process through which the special theory was abandoned can be described as follows. It was precisely the law concerning the interchangeability of mass and energy (a fundamental result of special relativity), which brought about the downfall of STR. Special relativity as it were 'succumbed to its inner contradictions', more particularly to the impossibility, in Einstein's view, of reconciling $E = mc^2$ with gravitation.

§8.3 The Heuristic Role of the Principle of Equivalence

In his 1912 article, Einstein wrote:

> In the immediate future our task is to create a relativistic scheme for expressing the equivalence between gravitational and inertial masses. I made a first, albeit modest, attempt towards attaining this aim in my work on the static gravitational field. In this I started from the most obvious viewpoint, namely that the equivalence between inertial and gravitational masses should be explained in terms of a fundamental identity between these two primitive qualities of matter, viz. energy: from a physical standpoint, the presence of a static gravitational field should be considered as essentially identical with an acceleration of the reference frame. I have to confess that I was able consistently to push through this interpretation only for infinitely small regions, a state of affairs of which I can give no satisfactory account. (Einstein 1912, p. 1063; my translation)

[2] For this (second) point, I am indebted to Professor C. Kilmister.

The most striking word in this passage is '*Wesensgleichheit*', 'identity of essence', which reveals a strong essentialist streak in Einstein's scientific character. As already mentioned, a method often used by him was to question well-known 'facts' and so isolate certain features which are unsatisfactorily explained by extant hypotheses. Newtonian theory accounts for the 'fact' that all bodies fall with the same acceleration by simply *postulating* the equality of gravitational and inertial masses ($m = m'$). For Einstein, this equality should manifest a deep *essential* identity: bodies fall with the same acceleration because gravitational fields are created by local accelerations of the frame of reference.

The last two sentences in the above passage also indicate what difficulties Einstein faced in 1912. He could explain the behaviour of the field *in* infinitely small regions but was unable to construct a global theory of gravitation.

Consider a point P at which the field is represented by the acceleration $\vec{g}$. In an infinitely small neighbourhood N of P the field can be regarded as constant.

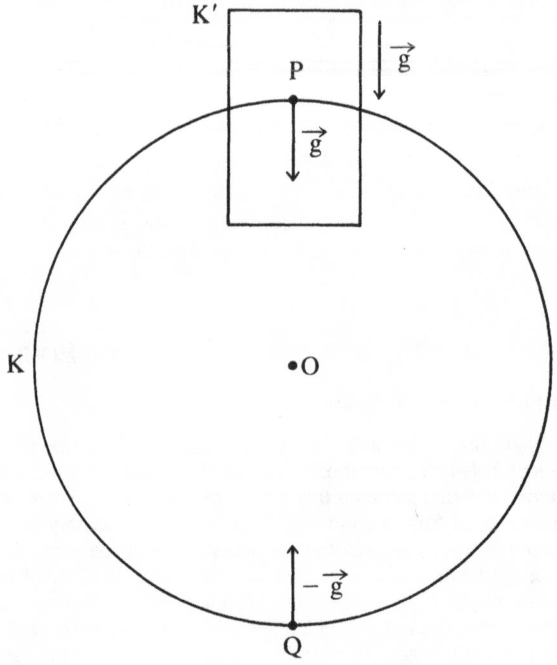

(Field created, in K, by a single mass point at O)

Figure 2.

Let K be our original frame and K′ a system whose acceleration with respect to K is equal to $\vec{g}$. The equivalence principle states that K′ can be considered as field-free[3] and the field in K as generated by the acceleration of K relatively to K′. However, this argument applies only to the immediate neighbourhood of P; for example, if the field is created by a single particle at O and Q is the point diametrically opposed to P with respect to O, then, within K′, the field at Q is $-2\vec{g}$. K′ is inertial or field-free only at P; in fact, transforming the field away at P entails piling it up somewhere else (at Q for example).

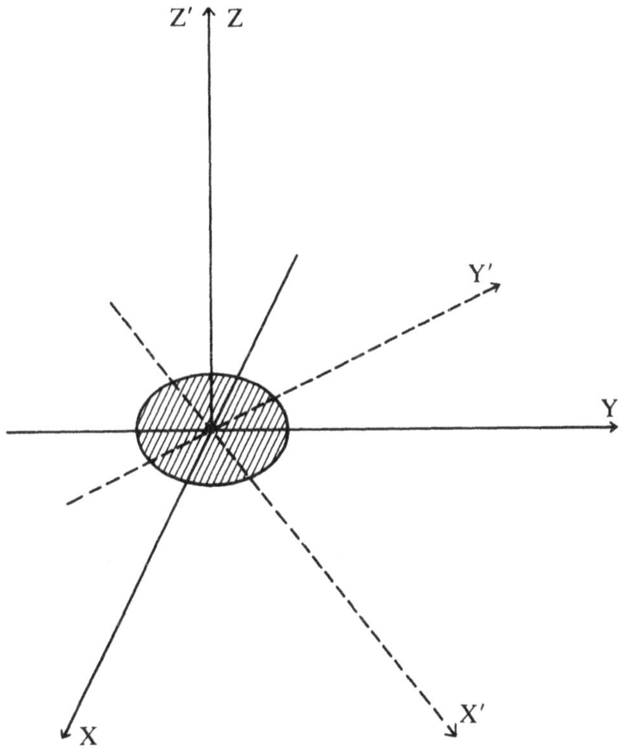

Figure 3.

Einstein thus found himself compelled to abandon Euclidean kinematics. An important thought-experiment which finally convinced him that Euclidean geometry had to be given up is the so-called rotating disc experiment

[3] We shall however see that this is not, strictly speaking, true.

(see Stachel 1979). Starting from STR, let us consider an inertial frame OXYZ with respect to which a material disc rotates in the XOY plane. It is important to look upon the disc as already rotating and not to wonder how its motion arose in the first place. Let an observer moving with the disc, i.e., at rest relatively to it, measure its circumference and its radius and then divide the results of these two measurements. He will obtain, not 2π, but $2\pi/\sqrt{1 - w^2r^2/c^2}$, where w is the angular velocity of the disc; for, by the Lorentz contraction, the rods he places along the circumference will be shortened by the factor $\sqrt{1 - w^2r^2/c^2}$, while the rods he uses for measuring the radius, being transverse to the motion, are unaffected by it. Since the result $2\pi/\sqrt{1 - w^2r^2/c^2}$ is independent of the composition both of the rod and of the disc, the observer will conclude that his (2-dimensional) geometry is non-Euclidean. Einstein thus faced the problem of providing an explanation of gravity in non-Euclidean terms. In 1913–14 he found a solution by turning the principle of equivalence into a powerful heuristic instrument. He looked at it from the following new angle.

Consider a classical inertial frame K and the (Newtonian) gravitational field created by a single mass point at O. Take K′ to be an accelerated frame with the same origin O. Finally let P be a material point having the (negligible) gravitational and inertial masses m′ and m respectively. By the second law of motion, the total force $\vec{F}$ acting on P is given by:

$$\vec{F} = m\frac{d^2}{dt^2}(\xi_i\vec{w}_i) = m(\ddot{\xi}_i\vec{w}_i + 2\dot{\xi}_i\dot{\vec{w}}_i + \xi_i\ddot{\vec{w}}_i).$$

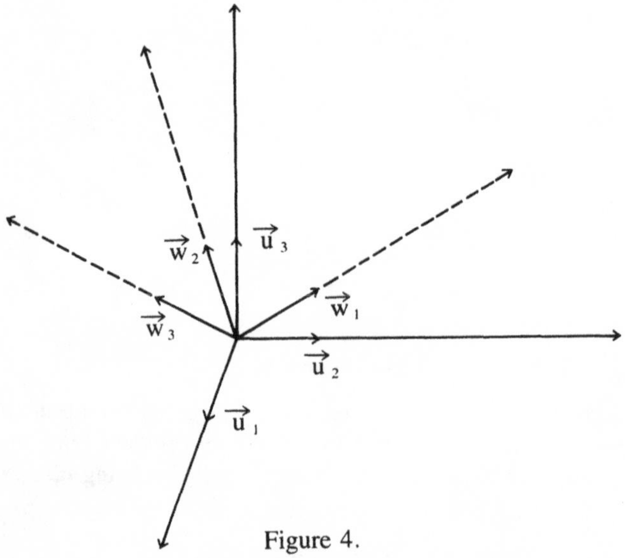

Figure 4.

where (ξ_i) are the components of $\vec{OP}$ in K′ and the dots represent differentiations with respect to time. $\ddot{\xi}_i\vec{w}_i$ is therefore the acceleration $\vec{\alpha}$ of P in K′. Putting $\vec{I} = -(2\dot{\xi}_i\dot{\vec{w}}_i + \xi_i\ddot{\vec{w}}_i)$, we obtain: $\vec{F} = m\vec{\alpha} - m\vec{I}$; hence: $m'\vec{g} + \vec{H} = m\vec{\alpha} - m\vec{I}$, where $\vec{g}$ is the gravitational field at P and $\vec{H}$ the resultant of the remaining forces exerted on P. Since $m = m'$, the last equation can be rewritten as:

(1) $\vec{H} + m(\vec{g} + \vec{I}) = m\vec{\alpha}$

Note that the so-called inertial force $\vec{I}$ is independent of the composition of P. Everything happens in K′ as if an extra inertial field $\vec{I}$ had been added, where $\vec{I}$, exactly like Newtonian gravitation, acts on P proportionally to P's mass (inertial or gravitational). In classical physics, we regard $m\vec{I}$ as an artificial force which can be transformed away by choosing K′ to be an inertial frame, e.g. by making K′ coincide with K. But this option is no longer open to a physicist who has given up the idea of an a priori privileged set of inertial systems. He has now to look upon K′ as a legitimate frame and upon $\vec{I}$ as a real physical field. Note however that, in K′, the fields $\vec{g}$ and $\vec{I}$ cannot be experimentally told apart. By equation (1):

$$\vec{g} + \vec{I} = \vec{\alpha} - \frac{1}{m}\vec{H}.$$

In whatever way we vary the test-particle at P, we shall be able to solve this equation only for the sum ($\vec{g} + \vec{I}$) and not for $\vec{g}$ and $\vec{I}$ separately. If, instead of a mass at O, we had an electric charge creating an electrostatic field $\vec{E}$, then equation (1) would become $\vec{H} + e\vec{E} + m\vec{I} = m\vec{\alpha}$, where e is the charge carried by P. Hence:

(2) $\frac{e}{m}\vec{E} + \vec{I} = \vec{\alpha} - \frac{1}{m}\vec{H}.$

By varying (e/m) through taking different test-particles in turn, we can solve two equations of type (2) in the two unknowns $\vec{E}$ and $\vec{I}$. Thus, $\vec{E}$ and $\vec{I}$ can be told apart in a way in which $\vec{g}$ and $\vec{I}$ cannot.

Since gravitational and inertial fields always occur indissolubly mixed, Einstein typically refused to regard this 'fact' as a mere accident. The division between inertial and strictly gravitational components is artificial; there exists only one global G-field, namely $\vec{G} = \vec{g} + \vec{I}$, which a new theory would have to explain. In view of $\vec{G} = \vec{g} + \vec{I}$, $\vec{I}$ is now a special case of a G-field, namely that variety which can be completely transformed away by a single change of coordinates. In the above example, this corresponds to the case where the mass at O vanishes and the field in

K′ is entirely due to the acceleration of the frame; the field vanishes when we go over from K′ to K.

By a stroke of genius, Einstein put this result to heuristic use as follows. Reducible G-fields can be generated at will by accelerating frames of reference relatively to Minkowski spaces, in which both the equation:

$$\begin{aligned} ds^2 &= (dx^0)^2 - (dx^1)^2 - (dx^2)^2 - (dx^3)^2 \\ &= c^2dt^2 - dx^2 - dy^2 - dz^2 \; [x^0 \underset{\text{Def}}{=} ct] \end{aligned}$$

and the law of inertia hold. In the accelerating frame K′, ds^2 will assume the form: $ds^2 = g_{\mu\nu}dx^\mu dx^\nu$, where the matrix $(g_{\mu\nu})$ is reducible to

$$\begin{bmatrix} 1 & & & 0 \\ & -1 & & \\ & & -1 & \\ 0 & & & -1 \end{bmatrix}$$

by a global transformation of the coordinates. In K′, we can first study behaviour of the G-field generated by the acceleration of the frame, then extend our results to the irreducible case by lifting the restriction initially imposed on the $g_{\mu\nu}$'s [namely the restriction that $(g_{\mu\nu})$ should be reducible to a constant matrix by one global transformation]. But, in order to carry out such a programme, *our results in K′ have to be expressed directly in terms of the $g_{\mu\nu}$'s, i.e. without reference to any special properties which the $g_{\mu\nu}$'s may initially have possessed. This is where mathematics, more particularly the absolute differential calculus, played a crucially important role.*

Using this method, Einstein determined the path followed by a particle falling freely in a gravitational field. In a frame K where no G-field is present, and where $ds^2 = (dx^0)^2 - (dx^1)^2 - (dx^2)^2 - (dx^3)^2$, the trajectory of a free particle is a straight line, whose equations are:

$$d^2x^i/(dx^0)^2 = 0, \text{ for } i = 1, 2, 3.$$

Since our aim is to look at the same particle from an accelerated system, it will be useful to characterise the path in an invariant way. Let us therefore use, instead of x_0, the intrinsic parameter s as independent variable. Since $d^2x^i/(dx^0)^2 = 0$, we have $dx^i/dx^0 =$ constant, hence:

$$\left[\frac{dx^1}{dx^0}\right]^2 + \left[\frac{dx^2}{dx^0}\right]^2 + \left[\frac{dx^3}{dx^0}\right]^2 = k^2 \;(k \text{ being a constant})$$

$$\text{i.e. } (dx^1)^2 + (dx^2)^2 + (dx^3)^2 = k^2(dx^0)^2.$$

$$\text{Thus: } ds^2 = (dx^0)^2 - (dx^1)^2 - (dx^2)^2 - (dx^3)^2 = [1 - k^2]\,(dx^0)^2.$$

Therefore: $ds = h \cdot dx^0$, where $h = \pm\sqrt{1 - k^2}$ = constant.

Since $\frac{d^2x^i}{(dx^0)^2} = 0$, it follows that: $\frac{d^2x^i}{ds^2} = 0$ for all i = 1, 2, 3.

Moreover, $\frac{d^2x^0}{ds^2} = 0$, in view of $ds = h \cdot dx^0$. Hence $\frac{d^2x^\mu}{ds^2} = 0$ for all $\mu = 0, 1, 2, 3$.

$$(3)\quad \begin{cases} \text{The equation } \dfrac{d^2x^\mu}{ds^2} = 0 \text{ gives a stationary value to the integral} \\ \int ds, \text{ because } \delta(\int ds) = 0. \end{cases}$$

Since ds is an invariant, this result applies to all frames of reference. Therefore, in K′, the path of the same particle will be given by $\delta(\int ds) = 0$. This means that in K′, as in any other frame of reference, the path will be a geodesic; provided of course we take the infinitesimal distance to be given by:

(4) $ds^2 = g_{\mu\nu}dx^\mu dx^\nu$.

From the absolute differential calculus we know that a geodesic satisfies the following set of differential equations (see appendix five):

$$(5)\quad \begin{cases} \dfrac{d^2x^\mu}{ds^2} + \left\{ {\mu \atop \alpha\beta} \right\} \cdot \dfrac{dx^\alpha}{ds} \cdot \dfrac{dx^\beta}{ds} = 0, \text{ where} \\ \left\{ {\mu \atop \alpha\beta} \right\} \underset{\text{Def}}{=} g^{\mu\sigma} \cdot [\alpha\beta,\sigma] \underset{\text{Def}}{=} \tfrac{1}{2}\, g^{\mu\sigma}\left[\dfrac{\partial g_{\alpha\sigma}}{\partial x^\beta} + \dfrac{\partial g_{\beta\sigma}}{\partial x^\alpha} - \dfrac{\partial g_{\alpha\beta}}{\partial x^\sigma}\right] \end{cases}$$

Comparing (3) and (5), we conclude that in the accelerating frame the trajectory of the particle is *no longer straight in the ordinary sense of the word* and that $\left\{ {\mu \atop \alpha\beta} \right\}$ is the quantity which deflects the particle from a straight path. Since $\left\{ {\mu \atop \alpha\beta} \right\}$ is a function of the $g_{\alpha\beta}$'s and since we ascribe the deviation from a straight trajectory to the action of a G-field, the latter is represented by the $g\alpha\beta$'s, or rather by the partial derivatives $\frac{\partial g_{\alpha\beta}}{\partial x^\sigma}$; for the quantities $\left\{ {\mu \atop \alpha\beta} \right\}$ vanish if the $g\alpha\beta$'s are constant. The $g\alpha\beta$'s can thus be regarded as gravitational potentials.

So far, we have assumed the G-field to be reducible (inertial, in the old terminology). Applying the method described above, we now generalise our results by extending them to cases where there need not exist a global transformation making the $g_{\alpha\beta}$'s constant throughout the manifold. Thus, even if the field is irreducible, the path of a free particle will be a geodesic and the metric tensor will still represent the gravitational potentials.

In this process of generalisation the importance of the absolute calculus cannot be overestimated. Right from the beginning the path was characterised, in an invariant way, as a curve for which $\delta(\int ds) = 0$. In the differential equation $\frac{d^2x^\mu}{ds^2} + \left\{ {\mu \atop \alpha\beta} \right\} \frac{dx^\alpha}{ds} \cdot \frac{dx^\beta}{ds} = 0$, the Christoffel quantities $\left\{ {\mu \atop \alpha\beta} \right\}$ directly relate to the $g_{\alpha\beta}$'s *without reference to any special properties which the* $g_{\alpha\beta}$'s might have. In other words: Einstein had at his disposal a calculus in which *the functional dependence of the affinities on the metric is exhibited independently of the frame chosen and also of the particular properties of the space under consideration*. Had we started from the equations $\frac{d^2x^i}{dt^2} = 0$, and tried to go over directly to an accelerated frame, it would have been virtually impossible for us to get away from flat space.

The same technique was applied in writing down Maxwell's equations in generally covariant form, then in extending them to cases where the G-field is irreducible (see appendix four). In a Lorentzian frame, the equations for the electromagnetic field are:

$$\text{(6)} \quad \frac{\partial F_{\mu\nu}}{\partial x^\sigma} + \frac{\partial F_{\nu\sigma}}{\partial x^\mu} + \frac{\partial F_{\sigma\mu}}{\partial x^\nu} = 0; \text{ i.e. } F_{[\mu\nu|\sigma]} = 0$$

and

$$\text{(7)} \quad \frac{\partial F^{\mu\nu}}{\partial x^\nu} = S^\mu; \text{ i.e. } F^{\mu\nu}{}_{|\nu} = S^\mu$$

Because $F_{\mu\nu}$ is antisymmetric, equation (6) is already in generally covariant form. As for (7), we notice that, since the $g_{\mu\nu}$'s are constant, the affinities $\left\{ {\mu \atop \alpha\beta} \right\}$ vanish; so the ordinary and covariant derivatives coincide. Hence (7) can be written as:

7'. $F^{\mu\nu}{}_{\|\nu} = S^\mu$ (two strokes denote covariant differentiation),

which is generally covariant. We now assume that, even when the space is curved, (6) and (7') still hold.

One last example illustrating the way in which the principle of equivalence can be heuristically exploited was given by Einstein in his *Die formale Grundlage der allgemeinen Relativitätstheorie* (1914).

If $T^{\mu\nu}$ is the total energy-tensor in flat space, then the conservation laws for energy and momentum can be written as $T^{\mu\nu}{}_{|\nu} = 0$. This equation is equivalent to $T^{\mu\nu}{}_{\|\nu} = 0$ since, in flat space, $\left\{ \begin{matrix} \mu \\ \alpha\beta \end{matrix} \right\} = 0$ for all μ, α, β. But $T^{\mu\nu}{}_{\|\nu} = 0$ is generally covariant; so it still holds when the frame of reference is accelerated. We now go over to the general case by postulating that $T^{\mu\nu}{}_{\|\nu}$ vanishes even when an irreducible field is present, i.e., even when the space ceases to be flat.

We now face the question whether it is really necessary to give a heuristic twist to the equivalence principle. Can it not be treated as an *ordinary* statement which entails an extension of special relativistic laws to cases where an irreducible G-field is present? This seems to be Eddington's position, according to which the transition from $T^{\mu\nu}{}_{|\nu} = 0$ to $T^{\mu\nu}{}_{\|\nu}$ involves an appeal to the principle of equivalence. Eddington's argument runs as follows: take any law like $T^{\mu\nu}{}_{\|\nu} = 0$ which holds in gravitation-free space; at an arbitrary point P of curved space, the field can be transformed away and a geodesic frame chosen in which all the affinities vanish; since the field has disappeared, the relation $T^{\mu\nu}{}_{|\nu} = 0$ holds at P and, because the affinities vanish, $T^{\mu\nu}{}_{|\nu} = T^{\mu\nu}{}_{\|\nu}$; hence $T^{\mu\nu}{}_{\|\nu} = 0$ and this equation, being generally covariant, is true in the original frame; P being an arbitrary point, $T^{\mu\nu}{}_{\|\nu} = 0$ holds throughout the manifold.

Eddington's argument obviously depends on the meaning given to the expression: 'transforming the field away'. We can of course choose a (geodesic) system such that, at P, $(g_{\mu\nu})$ assumes the value

$$\begin{bmatrix} 1 & & & 0 \\ & -1 & & \\ & & -1 & \\ 0 & & & -1 \end{bmatrix}$$

and $g_{\mu\nu|\sigma} = 0$ for all μ, ν, σ. The $g_{\mu\nu}$'s are thus quasi-constant in the neighbourhood of P. However, the *second* derivatives of the $g_{\mu\nu}$'s cannot all be made to vanish at an arbitrary point P; otherwise, the curvature tensor itself would go to zero and, being a tensor, it will vanish in every frame of reference. The manifold would thus have to be flat, which is impossible in the case of an irreducible field. Hence, we cannot say of a geodesic frame that it is strictly Lorentzian and so conclude that the special-relativistic laws must hold at P.

To sum up: by turning the equivalence principle into a heuristic instrument, Einstein solved the problem of constructing a global theory of gravitation. In this connection, two points can be made:

1. If taken to mean that the G-field can be completely transformed away at each point, the principle of equivalence is, strictly speaking, false. It would, if so interpreted, apply only to flat space.

2. However, the principle of equivalence can be heuristically exploited as follows: we start from a Lorentzian reference system, generate a reducible G-field by accelerating the frame, study the properties of such a particular field and finally go over to the case of an irreducible field. *The last step however involves a genuine generalisation.*

§8.4 The Field Equations

So far, we have regarded the $g_{\mu\nu}$'s as given and determined the effects of gravitation on other physical processes, such as the electromagnetic phenomena. The path of a free particle was found to be a geodesic, i.e., a path which 'extremises' the value of $\int ds = \int \sqrt{g_{\mu\nu} dx^\mu dx^\nu}$. The equation of a (non-null) geodesic is:

$$\frac{d^2x^\mu}{ds^2} + \left\{ \begin{matrix} \mu \\ \alpha\beta \end{matrix} \right\} \frac{dx^\alpha}{ds} \cdot \frac{dx^\beta}{ds} = 0$$

If an electromagnetic field is present, the trajectory of a charged particle deviates from a geodesic according to the following equation of motion:

$$-m_0 \left[\frac{d^2x^\mu}{ds^2} + \left\{ \begin{matrix} \mu \\ \alpha\beta \end{matrix} \right\} \frac{dx^\alpha}{ds} \cdot \frac{dx^\beta}{ds} \right] = k\rho_0 \, F^\mu_\nu \frac{dx^\nu}{ds}$$

(k = constant, ρ_0 = proper density of charge) (see Eddington 1923, § 80).

At the beginning of his 1914 paper, Einstein gave these results somewhat in the same way as Newton had proposed his equations of motion before proceeding to 'derive' his inverse-square law. Thus all that remained for Einstein to do was to lay down field equations for the $g_{\mu\nu}$'s; but this was precisely where most of the difficulties arose.

We have already seen how Einstein decided that the gravitational potentials were to coincide with the metric; this resulted from regarding gravitation as due to an acceleration field determined by the geometry of space-time. The principle of equivalence also tells us that the G-field in every frame of reference must be viewed as a real physical field. By Newtonian theory, such a field is determined by certain boundary conditions together with the distribution of mass throughout space. Finally, by special relativity, mass and energy are interchangeable. It follows from the conjunction of all these statements that the gravitational potential, i.e., the metric, must be related to the energy content of space which, in special relativity, is described by a symmetric tensor $T^{\mu\nu}$.

In 1913 Einstein and Grossmann published a joint paper to which the former contributed the first—physical—section and the latter the mathematical section. Einstein proposed to find field equations of the form: $H^{\mu\nu} = -kT^{\mu\nu}$ where $H^{\mu\nu}$ was to be a function of the $g_{\alpha\beta}$'s and of their derivatives. He was guided by the idea that Poisson's equation $\nabla^2\varphi = 4\pi G\rho$ must be yielded as a limiting case of his own solution. From Poisson's equation he drew the following two conclusions of which the second, because it was not explicitly stated but taken for granted, proved an obstacle to further progress:

(a) Like $\nabla^2\varphi$, $H^{\mu\nu}$ was to be of the second order of differentiation in the potential $g_{\mu\nu}$'s.

(b) $\nabla^2\varphi = 4\pi G\rho$ is a single independent equation for a single unknown function φ. Einstein therefore expected his own solution to consist of ten independent equations for the ten functions $g_{\alpha\beta}$; also, given suitable boundary conditions, the $g_{\alpha\beta}$'s were to be uniquely determined.

We shall not go into Einstein's proposed equations; they were later modified. For an understanding of subsequent developments, it is more interesting to concentrate on Einstein's concluding remarks and on an approach initiated by Grossmann. Towards the end of the paper Grossmann writes that, since Riemann and Christoffel, the geometrical properties of space had been known to be closely connected with the behaviour of the curvature tensor $B^{\alpha}{}_{\mu\nu\sigma}$. For example, the vanishing of this tensor constitutes a necessary and sufficient condition for the flatness of space. Consequently, $B^{\alpha}{}_{\mu\nu\sigma} = 0$ is too stringent a constraint in the case of an irreducible field. Looking at Einstein's law $H_{\mu\nu} = -kT_{\mu\nu}$, Grossmann proposes to substitute for $H_{\mu\nu}$ a symmetric tensor $R_{\mu\nu}$ constructed from the $B^{\alpha}{}_{\mu\nu\sigma}$'s alone. In the absolute calculus, a standard way of lowering the rank of tensors is through contraction of certain indices. Note that $B^{\alpha}{}_{\mu\alpha\nu} = -B^{\alpha}{}_{\mu\nu\alpha}$, since $B^{\alpha}{}_{\mu\nu\beta}$ is antisymmetric in α and β. Also, if we contract α and μ, we obtain an expression which vanishes identically (see appendix seven); for:

$$B^{\alpha}{}_{\alpha\nu\beta} = g^{\alpha\sigma}B_{\sigma\alpha\nu\beta} = g^{\sigma\alpha}B_{\sigma\alpha\nu\beta} = -g^{\sigma\alpha}B_{\alpha\sigma\nu\beta} \text{ (by antisymmetry of } B_{\sigma\alpha\nu\beta}$$
$$\text{in } \sigma \text{ and } \alpha) = -B^{\sigma}{}_{\sigma\nu\beta} = -B^{\alpha}{}_{\alpha\nu\beta}.$$

Hence $B^{\alpha}{}_{\alpha\nu\beta} = 0$. Thus, there exists essentially one way of contracting two indices in $B^{\alpha}{}_{\mu\nu\beta}$, namely the contraction of α with β. Hence, one form of the gravitational field equations which readily suggests itself is:

(1) $R_{\mu\nu} = -kT_{\mu\nu}$, where: $R_{\mu\nu} = B^{\alpha}{}_{\mu\nu\alpha}$.

This example illustrates the major heuristic role which mathematics can play in physical discovery (see §1.2). Grossmann starts from a vague physical principle, namely that geometry is connected with the energy content of space; this is then mapped into a mathematical system, namely Riemannian geometry; one thereby obtains the much stronger proposition: $R_{\mu\nu} = -kT_{\mu\nu}$. Paradoxically, Grossmann rejects this relation on the grounds that it does not admit Poisson's equation as limiting case.[4] With hindsight we realise that, although he rightly rejected this law, Grossmann narrowly missed the opportunity of proposing the correct field equations, namely $R_{\mu\nu} - \frac{1}{2} g_{\mu\nu}R = -kT_{\mu\nu}$. The irony is that, for empty space[5]: $(R_{\mu\nu} - \frac{1}{2} g_{\mu\nu}R = 0) \Leftrightarrow (R_{\mu\nu} = 0)$; i.e. $R_{\mu\nu} = 0$ holds, after all, for this special case.

At the end of the paper Einstein tries to establish, by *reductio ad absurdum*, the impossibility of the field being determined by a generally covariant law. In fact he showed, without being fully aware of it, the incompatibility of the following two statements:

(1) The field equations are generally covariant.

(2) Given suitable boundary conditions, the equations completely determine the potentials.

Since he took (2) for granted, he concluded that (1) must necessarily be false. His proof is, roughly, as follows:

Suppose the field equations to be of the form: $E_{\alpha\beta}(g_{\mu\nu}, g_{\mu\nu|\gamma}, g_{\mu\nu|\gamma\sigma}) = 0$, where the left-hand side is a tensor. If, in some frame $(x^1, \ldots, x^N)$, $g_{\mu\nu} = \varphi_{\mu\nu}(x)$ is a solution, then:

$$\bar{g}_{\mu\nu} = \bar{\varphi}_{\mu\nu}(\bar{x}) \underset{\text{Def}}{\equiv} \frac{\partial x^\alpha}{\partial \bar{x}^\mu} \cdot \frac{\partial x^\beta}{\partial \bar{x}^\nu} \cdot \varphi_{\alpha\beta}{}^{(x)}$$

must be the solution in another coordinate system $(\bar{x}^1, \ldots, \bar{x}^N)$, i.e. $[E_{\alpha\beta}(\varphi_{\mu\nu}(x), \varphi_{\mu\nu|\gamma}(x), \varphi_{\mu\nu|\gamma\sigma}(x)) = 0] \Rightarrow [E_{\alpha\beta}(\bar{\varphi}_{\mu\nu}{}^{(\bar{x})}, \bar{\varphi}_{\mu\nu|\gamma}{}^{(\bar{x})}, \bar{\varphi}_{\mu\nu|\gamma\sigma}{}^{(\bar{x})}) = 0]$. However, $E_{\alpha\beta}(\varphi_{\mu\nu}(x), \varphi_{\mu\nu|\gamma}(x), \varphi_{\mu\nu|\gamma\sigma}(x))$ vanishes *identically* for all values of the independent variables $x^1, \ldots, x^N$; so:

$$E_{\alpha\beta}(\varphi_{\mu\nu}{}^{(\bar{x})}, \varphi_{\mu\nu|\gamma}{}^{(\bar{x})}, \varphi_{\mu\nu|\gamma\sigma}{}^{(\bar{x})}) = 0$$

also holds. By the uniqueness of the solution, we must therefore have: $\varphi_{\mu\nu}(\bar{x}) = \bar{\varphi}_{\mu\nu}(\bar{x})$.

The boundary conditions have of course to be suitably chosen in order for the uniqueness condition to apply. Einstein considers a region Ω of space in which the energy tensor vanishes; the coordinate systems are such that

---

[4]In a letter to Sommerfeld, Einstein admits that this was a (technical) mistake (see Einstein 1915e.

[5]Suppose $R_{\mu\nu} - \frac{1}{2} g_{\mu\nu} \cdot R = 0$. Multiplying by $g^{\sigma\mu}$: $R^\sigma{}_\nu - \frac{1}{2} \delta^\sigma_\nu R = 0$. Contracting σ and ν: $0 = R - 2R$; i.e., $R = 0$. Hence, in view of $R_{\mu\nu} - \frac{1}{2} g_{\mu\nu}R = 0$, we have: $R_{\mu\nu} = 0$.

they coincide outside and on the boundary of Ω, but differ within Ω in a continuous way. We have now reached a contradiction for, in general, the two functions $\varphi_{\mu\nu}$ and $\overline{\varphi}_{\mu\nu}$ are different; i.e. $\varphi_{\mu\nu}(\overline{x}) \neq \overline{\varphi}_{\mu\nu}(\overline{x})$ for some $\overline{x}$.

Einstein expresses his, and Grossmann's, regret at having to renounce the covariance condition which they had set out to meet. However, Grossmann's efforts were later to prove very fruitful. He initiated a general method of approaching the field equations via the Riemann and Ricci tensors which are bound up with the geometrical properties of space. To repeat: we have here a situation in which mathematics played an essential heuristic role in the 'logic' of physical discovery. Having once and for all fixed the mathematical framework in which to formulate his theories, the physicist can then let the framework guide him in the choice of the physically relevant quantities and relations. He implicitly relies on the possibility that the mathematics has captured more of the physics than he had consciously intended; i.e. he lets the formalism do the work for him (see section 1 (i) above). In this particular case, after identifying the metric with the gravitational potentials, Grossman and Einstein rely on the mathematics, i.e., on the absolute differential calculus, to tell them which quantities are relevant from the geometrical, hence also from the physical, point of view. (Curvature tensor, Ricci tensor, parallel transport of vectors, etc.)

Between 1913 and 1916 the progress in Einstein's thinking about relativity can be viewed as a gradual and somewhat painful reclaiming of general covariance. Considering that he had no new 'facts' to guide him and had moreover reached a negative conclusion about the covariance principle, it was mainly faith in the correctness of Mach's position that sustained him; more particularly, his Machian belief that no set of reference frames has a privileged status. However, the covariance requirement which he effectively applied was far more stringent than the general relativity principle which proves very weak. He looked for covariant differential equations involving only the $g_{\mu\nu}$'s together with their first and second derivatives.

Qua *physicist*, Mach certainly exerted a considerable influence on Einstein but it was the equivalence principle, rather than Machism, which led Einstein to identify the metric with the field and hence harden the covariance requirement. It was also the use of Poisson's equation as a limiting case which made him search for a solution containing at most the second order derivatives of the $g_{\mu\nu}$'s. Later, we shall see in what sense these requirements uniquely determine the field equations. Hilbert pointed out that, in view of $G^{\mu\nu}{}_{\|\nu} = 0$,[6] the system of equations $R_{\mu\nu} = 0$ is underdetermined. If the field equations are initially satisfied, then the further development of the system is governed by six equations only; this allows for the required arbitrariness

[6] $G^{\mu\nu}$ is Einstein's tensor $R^{\mu\nu} - \frac{1}{2} g^{\mu\nu}R$ (see appendix eight).

in the choice of the coordinate system (four redundant equations corresponding to the choice of four coordinates). This situation is analogous to the case, in elementary algebra, where we have to solve linear equations whose left-hand sides are not independent; eg. $2x + 7y = a$ and $6x + 21y = b$. This system may be impossible (namely if $b \neq 3a$). However, if the system holds for one pair (x_0, y_0), then: $b = 6x_0 + 21y_0 = 3(2x_0 + 7y_0) = 3a$; consequently, one equation is redundant; so there exist infinitely many solutions all governed by $2x + 7y = a$. x can be arbitrarily chosen and y is then given the value $(a - 2x)/7$.

Thus, Einstein's general argument against covariance lost its force. However, there remained the problem of whether the specific equations envisaged by Grossmann, namely $R_{\mu\nu} = -kT_{\mu\nu}$, are *physically* acceptable. In 1915, Einstein published four papers (Einstein 1915a-d) which culminated with the field equations as we know them today; namely: $R_{\mu\nu} - \frac{1}{2} g_{\mu\nu}R = -kT_{\mu\nu}$, or $R_{\mu\nu} = -k(T_{\mu\nu} - \frac{1}{2} g_{\mu\nu}T)$. A key role in this development was played by the mathematical techniques of Riemannian geometry and by the special-relativistic equation: $E = mc^2$.

The first paper, published on November 4, contains a major revision of the basic concepts of Riemannian geometry. The class of allowable frames of reference is narrowed and the definition of covariant divergence altered. The paper is a little bewildering, for not until the end does the author give any clear reason for the drastic steps he takes. Instead of following Einstein's rather confusing order of presentation, I shall carry out a reconstruction which remains nonetheless faithful to the original work.

The main difficulty stems from the fact that, in $R^{\mu\nu} = -kT^{\mu\nu}$, the divergence of $T^{\mu\nu}$ but, in general, not that of $R^{\mu\nu}$ vanishes. $R^{\mu\nu} = -kT^{\mu\nu}$ can alternatively be written as $R^{\mu}_{\nu} = -kT^{\mu}_{\nu}$, or as $R_{\mu\nu} = -kT_{\mu\nu}$.

Contracting μ with ν, we obtain: $R = -kT$. Hence, $(R^{\mu\nu} - \frac{1}{2}g^{\mu\nu}R) = -k(T^{\mu\nu} - \frac{1}{2}g^{\mu\nu}T)$. Taking divergences on both sides, then noting that the divergence of $(R^{\mu\nu} - \frac{1}{2}g^{\mu\nu}R)$ vanishes identically (see appendix eight) while $T^{\mu\nu}_{\ \ \|\nu} = 0$, we have:

$$(g^{\mu\nu}T)_{\|\nu} = 0; \text{ i.e. } g^{\mu\nu}\frac{\partial T}{\partial x^{\nu}} = 0.$$

Multiplying by $g_{\sigma\mu}$:

$$\delta^{\nu}_{\sigma}\frac{\partial T}{\partial x^{\nu}} = 0; \text{ i.e. } \frac{\partial T}{\partial x^{\sigma}} = 0.$$

The last equation holds for all $\sigma = 0, 1, 2, 3$. T must consequently be constant. Einstein knew this was not so in the general case. The equation

$R_{\mu\nu} = -kT_{\mu\nu}$ had therefore to be modified, for a mere limitation on the number of allowable reference frames does not affect the above argument. In the paper, Einstein offers a drastic solution covariant with respect only to a proper subgroup of the set of all possible transformations. $R_{\mu\nu}$ is written as:

(2) $R_{\mu\nu} = K_{\mu\nu} + H_{\mu\nu};$

where:

$$K_{\mu\nu} = \begin{Bmatrix} \alpha \\ \mu\beta \end{Bmatrix} \cdot \begin{Bmatrix} \beta \\ \nu\alpha \end{Bmatrix} - \frac{\partial}{\partial x^{\alpha}}\left[\begin{Bmatrix} \alpha \\ \mu\nu \end{Bmatrix}\right]$$

and:

$$H_{\mu\nu} = \frac{\partial^2}{\partial x^{\mu}\partial x^{\nu}}(\log\sqrt{-g}) - \begin{Bmatrix} \alpha \\ \mu\nu \end{Bmatrix} \cdot \frac{\partial}{\partial x^{\alpha}}(\log\sqrt{-g})$$

Consider the group G of all transformations $\bar{x} = \bar{x}(x_1, \ldots, x^N)$ such that

$$D \underset{\text{Def}}{=} \frac{\partial(\bar{x}^1, \bar{x}^2, \ldots, \bar{x}^N)}{\partial(x^1, x^2, \ldots, x^N)} = 1; \text{ hence:}$$

$$\bar{g} \underset{\text{Def}}{=} det(\bar{g}_{\mu\nu}) = det\left[\frac{\partial x^{\alpha}}{\partial \bar{x}^{\mu}} \cdot \frac{\partial x^{\beta}}{\partial \bar{x}^{\nu}} g_{\alpha\beta}\right] = D^{-2} \cdot det(g_{\alpha\beta}) = det(g_{\alpha\beta}) \underset{\text{Def}}{=} g.$$

It follows that g, hence also $\sqrt{-g}$, are invariants. $\partial(\log\sqrt{-g})/\partial x^{\nu}$ is now a covariant vector and:

$$\left[\frac{\partial \log\sqrt{-g}}{\partial x^{\nu}}\right]_{\|\mu} = \frac{\partial^2}{\partial x^{\mu} \cdot \partial x^{\nu}}(\log\sqrt{-g}) - \begin{Bmatrix} \alpha \\ \mu\nu \end{Bmatrix} \cdot \frac{\partial(\log\sqrt{-g})}{\partial x^{\alpha}} = H_{\mu\nu}$$

is a second order tensor. Since, by (2): $K_{\mu\nu} = R_{\mu\nu} - H_{\mu\nu}$, $K_{\mu\nu}$ is also a tensor, so the relations:

(3) $K_{\mu\nu} = -kT_{\mu\nu},$

which Einstein chose as his field equations, are covariant with respect to G. Because g is an invariant, the definition of the covariant divergence can be simplified as follows: let $A^{\mu\nu}$ be an arbitrary contravariant tensor of the second order.

$$A^{\mu\nu}_{\|\sigma} = \frac{\partial A^{\mu\nu}}{\partial x^{\sigma}} + \begin{Bmatrix} \mu \\ \alpha\sigma \end{Bmatrix} \cdot A^{\alpha\nu} + \begin{Bmatrix} \nu \\ \alpha\sigma \end{Bmatrix} \cdot A^{\mu\alpha}$$

Therefore:

$$A^{\mu\nu}_{\|\nu} = \frac{\partial A^{\mu\nu}}{\partial x\nu} + \begin{Bmatrix} \mu \\ \alpha\nu \end{Bmatrix} \cdot A^{\alpha\nu} + \begin{Bmatrix} \nu \\ \alpha\nu \end{Bmatrix} \cdot A^{\mu\alpha}$$

$$= \frac{\partial A^{\mu\nu}}{\partial x^{\nu}} + \begin{Bmatrix} \mu \\ \alpha\nu \end{Bmatrix} \cdot A^{\alpha\nu} + \frac{\partial \log\sqrt{-g}}{\partial x^{\alpha}} \cdot A^{\mu\alpha}$$

Hence:

$$A^{\mu\nu}_{\|\nu} - \frac{\partial \log\sqrt{-g}}{\partial x^{\alpha}} \cdot A^{\mu\alpha} = \frac{\partial A^{\mu\nu}}{\partial x^{\nu}} + \begin{Bmatrix} \mu \\ \alpha\nu \end{Bmatrix} \cdot A^{\alpha\nu}$$

Since $\frac{\partial \log\sqrt{-g}}{\partial x^{\mu}}$ is a covariant vector, $\frac{\partial \log\sqrt{-g}}{\partial x^{\alpha}} \cdot A^{\mu\alpha}$ is a contravariant one; consequently both sides of the last equation are contravariant vectors. We can therefore define a new covariant divergence $A^{\mu\nu}_{:\nu}$ as follows:

$$A^{\mu\nu}_{:\nu} \underset{\text{Def}}{=} \frac{\partial A^{\mu\nu}}{\partial x^{\nu}} + \begin{Bmatrix} \mu \\ \alpha\nu \end{Bmatrix} \cdot A^{\alpha\nu}$$

It is important to note that the set of reference frames must be such that g assumes a constant value in *none* of them. Otherwise, $\frac{\partial(\log\sqrt{-g})}{\partial x^{\alpha}}$ vanishes and so does $H_{\mu\nu}$, by equation (2); thus the relation $R_{\mu\nu} = K_{\mu\nu} + H_{\mu\nu} = -kT_{\mu\nu}$ collapses into $K_{\mu\nu} = -kT_{\mu\nu}$; and Einstein had already found the first relation to be unacceptable.

Einstein was obviously dissatisfied with this first desperate solution in which the geometrical significance of the matrix $K_{\mu\nu}$ remains unclear. An alternative solution, published on November 11, 1915, consisted in trying to get round the difficulty presented by the non-constant character of $T = T^{\nu}_{\nu}$. One important case in which the trace T vanishes identically is that of the electromagnetic field, where:

$$T^{\mu}_{\nu} = -F^{\mu\sigma}F_{\nu\sigma} + \tfrac{1}{4}\delta^{\mu}_{\nu}F^{\alpha\sigma}F_{\alpha\sigma}$$

(In the case of incoherent matter on which no external forces act, we have:

$$T^{\mu\nu} = \rho_0 \frac{dx^{\mu}}{ds} \cdot \frac{dx^{\nu}}{ds} \text{ and}$$

$$T \underset{\text{Def}}{=} g_{\mu\nu}T^{\mu\nu} = \rho_0, \text{ since } g_{\mu\nu} \cdot dx^{\mu} \cdot dx^{\nu} = ds^2;$$

thus T will, in general, be nonzero.)

At this point, Einstein pinned his hopes on a complete reduction of physics to some—possibly extended—electromagnetic theory in which the trace T of T^{μ}_{ν} would still vanish. Matter would of course have to be explained in purely electromagnetic terms. In other words, this second solution of November 11 consisted in going back to general covariance and accepting the equation $R_{\mu\nu} = -kT_{\mu\nu}$, while requiring that T should vanish. At this stage, Einstein's position closely resembled that of Lorentz who had tried to reduce matter to electromagnetic inertia and thus deprived the electron of its material mass (see chapter two). This corresponds to putting $\rho_0 = 0$ in the energy tensor of incoherent matter. In chapter seven, I argued that one advantage of Einstein over Lorentz was the former's agnosticism about the ultimate nature both of inertia and of the molecular forces. Einstein must have realised how risky it was to tie general relativity to any specific theory about the nature of matter; for on November 25, 1915, he published the paper in which the final breakthrough was achieved. In fact, the article of November 4 already contained the kernel of the definitive solution; and the first hints go back to Einstein's 1914 paper. There, Einstein considered the special relativistic conservation law $\partial T^{\mu\nu}/\partial x^{\nu} = T^{\mu\nu}_{\ |\nu} = 0$ which, on going over to curved space, had to be replaced by $T^{\mu\nu}_{\ \|\nu} = 0$ (or equivalently, $T^{\nu}_{\mu\|\nu} = 0$). He remarked that this last equation no longer constitutes a strict conservation law; in it the gravitational energy does not explicitly appear because it is imbedded in the underlying geometry. Hence arises the problem of determining a gravitational energy matrix t^{ν}_{μ} in such a way that T^{ν}_{μ} and t^{ν}_{μ}, *taken together,* are conserved; i.e. such that:

$$\frac{\partial}{\partial x^{\nu}}(kT^{\nu}_{\mu} + t^{\nu}_{\mu}) = 0.$$

In his paper of November 4, 1915, Einstein used variational methods in order to derive the field equations and thereby identified the quantity t^{ν}_{μ} as (cf. appendix nine):

$$\textbf{(4)}\quad t^{\nu}_{\mu} = \tfrac{1}{2}\delta^{\nu}_{\mu} \cdot g^{\lambda\sigma} \cdot \Gamma^{\alpha}_{\lambda\beta} \cdot \Gamma^{\beta}_{\sigma\alpha} - g^{\lambda\sigma} \cdot \Gamma^{\alpha}_{\lambda\mu} \cdot \Gamma^{\nu}_{\sigma\alpha} \quad \left(\Gamma^{\alpha}_{\beta\gamma} = \left\{ {\alpha \atop \beta\gamma} \right\}\right)$$

The field equations could then be written as:

$$\textbf{(5)}\quad \frac{\partial}{\partial x^{\alpha}}(g^{\sigma\nu} \cdot \Gamma^{\alpha}_{\mu\sigma}) + \tfrac{1}{2}\delta^{\nu}_{\mu} \cdot g^{\gamma\sigma} \cdot \Gamma^{\alpha}_{\gamma\beta} \cdot \Gamma^{\beta}_{\sigma\alpha} = kT^{\nu}_{\mu} + t^{\nu}_{\mu}.$$

Let us now turn to the last of Einstein's 1915 papers (1915d). Again, all frames of reference are to be allowed; so we can choose a coordinate system in which $g = -1$ at all points. Einstein takes $R_{\mu\nu} = 0$ to be the correct law for empty space, because these equations had enabled him, earlier that year, to explain the precession of Mercury's perihelion (Einstein 1915e). Since he no longer wanted to assume $T = 0$, he set out to modify the relation $R_{\mu\nu} = -kT_{\mu\nu}$ but had first to identify where the error lay. Let us note that, since g is constant, $\partial(\log\sqrt{-g})/\partial x^{\alpha} = 0$; hence $H_{\mu\nu} = 0$ and $R_{\mu\nu} = K_{\mu\nu}$ (by equation (2)). Thus the field equation $R_{\mu\nu} = -kT_{\mu\nu}$ reduces to $K_{\mu\nu} = -kT_{\mu\nu}$. This judicious choice of coordinate system means that, from a strictly *technical* viewpoint, we have to deal with the same mathematical relations as in Einstein's (1915a); so the results achieved there can be brought to bear on the present problem.

With some rational reconstruction, Einstein's reasoning seems to have been as follows. The transition from $R_{\mu\nu} = 0$ to $R_{\mu\nu} = -kT_{\mu\nu}$ was made on the assumption that we go from zero to positive energy level. This is mistaken, in view of the law identifying mass with energy ($E = mc^2$). $R_{\mu\nu} = 0$ applies to space empty of all *but* gravitational energy. However, since gravitational energy is present, it contributes to inertial mass and hence, by the equivalence principle, also to gravitational mass. We can roughly say that the field is self-generating in the sense that its energy acts on one of its sources. (This explains the non-linear character of the field equations.) Einstein already had a gravitational energy matrix t^{ν}_{μ}. t^{ν}_{μ} was obtained by means of a standard mathematical method which was classically known to yield conservation laws, namely the variational method (see appendix nine). Einstein considered a frame throughout which $\sqrt{-g} = 1$, i.e. $g = -1$. He then required: $\delta(\int H \cdot dw) = 0$, where: $dw = \sqrt{-g} \cdot dx^0 \cdot dx^1 \cdot dx^2 \cdot dx^3 = dx^0 \cdot dx^1 \cdot dx^2 \cdot dx^3$ and $H = g^{im} \cdot \Gamma^{\nu}_{i\mu} \cdot \Gamma^{\mu}_{m\nu}$. In this way he obtained both the field equations for empty space:

$$-\frac{\partial \Gamma^{\sigma}_{\mu\nu}}{\partial x^{\sigma}} + \Gamma^{j}_{\mu i} \cdot \Gamma^{i}_{\nu j} = \frac{\partial}{\partial x^{\sigma}}\left(\frac{\partial H}{\partial g^{\mu\nu}_{\sigma}}\right) - \frac{\partial H}{\partial g^{\mu\nu}} = 0 \left(g^{\mu\nu}_{\sigma} \underset{\text{Def}}{=} \frac{\partial g^{\mu\nu}}{\partial x^{\sigma}}\right)$$

and the conservation law $t^{\nu}_{\mu|\nu} = 0$ (see appendix nine).

Because of this conservation law, i.e. because of a mathematical equation which usually expresses the constancy of a substance through time, he interpreted the matrix t^{ν}_{μ} as expressing gravitational energy; this, despite the fact that t^{ν}_{μ} is frame-dependent and should therefore represent nothing physically significant in relativity theory.

Let me now describe how Einstein arrived at field equations for the case where non-gravitational forms of energy are present. One method of obtaining new equations is to rewrite $R_{\mu\nu} = 0$ in a way which clearly exhibits the dependence of the field on its own energy, i.e. on t^{ν}_{μ}. We can then generalise the equation by adding kT^{ν}_{μ} to t^{ν}_{μ} (and kT to t); kT^{ν}_{μ} and t^{ν}_{μ} ought to play symmetric roles, which is certainly not the case if we adopt $R_{\mu\nu} = -kT_{\mu\nu}$ as our field equations. This can be seen by contracting ν with μ in (4).

We obtain:

$$t \underset{\text{Def}}{=} t^{\mu}_{\mu} = {}^{4}\!/_{2}\, g^{\lambda\sigma}\Gamma^{\alpha}_{\lambda\beta} \cdot \Gamma^{\beta}_{\sigma\alpha} - g^{\lambda\sigma} \cdot \Gamma^{\alpha}_{\lambda\mu} \cdot \Gamma^{\mu}_{\sigma\alpha} = g^{\lambda\sigma} \cdot \Gamma^{\alpha}_{\lambda\beta} \cdot \Gamma^{\beta}_{\sigma\alpha}.$$

Substituting in (5):

$$\textbf{(5')} \quad \frac{\partial}{\partial x^{\alpha}}(g^{\sigma\nu}\Gamma^{\alpha}_{\mu\sigma}) + {}^{1}\!/_{2}\delta^{\nu}_{\mu} \cdot t = kT^{\nu}_{\mu} + t^{\nu}_{\mu}$$

It is obvious that, in (5′), kT^{ν}_{μ} and t^{ν}_{μ} play totally asymmetric roles. Dissatisfaction with this anomaly started Einstein on his search for different field equations.

In fact, the solution is near at hand if we remember that (5′) is supposed to hold for empty space, i.e., for the case where $T^{\nu}_{\mu} = 0$. Hence:

$$\textbf{(6)} \quad \frac{\partial}{\partial x^{\alpha}}(g^{\sigma\nu} \cdot \Gamma^{\alpha}_{\mu\sigma}) + {}^{1}\!/_{2}\delta^{\nu}_{\mu} \cdot t = t^{\nu}_{\mu}; \text{ i.e.}$$

$$\frac{\partial}{\partial x^{\alpha}}(g^{\sigma\nu} \cdot \Gamma^{\alpha}_{\mu\sigma}) = t^{\nu}_{\mu} - {}^{1}\!/_{2}\delta^{\nu}_{\mu} \cdot t.$$

In this last equation, which is equivalent to $R_{\mu\nu} = 0$, the dependence of the field on gravitational energy is clearly exhibited. For the case where the extra energy represented by T^{ν}_{μ} is present, we generalise (6) by adding kT^{ν}_{μ} to t^{ν}_{μ} and kT to kt, where k is some constant. Thus we obtain:

$$\frac{\partial}{\partial x^{\alpha}}(g^{\sigma\nu} \cdot \Gamma^{\alpha}_{\mu\sigma}) = (t^{\nu}_{\mu} + kT^{\nu}_{\mu}) - {}^{1}\!/_{2}\delta^{\nu}_{\mu}(t + kT).$$

Working back from these equations we arrive at (see appendix nine):

$$\textbf{(7)} \quad \begin{cases} R^{\nu}_{\mu} = -k(T^{\nu}_{\mu} - {}^{1}\!/_{2}\delta^{\nu}_{\mu}T); \text{ i.e.} \\ R_{\mu\nu} = -k(T_{\mu\nu} - {}^{1}\!/_{2}\, g_{\mu\nu}T); \text{ i.e.} \\ R_{\mu\nu} - {}^{1}\!/_{2}g_{\mu\nu}R = -kT_{\mu\nu} \end{cases}$$

These are the field equations as they are generally accepted today.

Two philosophical points can be made. First, this process of discovery illustrates, once again, the crucial role played by '$E = mc^2$' in the genesis of GTR, thus underlining the continuity between the special and the general theories. The second point is that the heuristic role of mathematics, at any rate in the present case of physical discovery, can hardly be exaggerated; the most important step consisted in isolating a gravitational energy matrix

t_μ^ν. Although 'energy' is a physical concept, the fact that t_μ^ν is not a tensor and hence has no invariance properties, makes its physical interpretation highly problematic; its main right to being called an *energy* pseudo-tensor stems from the vanishing of the ordinary divergence $\partial(kT_\mu^\nu + t_\mu^\nu)/\partial x^\nu$, i.e., from a strictly mathematical property. Once t_μ^ν had been singled out, there remained the *mathematical* problem of rewriting $R_{\mu\nu} = 0$ so as to bring out the energy term t_μ^ν; then of generalising the relation thus obtained, by adding kT_μ^ν to t_μ^ν (and kT to t).

An important qualification has to be made about the role of mathematics in scientific discovery. Einstein *could,* after all, have proceeded in the way which many textbooks use for purposes of exposition: realising that the divergence of $R^{\mu\nu}$ does not vanish identically, he might have tried to add extra terms to the left-hand side of $R^{\mu\nu} = -kT^{\mu\nu}$ so as to obtain a divergenceless tensor. Such a method would in fact have been mathematically sound, since $(R^{\mu\nu} - \frac{1}{2}g^{\mu\nu}R + \lambda g^{\mu\nu})$ is the most general divergenceless tensor satisfying the condition of being a function of the $g_{\mu\nu}$'s and of their first and second derivatives alone.[7] That Einstein did not resort to such manipulative tricks confirms the view that purely mathematical gimmickry played no significant part in the development of the relativity programme. Einstein was concerned with the physics and not with the mathematical machinery.

However, having said this, I still maintain that one can identify situations in which mathematics, rather than physical intuition, played a dominant heuristic role. An example will illustrate this distinction. In classical mechanics, Descartes had an intuitive notion of the momentum as a quantity proportional both to velocity and to quantity of matter. Newton, using the same concept of momentum, laid down three postulates from which the conservation law of momentum was deduced. The mathematical formalism captured concepts and relations which could be intuitively grasped; the conservation laws then follow from the axioms connecting the basic concepts and relations. At the present stage of Einstein's programme however, exactly the reverse process took place. Einstein was already in possession of a mathematical system; the system guided him in isolating relevant physical quantities like t_μ^ν, which are *mathematically* (not intuitively) characterised by a conservation law. This example illustrates the second heuristic role which mathematics can play in physical discovery (see §1.2.2). Through physically interpreting the hitherto uninterpreted, or rather the seemingly uninterpretable, mathematical entity t_μ^ν, Einstein was led to a physical discovery. Einstein's position is analogous to Lorentz's, who was led by the mathematical structure of Maxwell's equations to the Lorentz transformation and then to the contraction hypothesis (see chapter two above).

[7]Cartan first established this result for the case where the tensor is linear in the second derivatives. Professor J. Stachel pointed out to me that the linearity can be omitted.

§8.5 THE NEGATIVE HEURISTIC

We have seen that the covariance principle—construed as a metastatement—forms part of the hard core of the relativity programme. There arises the question whether there exists some empirical law also belonging to this hard core. One proposition running through all the later stages of the programme is the law about the interchangeability of mass and energy ($E = mc^2$). We have seen how this relation forced Einstein to abandon special relativity and then, much later, enabled him to alter the equation $R_{\mu\nu} = -kT_{\mu\nu}$ into $R_{\mu\nu} = -k(T_{\mu\nu} - \frac{1}{2}g_{\mu\nu} \cdot T)$. Apart from its technical importance, this law possesses, as already mentioned, a deep metaphysical significance: traditionally, matter had been considered the underlying substance of the world and energy a mode of existence of matter. '$E = mc^2$' shows that matter and energy are continuous with each other, or rather that they constitute one and the same thing. With this new law, we have a complete revision of our way of looking at the physical world. The reason why I hesitate about taking this law to belong to the hard core of the programme is that its importance was fully appreciated only *with hindsight*. Unlike the covariance principle, it was not a central tenet of Einstein's system as presented in the EMB of 1905. Later on in 1905, Einstein realised that the inertia of a body may depend on its energy content and, in 1907, he gave another 'proof' of $E = mc^2$ (see §7.3). It seems therefore that the full significance of this equation was only gradually realised. However, if we allow the negative heuristic to emerge gradually as the programme unfolds, we can look upon $E = mc^2$ as the empirical hard core of relativity; or rather, instead of this precise equation, we take the more qualitative proposition that mass and energy are interchangeable.

§8.6 WEYL, EDDINGTON AND SCHRÖDINGER

Let me now briefly examine various ways in which Einstein's programme was further developed. The following examples are designed to show that the attempt to imbed ever greater parts of physics into 'natural geometry', to construct unified theories, was characteristic both of Einstein's approach and of that of other workers in the programme. This is precisely why one can legitimately refer to a single *programme*. Let us also note that one speaks of 'unified' and not of 'simple' field theories. I have tried above to explain why 'unified' and 'simple' have, to my mind, very different meanings.

Einstein's and Maxwell's equations, when written in generally covariant form, are:

(1) $R_{\mu\nu} - \frac{1}{2}g_{\mu\nu}R = -kT_{\mu\nu}$

$$(2)\quad \begin{cases} F_{[\mu\nu|\sigma]} \underset{\text{Def}}{=} \frac{1}{3}(F_{\mu\nu|\sigma} + F_{\nu\sigma|\mu} + F_{\sigma\mu|\nu}) = 0 \\ F^{\mu\nu}_{\ \ \|\nu} = S^{\mu} \end{cases}$$

where: $R_{\mu\nu}$ is the Ricci tensor, $g_{\mu\nu}$ the fundamental tensor, $T_{\mu\nu}$ the energy-momentum tensor, $F_{\mu\nu}$ the (anti-symmetric) electromagnetic tensor and S^{μ} the source vector.

In this Einstein-Maxwell theory the gravitational and the electromagnetic fields play essentially asymmetric roles. The gravitational potentials constitute the space-time metric. If electromagnetic energy is present, one term occurring in T^{ν}_{μ} is some multiple of ($\frac{1}{4}\delta^{\nu}_{\mu} \cdot F^{\alpha\sigma} \cdot F_{\alpha\sigma} - F^{\nu\sigma} \cdot F_{\mu\sigma}$). Thus, Einstein's equations for the case where only electromagnetic energy is present are obtained by applying a variational principle to the integral

$$(3)\ I = \int [R + \tfrac{1}{2}h\, F^{\alpha\sigma} \cdot F_{\alpha\sigma}] \sqrt{-g} \cdot dx,$$

where: $dx = dx^0 \cdot dx^1 \cdot dx^2 \cdot dx^3$, h = constant, and $R = g^{\alpha\sigma}R_{\alpha\sigma}$.

In view of (1), the electromagnetic field contributes to the geometry but cannot, without further assumptions, be extracted from it. The electromagnetic field remains extraneous to the structure of space-time. Hermann Weyl tried to remedy this defect by constructing a geometry more general than Riemann's. Let us note that the first equation in (2) is equivalent to the existence of a vector potential K_{μ} such that:

$$F_{\mu\nu} = K_{\mu|\nu} - K_{\nu|\mu}.$$

Thus four functions, namely K_0, K_1, K_2 and K_3, completely determine the electromagnetic field. Weyl's solution consists in extending Riemannian geometry so as to include four additional quantities. In Einstein's theory, the gravitational field manifests its presence in that a vector A^{μ}, when taken round an infinitesimal circuit C, changes direction in accordance with the formula (see appendix six):

$$\Delta A^{\mu} = \tfrac{1}{2}B^{\mu}{}_{\alpha\beta\gamma} \cdot A^{\alpha} \cdot H^{\beta\gamma},$$

where: $B^{\mu}{}_{\alpha\beta\gamma}$ = curvature tensor, and

$$H^{\beta\gamma} = \tfrac{1}{2} \oint_C (x^{\gamma} \cdot dx^{\beta} - x^{\beta} \cdot dx^{\gamma}).$$

However, the length of A^{μ} is not altered by the transport. Weyl constructs a new geometry which allows for a change of length given by:

$$(4)\ \frac{\delta\ell}{\ell} = K^*_{\mu} \cdot dx^{\mu};\ \text{i.e. } \delta\ell = \ell \cdot K^*_{\mu} \cdot dx^{\mu},\ \text{where: } \ell^2 = g_{\alpha\beta} \cdot A^{\alpha} \cdot A^{\beta}$$

and K^*_{μ} is some covariant vector which is later identified with the electromagnetic potential. Thus, the change in the length of A^{μ} manifests the presence of the electromagnetic field. By (4), $\delta\ell$ is clearly proportional to ℓ while being independent of the direction of A^{μ}. These restrictions give us exactly the number of extra entities we need, namely the four functions K_i^*, i = 0, 1, 2, 3. The affinity $\Gamma^{\mu}_{\lambda\sigma}$ is now determined by the formula:

(5) $\delta A^{\mu} = -\Gamma^{\mu}_{\lambda\sigma} \cdot A^{\lambda} \cdot dx^{\sigma}$

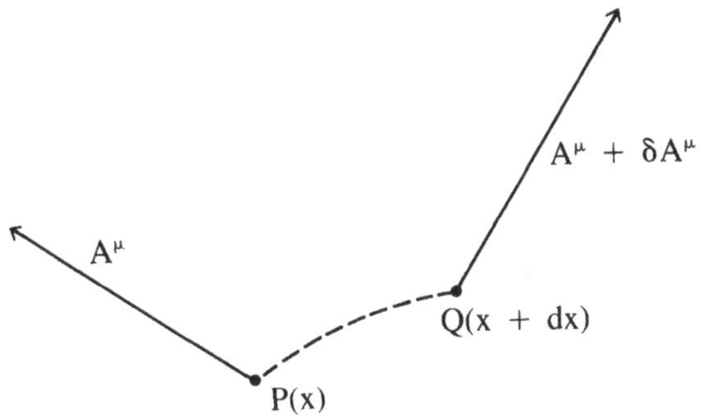

Consider the change in length which occurs when we transport a vector A^{μ} parallel to itself from a point P(x) to a neighbouring point Q(x + dx). Then:

(6) $$\delta\ell^2 = [g_{\alpha\beta}(x + dx) \cdot (A^{\alpha} + \delta A^{\alpha}) \cdot (A^{\beta} + \delta A^{\beta}) - g_{\alpha\beta} \cdot A^{\alpha} \cdot A^{\beta}]$$

$$\doteqdot [(g_{\alpha\beta} + g_{\alpha\beta|\sigma} \cdot dx^{\sigma}) \cdot (A^{\alpha} - \Gamma^{\alpha}_{\mu\nu} \cdot A^{\mu} \cdot dx^{\nu}) \cdot (A^{\beta} - \Gamma^{\beta}_{\sigma\gamma} \cdot A^{\sigma} dx^{\gamma}) - g_{\alpha\beta} \cdot A^{\alpha} \cdot A^{\beta}]$$

$$\doteqdot [g_{\alpha\beta|\gamma} \cdot A^{\alpha} \cdot A^{\beta} \cdot dx^{\gamma} - g_{\sigma\beta}\, \Gamma^{\sigma}_{\alpha\gamma} \cdot A^{\alpha} \cdot A^{\beta} \cdot dx^{\gamma} - g_{\sigma\alpha} \cdot \Gamma^{\sigma}_{\beta\gamma} \cdot A^{\alpha} \cdot A^{\beta} \cdot dx^{\gamma}]$$

(by interchange of dummies and by the symmetry of $g_{\mu\nu}$).

It follows that if = $\Gamma^{\sigma}_{\mu\nu} = \{^{\sigma}_{\mu\nu}\}$, the last member of (6) vanishes, i.e. the vector length is unaltered. This is why Weyl distinguished between the affinity $\Gamma^{\sigma}_{\mu\nu}$ and the Christoffel brackets $\{^{\sigma}_{\mu\nu}\}$.

By (4), $\delta\ell^2$ can also be expressed as follows:

(7) $$\delta\ell^2 = 2\ell \cdot \delta\ell = 2\ell \cdot \ell \cdot K^*_{\gamma} dx^{\gamma} = 2\ell^2 \cdot K^*_{\gamma} \cdot dx^{\gamma}$$

$$= 2g_{\alpha\beta} \cdot A^{\alpha} \cdot A^{\beta} \cdot K^*_{\gamma} \cdot dx^{\gamma}.$$

(Remember: $\ell^2 \underset{\text{Def}}{=} g_{\alpha\beta} A^{\alpha} A^{\beta}$)

Comparing the two expressions for $\delta\ell^2$ and noting that the equality must hold for arbitrary values of A^{μ} and dx^{γ}, we conclude:

(8) $(g_{\alpha\beta|\gamma} - 2g_{\alpha\beta} \cdot K^*_\gamma) - g_{\sigma\beta} \cdot \Gamma^\sigma_{\alpha\gamma} - g_{\sigma\alpha} \cdot \Gamma^\sigma_{\beta\gamma} = 0$

By circular permutation of α, β, γ, we obtain two further equations:

(9) $(g_{\beta\gamma|\alpha} - 2g_{\beta\gamma} \cdot K^*_\alpha) - g_{\sigma\gamma} \cdot \Gamma^\sigma_{\beta\alpha} - g_{\sigma\beta} \cdot \Gamma^\sigma_{\gamma\alpha} = 0$

(10) $(g_{\gamma\alpha|\beta} - 2g_{\gamma\alpha} \cdot K^*_\beta) - g_{\sigma\alpha} \cdot \Gamma^\sigma_{\gamma\beta} - g_{\sigma\gamma} \cdot \Gamma^\sigma_{\alpha\beta} = 0.$

Now take $\Gamma^\alpha_{\beta\gamma}$ to be symmetric in the indices β and γ. Subtracting the last 2 equations from the first one, we have:

$$g_{\sigma\gamma} \cdot \Gamma^\sigma_{\alpha\beta} = [\alpha\beta,\gamma] - (g_{\gamma\alpha} \cdot K^*_\beta + g_{\gamma\beta} \cdot K^*_\alpha - g_{\alpha\beta} \cdot K^*_\gamma)$$

where $[\alpha\beta,\gamma] = \frac{1}{2}(g_{\gamma\alpha|\beta} + g_{\gamma\beta|\alpha} - g_{\alpha\beta|\gamma})$ is the Christoffel symbol of the first kind. Multiplying by $g^{\mu\gamma}$:

(11) $\Gamma^\mu_{\alpha\beta} = \{^{\mu}_{\alpha\beta}\} - g^{\mu\gamma}(g_{\gamma\alpha} \cdot K^*_\beta + g_{\gamma\beta} \cdot K^*_\alpha - g_{\alpha\beta} \cdot K^*_\gamma).$

The affinity is thus uniquely determined and differs from the usual Christoffel quantity $\{^{\mu}_{\alpha\beta}\}$ by the last term. This additional term is responsible for the change in length of a vector transported parallel to itself from one point of the manifold to another.

We now identify K^*_μ with the potential K_μ and claim that the electromagnetic field manifests its presence through the change in length experienced by a vector taken round a circuit. The electromagnetic field is thus imbedded in an extended geometric structure. However, Weyl's solution is a little contrived. The two fields are not properly unified but forcibly brought together. Through conveniently assuming that $\delta\ell/\ell$ is independent of the direction of the transported vector, Weyl extends Riemannian geometry just far enough to obtain the four extra quantities he wants. The unsatisfactory character of this unification of electromagnetism and gravitation is also brought out by Weyl's method of obtaining the field equations, namely through varying the integral:

$$\textbf{(12)}\begin{cases} J = \int(R^2 + kF^{\alpha\gamma}F_{\alpha\gamma}) \cdot \sqrt{-g} \cdot dx, \text{ where: } k = \text{constant,} \\ R = g^{\alpha\gamma}R_{\alpha\gamma},\ F_{\alpha\gamma} = K_{\alpha|\gamma} - K_{\gamma|\alpha} \end{cases}$$

and $R_{\alpha\gamma}$ is the Ricci tensor built out of the affine connections $\Gamma^\mu_{\alpha\gamma}$. Except for the fact that R is replaced by R^2, the two integrals (3) and (12) have identical forms. Before commenting on the choice of the integrand in (12), it is necessary to explain the notion of gauge-invariance which plays a central role in Weyl's theory.

A gauge-transformation consists in a change of the unit of length at each point of space-time. This leads to the equations:

(13) $d\bar{s}^2 = f \cdot ds^2$, where f is some function of the coordinates.

(14) Hence, $\hat{\ell} = \sqrt{f} \cdot \ell$.

It follows from (13) that:

(15) $\hat{g}_{\mu\nu} \cdot dx^\mu \cdot dx^\nu = d\hat{s}^2 = f \cdot ds^2 = f \cdot g^{\mu\nu} \cdot dx^\mu \cdot dx^\nu$.

Thus:

(16) $\hat{g}_{\mu\nu} = f \cdot g_{\mu\nu}$

But we also have, by (4): $\frac{\delta\ell}{\ell} = K_\mu \cdot dx^\mu$, i.e. $\delta(\log\ell) = K_\mu dx^\mu$. Similarly: $\delta(\log\hat{\ell}) = \hat{K}_\mu \cdot dx^\mu$. By (14): $\hat{\ell} = \sqrt{f} \cdot \ell$. Thus:

(17) $\hat{K}_\mu \cdot dx^\mu = \delta(\log\hat{\ell}) = \delta(\log(\ell\sqrt{f})) = \frac{1}{2}\delta(\log f) + \delta(\log\ell)$

$$= \frac{1}{2}\frac{\partial \log f}{\partial x^\mu} \cdot dx^\mu + K_\mu \cdot dx^\mu.$$

Thus:

(18) $\hat{K}_\mu = \frac{1}{2}\frac{\partial \log f}{\partial x^\mu} + K_\mu$.

By (16) and (18), we have that a gauge-transformation consists of:

(19) $\hat{g}_{\mu\nu} = f \cdot g_{\mu\nu}$ and $\hat{K}_\mu = K_\mu + \frac{1}{2}\frac{\partial \log f}{\partial x^\mu}$,

where f is some function of the coordinates. Substituting in (11), we easily verify that:

(20) $\hat{\Gamma}^\mu_{\alpha\beta} = \Gamma^\mu_{\alpha\beta}$

We say of a function that it is gauge-invariant if it is unaffected by any transformation of the form (19). Thus, $\Gamma^\mu_{\alpha\beta}$ is gauge-invariant; this obviously also applies to every entity built out of the $\Gamma^\mu_{\alpha\beta}$'s and of the coordinates. One important problem in Weyl's theory is to determine quantities having invariance properties with respect to both gauge and coordinate transformations; eg. gauge-invariant tensors. With regard to equations (3) and (12), R^2 replaces R because $(R^2 + kF^{\alpha\gamma} F_{\alpha\gamma}) \sqrt{-g}$, but not $(R + \frac{1}{2}h F^{\alpha\gamma} F_{\alpha\gamma}) \cdot \sqrt{-g}$, is gauge-invariant. By the way, it should also be noted that $(R^2 + kF^{\alpha\gamma} F_{\alpha\gamma}) \sqrt{-g}$ is gauge-invariant *only* in a 4-dimensional space. Thus Weyl's theory singles out four-dimensionality; or, more accurately, Weyl's theory explains why our world is four-dimensional, *provided* some rationale is given for the choice of $(R^2 + kF^{\alpha\gamma} F_{\alpha\gamma}) \cdot \sqrt{-g}$ as the appropriate tensor density. Eddington strongly doubted the existence of such a rationale. He remarked that, apart from the analogy between (12) and (3), there is no reason for stringing together two such heterogenous functions as R^2 and $k \cdot F^{\alpha\gamma} F_{\alpha\gamma}$. Thus the choice of $(R^2 + kF^{\alpha\gamma} F_{\alpha\gamma}) \sqrt{-g}$ defeats Weyl's purpose, namely to explain why gravity which, in Einstein's case, is represented by $R_{\alpha\beta}$ should naturally coalesce with electromagnetism, which is represented by $F_{\alpha\beta}$. Eddington was, to my mind, right when he wrote:

> But the connection, though reduced to simpler terms, is not explained in any way by Weyl's action-principle. It is obvious that his action as it stands has no deep significance; it is a mere stringing together of two in-invariants of different forms. To subtract $F_{\mu\nu}F^{\mu\nu}$ from $*G^2$ [i.e. from R^2] is a fantastic procedure which has no more theoretical justification than subtracting E_μ^ν from T_μ^ν. [$E_\mu^\nu = (\frac{1}{4}\delta_\mu^\nu F^{\alpha\beta}F_{\alpha\beta} - F^{\nu\sigma}F_{\mu\sigma})$ and $T_\mu^\nu = -\frac{1}{8}\pi(R_\mu^\nu - \frac{1}{2}\delta_\mu^\nu R + \lambda\delta_\mu^\nu)$]. At the moment we can only regard the assumed form of action A as a step towards some more natural combination of electromagnetic and gravitational variables. (Eddington 1923, p. 212)

Following Weyl, Eddington starts from the postulate that "parallel displacement has some significance in regard to the ultimate structure of the world—it does not much matter what significance" (Eddington 1923, p. 213). He then goes on to set up an affine geometry which *naturally* bifurcates into a gravitational part and an electromagnetic one. What 'naturalness' means in this context will become clear as we go along.

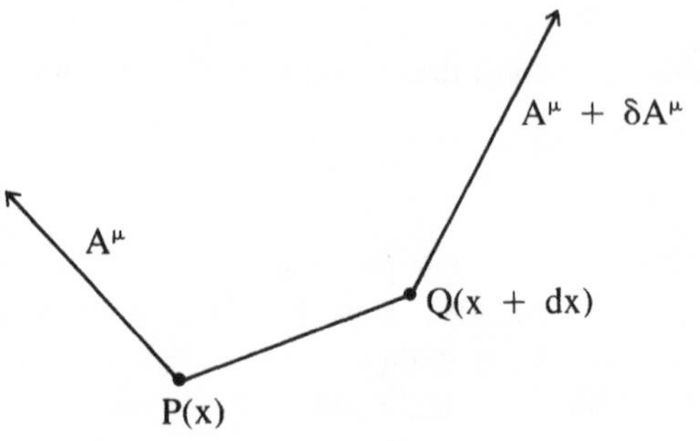

Let us again transport a contravariant vector A^μ from P(x) to Q(x + dx), while constantly keeping A^μ parallel to itself. At Q we thus obtain the vector $A^\mu + \delta A^\mu$, where δA^μ is bilinear and homogenous in A^α and dx^β. Hence: $\delta A^\mu = -\Gamma^\mu_{\alpha\beta} \cdot A^\alpha \cdot dx^\beta$, the $\Gamma^\mu_{\alpha\beta}$'s being functions defined at each point of the manifold.

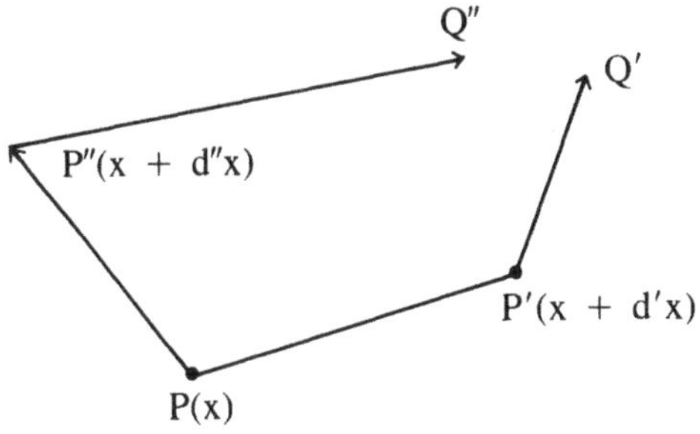

Consider three neighbouring points: P(x), P′(x + d′x) and P″(x + d″x). Let $\overrightarrow{P'Q'}$ and $\overrightarrow{P''Q''}$ be the vectors obtained through parallel transport of $\overrightarrow{PP''}$ and $\overrightarrow{PP'}$ respectively. The components of $\overrightarrow{P''Q''}$ are: $(d'x^{\mu} - \Gamma^{\mu}_{\alpha\beta} \cdot d'x^{\alpha} \cdot d''x^{\beta})$; the coordinates of Q″ relatively to P will thus be: $(d''x^{\mu} + d'x^{\mu} - \Gamma^{\mu}_{\alpha\beta} \cdot d'x^{\alpha} \cdot d''x^{\beta})$. Similarly, the coordinates of Q′ with respect to P are: $(d'x^{\mu} + d''x^{\mu} - \Gamma^{\mu}_{\alpha\beta} d''x^{\alpha} \cdot d'x^{\beta})$. If the parallelogramme law is to hold, Q′ and Q″ must coincide: i.e. $\Gamma^{\mu}_{\alpha\beta} \cdot d'x^{\alpha} \cdot d''x^{\beta} = \Gamma^{\mu}_{\alpha\beta} \cdot d''x^{\alpha} \cdot d'x^{\beta} = \Gamma^{\mu}_{\beta\alpha} \cdot d'x^{\alpha} \cdot d''x^{\beta}$. Since d″x and d′x are arbitrary infinitesimal displacements, $\Gamma^{\mu}_{\alpha\beta} = \Gamma^{\mu}_{\beta\alpha}$; i.e. the affinity must be symmetric in its lower indices. Eddington simply *assumes* the parallelogramme law, pointing out however that such an assumption is by no means physically necessary. This underlines, once again, the importance of mathematical considerations in the construction of physical theories.

Given the significance of parallel transport for the structure of the world, Eddington considers the total variation undergone by a vector A^{μ} as it is taken round a small closed curve about some point P. He finds that:

$$\text{(21)} \quad \begin{cases} \Delta A^{\mu} = \tfrac{1}{2} B^{\mu}{}_{\lambda\sigma\nu} \cdot A^{\lambda} \cdot H^{\sigma\nu}, \text{ where:} \\ \qquad H^{\sigma\nu} = \tfrac{1}{2} \oint (x^{\nu} \cdot dx^{\sigma} - x^{\sigma} \cdot dx^{\nu}), \\ \text{and: } B^{\mu}{}_{\lambda\sigma\nu} = \text{Curvature tensor} \underset{\text{Def}}{=} (-\Gamma^{\mu}_{\lambda\sigma|\nu} - \Gamma^{\tau}_{\lambda\sigma} \cdot \Gamma^{\mu}_{\tau\nu} + \Gamma^{\mu}_{\lambda\nu|\sigma} + \\ \qquad \Gamma^{\tau}_{\lambda\nu} \cdot \Gamma^{\mu}_{\tau\sigma}). \end{cases}$$

In view of (21), the Riemann-Christoffel tensor can be expected to play a fundamental role in general relativistic theories. It is also well known that the properties of a Riemannian space are closely connected with those of $B^{\mu}{}_{\lambda\sigma\nu}$; for example, the vanishing of this tensor is the necessary and sufficient condition for the space to be flat. This is why we look for second-rank tensors built out of the curvature tensor.

The simplest method for finding such quantities consists in contracting μ with one of the indices λ, σ or ν. Since ${B^{\mu}}_{\lambda\sigma\nu}$ is anti-symmetric in σ and ν, ${B^{\mu}}_{\lambda\sigma\nu} = -{B^{\mu}}_{\lambda\nu\sigma}$, so ${B^{\mu}}_{\lambda\sigma\mu} = -{B^{\mu}}_{\lambda\mu\sigma}$. Hence we need only consider ${B^{\mu}}_{\mu\sigma\nu}$ and ${B^{\mu}}_{\lambda\sigma\mu}$. Let ${B^{\mu}}_{\lambda\sigma\mu} \underset{\text{Def}}{=} R_{\lambda\sigma}$ and ${B^{\mu}}_{\mu\sigma\nu} \underset{\text{Def}}{=} H_{\sigma\nu}$. Thus

$$(22)\quad \begin{cases} R_{\lambda\sigma} = B^{\mu}_{\lambda\sigma\mu} = -\Gamma^{\mu}_{\lambda\sigma|\mu} + \Gamma^{\mu}_{\lambda\mu|\sigma} - \Gamma^{\tau}_{\lambda\sigma} \cdot \Gamma^{\mu}_{\tau\mu} + \Gamma^{\tau}_{\lambda\mu} \cdot \Gamma^{\mu}_{\tau\sigma} \text{ and} \\ H_{\sigma\nu} = {B^{\mu}}_{\mu\sigma\nu} = -\Gamma^{\mu}_{\mu\sigma|\nu} + \Gamma^{\mu}_{\mu\nu|\sigma} - \Gamma^{\tau}_{\mu\sigma} \cdot \Gamma^{\mu}_{\tau\nu} + \\ \qquad\qquad \Gamma^{\tau}_{\mu\nu} \cdot \Gamma^{\mu}_{\tau\sigma} = \Gamma^{\mu}_{\mu\nu|\sigma} - \Gamma^{\mu}_{\mu\sigma|\nu} \\ \qquad = \Gamma_{\nu|\sigma} - \Gamma_{\sigma|\nu}, \text{ where} \\ \Gamma_{\alpha} \underset{\text{Def}}{=} \Gamma^{\mu}_{\mu\alpha} = \Gamma^{\mu}_{\alpha\mu}. \end{cases}$$

Now note that, in view of the symmetry of $\Gamma^{\alpha}_{\beta\alpha}$ in β and α, equation (22) entails:

$$(23)\quad R_{\lambda\sigma} - R_{\sigma\lambda} = \Gamma_{\lambda|\sigma} - \Gamma_{\sigma|\lambda}$$

(24) Therefore: $H_{\lambda\sigma} = R_{\sigma\lambda} - R_{\lambda\sigma}$, i.e., the second mode of contraction merely yields the anti-symmetric part of $-2R_{\lambda\sigma}$. We are therefore *mathematically* led to write:

$$(25)\quad \begin{cases} R_{\lambda\sigma} = g^{*}_{\lambda\sigma} + F^{*}_{\lambda\sigma}, \text{ where} \\ g^{*}_{\lambda\sigma} \underset{\text{Def}}{=} \tfrac{1}{2}(R_{\lambda\sigma} + R_{\sigma\lambda}) \text{ and} \\ F^{*}_{\lambda\sigma} \underset{\text{Def}}{=} \tfrac{1}{2}(R_{\lambda\sigma} - R_{\sigma\lambda}). \end{cases}$$

Remembering that the gravitational and electromagnetic fields are described by a symmetric and an antisymmetric tensor ($g_{\lambda\sigma}$ & $F_{\lambda\sigma}$) respectively, we are driven to identify $g^{*}_{\lambda\sigma}$ with $g_{\lambda\sigma}$ and $F^{*}_{\lambda\sigma}$ with $F_{\lambda\sigma}$.

Commenting on Weyl's action-principle, Eddington remarked that a natural invariant density, which does not however single out four-dimensionality, is $\sqrt{-\det[R_{\mu\nu}]}$ (Eddington 1923, p. 206). Schrödinger took up this Eddingtonian suggestion, but he dropped the symmetry condition on $\Gamma^{\alpha}_{\mu\nu}$; he thereby gave himself more latitude in that he now had 64 independent functions at his disposal (Schrödinger 1950, p. 112).

Schrödinger wrote down his field equations in the form:

$$(26)\quad \delta\left(\int\sqrt{-\det[R_{\mu\nu}]} \cdot dx\right) = 0,$$

which is equivalent to:

$$\begin{cases} R_{\mu\nu|\sigma} - R_{\lambda\nu} \cdot {}^*\Gamma^{\lambda}_{\mu\sigma} - R_{\mu\lambda} \cdot {}^*\Gamma^{\lambda}_{\sigma\nu} = 0, \text{ where} \\ {}^*\Gamma^{\alpha}_{\mu\nu} \underset{\text{Def}}{=} \Gamma^{\alpha}_{\mu\nu} + \dfrac{1}{(N-1)}\delta^{\alpha}_{\mu}(\Gamma^{\tau}_{\nu\tau} - \Gamma^{\tau}_{\tau\nu}) \text{ and} \\ N = \text{dimension of space.} \end{cases}$$

Thus, like Eddington, Schrödinger tried to account both for gravitational and for electromagnetic phenomena in terms of a single affine geometry.

The above derivations were given in some detail in order to show that each step in these developments, which covered a period of almost forty years, was largely determined by a priori mathematical considerations.

§8.7 CONCLUSION

Considering the sequence: Newton, Einstein, Weyl, Eddington, Schrödinger and, once again, Einstein, we see that each stage yields a more unified, though not simpler, theory than the previous ones. In Newton's case the background metric is independent of the field. In Einstein's 1916 theory the metric and the gravitational potentials are identified; the electromagnetic field occupies an intermediate position in that it affects the metric but is not imbedded in it. Weyl extends the geometry in order to accommodate both fields; then Eddington sketches the outline of a theory in which an affine space unifies the two fields in a natural way. Finally, Schrödinger starts where Eddington left off. This sequence is intended merely to indicate a movement towards greater integration within the relativistic programme. The impression should not be given that each theory in the series was empirically preferable to the previous ones. In fact, both Weyl's hypothesis and Eddington's approach proved empirically unsuccessful. Moreover, Eddington's and Schrödinger's theories have a disturbing aspect. Einstein wanted both to geometrise physics and to turn geometry into an empirical science. In Weyl's theory, the geometry and the two physical fields are co-extensive in that both are described by fourteen quantities at each point. In Eddington's and Schrödinger's cases we have, respectively, forty and sixty-four independent functions at our disposal, but only fourteen 'physical' variables to account for. It is true that the theories are still empirical in the sense of being experimentally testable; but the distance between hypothesis and observational statements has grown. This, in itself, is not a defect but it makes the independent testability of unified field theories imperative.

The following historical and philosophical conclusions can be drawn:

1. The general covariance principle, together with a compactness requirement, was part of the heuristic which guided the relativity programme after 1908.

2. Since the Lorentz-covariance principle played a similar heuristic role before 1908, an important element of continuity appears between the special and the general theories. It is a dynamic type of continuity, which has to do with a common approach to theory-construction rather than with a fixed set of shared assumptions.

3. 'Compactness', or 'degree of unification', is to be substituted for 'simplicity'. For example, Weyl's theory is not simpler but more unified than the conjunction of Einstein's and Maxwell's field equations.

The following diagram indicates the way in which the development of relativity theory can be accounted for by the methodology of scientific research programmes.

NEGATIVE HEURISTIC (Hard Core)

1. Relativity principle (covariance principle)
2. Interchangeability of mass and energy ($E = mc^2$)

POSITIVE HEURISTIC

STR	GTR
(1) *Lorentz-covariance*. (Einstein applies Lorentz-covariance to Maxwell's equations; Planck applies it to Newton's second law; Einstein to the law of conservation of energy; Lewis and Tolman to the laws of conservation of momentum and energy.)	(1′) *General covariance, together with a unity requirement*. (Einstein and the 'minimal' coupling principle; Weyl's, Eddington's, Schrödinger's and Einstein's attempts to construct unified field theories.)
(2) *Correspondence Principle*. Use of classical laws as limiting cases of new relativistic laws (e.g. use of Newton's second law, of the conservation laws and of Maxwell's	(2′) *Correspondence Principle*. Use of old laws as limiting cases (e.g. Poisson's equation and special relativity).

equations. Maxwell's equations are retained).

(3) Analysis of the well-known 'fact' that electromagnetic induction depends on relative and not on absolute motion.

(3′) Analysis of the well-known 'fact' that all bodies fall with the same acceleration.

(4′) Heuristic function of the principle of equivalence. (An inertial field is nothing but a special, i.e., reducible, case of a gravitational field.)

APPENDIX ONE

TRANSFORMATION EQUATIONS FOR SPECIAL RELATIVISTIC VELOCITY, ACCELERATION AND FORCE

Consider the following Lorentz transformation:

$$(1)\quad \begin{cases} x' = \beta(x - ut),\ y' = y,\ z' = z, t' = \\ \qquad \beta(t - ux/c^2), \text{ where } \beta = (1 - u^2/c^2)^{-1/2}. \end{cases}$$

Therefore:

$$(2)\quad \begin{cases} dx' = \beta(dx - udt),\ dy' = dy,\ dz' = dz \\ \qquad \text{and } dt' = \beta(dt - udx/c^2) \end{cases}$$

Let $\vec{v} = (v_1, v_2, v_3)$ and $\vec{v}' = (v'_1, v'_2, v'_3)$ be the velocities of the same particle referred to the 'stationary' frame K and to the 'moving' frame K′ respectively. Thus:

$$(3)\quad \begin{cases} v_1 = dx/dt,\ v_2 = dy/dt,\ v_3 = dz/dt \text{ and} \\ v'_1 = dx'/dt',\ v'_2 = dy'/dt',\ v'_3 = dz'/dt'. \end{cases}$$

Let $\vec{a} = (a_1, a_2, a_3)$ and $\vec{a}' = (a'_1, a'_2, a'_3)$ be the accelerations of the same particle in K and K′ respectively; i.e.:

$$(4)\quad \begin{cases} a_1 = dv_1/dt,\ a_2 = dv_2/dt,\ a_3 = dv_3/dt \text{ and:} \\ a'_1 = dv'_1/dt',\ a'_2 = dv'_2/dt',\ a'_3 = dv'_3/dt' \end{cases}$$

Let $d\tau^2 = dt^2 - (dx^2 + dy^2 + dz^2)/c^2 = dt^2(1 - v^2/c^2)$; i.e. $d\tau = dt\sqrt{1 - v^2/c^2}$. We know that $d\tau$ is the basic invariant. Hence:

$$(5)\quad d\tau = dt\sqrt{1 - v^2/c^2} = dt'\sqrt{1 - v'^2/c^2}.$$

Thus:

$$(dt'/dt)\sqrt{1 - v'^2/c^2} = \sqrt{1 - v^2/c^2}.$$

By (2) and (3):

$$dt'/dt = \beta\left(1 - \frac{u}{c^2}\frac{dx}{dt}\right) = \beta(1 - uv_1/c^2).$$

Therefore:

$$(6)\begin{cases}\beta(1 - uv_1/c^2).\sqrt{1 - v'^2/c^2} = \sqrt{1 - v^2/c^2};\ \text{i.e.}\ \sqrt{1 - v'^2/c^2} = \\ \dfrac{\sqrt{1 - v^2/c^2}}{\beta(1 - uv_1/c^2)} = \dfrac{\sqrt{1 - u^2/c^2}.\sqrt{1 - v^2/c^2}}{(1 - uv_1/c^2)}\end{cases}$$

(I)
Transformation Equations for the Velocity

$$v'_1 = \frac{dx'}{dt'} = \frac{\beta(dx - udt)}{\beta\left(dt - \frac{u}{c^2}dx\right)} = \frac{\beta((dx/dt) - u)}{\beta\left(1 - \frac{u}{c^2}\frac{dx}{dt}\right)} =$$

$$\frac{v_1 - u}{1 - uv_1/c^2}\ \text{(by (2) \& (3))}$$

Similarly $v'_2 = \dfrac{dy'}{dt'} = \dfrac{dy}{\beta(dt - udx/c^2)} = \dfrac{v_2}{\beta(1 - uv_1/c^2)}$ and $v'_3 =$

$\dfrac{v_3}{\beta(1 - uv_1/c^2)}$. Hence:

$$(7)\ v'_1 = \frac{v_1 - u}{1 - uv_1/c^2};\ v'_2 = \frac{v_2}{\beta(1 - uv_1/c^2};\ v'_3 = \frac{v_3}{\beta(1 - uv_1/c^2)}$$

(II)
Transformation Equations for the Acceleration

Differentiating $v'_1 = (v_1 - u)/(1 - uv_1/c^2)$:

$$dv'_1 = (1 - u^2/c^2)\,dv_1/(1 - uv_1/c^2)^2 = dv_1/\beta^2(1 - uv_1/c^2)^2$$

Dividing by $dt' = \beta(dt - udx/c^2)$ and noting that $dv'_1/dt' = a'_1$, $dv_1/dt = a_1$ and $dx/dt = v_1$, we obtain:

$$a'_1 = a_1/\beta^3(1 - uv_1/c^2)^3.$$

Similarly, differentiating $v'_2 = v_2/\beta(1 - uv_1/c^2)$, we obtain:

$$dv'_2 = \frac{1}{\beta}\left[\frac{dv_2(1 - uv_1/c^2) + v_2udv_1/c^2}{(1 - uv_1/c^2)^2}\right].$$

Dividing by $dt' = \beta(dt - udx/c^2)$:

$$a'_2 = \frac{1}{\beta}\left[\frac{a_2(1 - uv_1/c^2) + v_2ua_1/c^2}{\beta(1 - uv_1/c^2)^3}\right]$$

$$= \frac{1}{\beta^2(1 - uv_1/c^2)^2}\left[a_2 + \frac{a_1uv_2/c^2}{(1 - uv_1/c^2)}\right]. \text{ Therefore:}$$

(8)
$$\begin{cases} a'_1 = \dfrac{a_1}{\beta^3(1 - uv_1/c_2)^3};\ a'_2 = \dfrac{1}{\beta^2(1 - uv_1/c^2)^2}\left[a_2 + \dfrac{a_1uv_2/c^2}{(1 - uv_1/c^2)}\right]; \\ a'_3 = \dfrac{1}{\beta^2(1 - uv_1/c^2)^2}\left[a_3 + \dfrac{a_1uv_3/c^2}{(1 - uv_1/c^2)}\right] \end{cases}$$

(III)
TRANSFORMATION EQUATIONS FOR THE FORCE

We know that $\vec{f} = \dfrac{d\vec{p}}{dt}$ and $\vec{f}' = \dfrac{d\vec{p}'}{dt'}$, where: $\vec{p} = m\vec{v}/\sqrt{1 - v^2/c^2}$, $\vec{p}' = m\vec{v}'/\sqrt{1 - v'^2/c^2}$ and m is the rest mass.

We start by establishing a preliminary result, namely $\vec{f}.\vec{v} = \dfrac{d}{dt}\left(\dfrac{mc^2}{\sqrt{1 - v^2/c^2}}\right)$. In this we assume that m remains constant.

More generally (i.e. when m varies with time)

$$\frac{dE}{dt} = \frac{d}{dt}\left(\frac{mc^2}{\sqrt{1 - v^2/c^2}}\right) = \vec{f}.\vec{v} + c^2\sqrt{1 - v^2/c^2}.\frac{dm}{dt},$$

where: $E = mc^2/\sqrt{1 - v^2/c^2}$.

Remember that $\vec{f} = d[m\vec{v}/\sqrt{1 - v^2/c^2}]/dt$. Integrating twice by parts:

$$\int \vec{f}.\vec{v}.dt = \int \frac{d}{dt}\left(\frac{m\vec{v}}{\sqrt{1 - v^2/c^2}}\right)\cdot \vec{v}\,dt$$

$$= \left[\frac{m\vec{v}.\vec{v}}{\sqrt{1 - v^2/c^2}} - \int \frac{m\vec{v}}{\sqrt{1 - v^2/c^2}}\cdot\frac{d\vec{v}}{dt}\cdot dt\right]$$

$$= \left[\frac{mv^2}{\sqrt{1 - v^2/c^2}} + \frac{c^2}{2}\int \frac{-2\vec{v}.(d\vec{v}/dt)/c^2.m.dt}{\sqrt{1 - v^2/c^2}}\right]$$

$$= \frac{mv^2}{\sqrt{1 - v^2/c^2}} + \frac{c^2}{2}\left(\frac{(1 - v^2/c^2)^{1/2}}{(1/2)} \cdot m - \int \frac{(1 - v^2/c^2)^{1/2}}{(1/2)} \cdot \frac{dm}{dt} \cdot dt\right)$$

$$= \cdot \frac{mc^2}{\sqrt{1 - v^2/c^2}} - c^2 \int \sqrt{1 - v^2/c^2} \cdot \frac{dm}{dt} \cdot dt.$$

Differentiating with respect to t:

$$\vec{f}.\vec{v} = \frac{d}{dt}\left(\frac{mc^2}{\sqrt{1 - v^2/c^2}}\right) - c^2.\sqrt{1 - v^2/c^2} \cdot \frac{dm}{dt}; \text{ i.e.}$$

(9) $$\sigma = \frac{dE}{dt} = \frac{d}{dt}\left(\frac{mc^2}{\sqrt{1 - v^2/c^2}}\right) = \vec{f}.\vec{v} + c^2.\sqrt{1 - v^2/c^2} \cdot \frac{dm}{dt},$$

where σ simply stands for dE/dt. If the rest-mass m remains constant, i.e. if dm/dt = 0, then (9) reduces to

(9′) $$\sigma = \frac{dE}{dt} = \frac{d}{dt}\left(\frac{mc^2}{\sqrt{1 - v^2/c^2}}\right) = \vec{f}.\vec{v} \cdot (m = \text{constant}).$$

Since $d\tau$ is an invariant, $\left(\frac{dt}{d\tau}, \frac{dx}{d\tau}, \frac{dy}{d\tau}, \frac{dz}{d\tau}\right)$ is a four-vector. By (5)

$$\frac{dt}{d\tau} = \frac{1}{\sqrt{1 - v^2/c^2}}; \frac{dx}{d\tau} = \frac{dx}{dt} \cdot \frac{dt}{d\tau} = \frac{v_1}{\sqrt{1 - v^2/c^2}};$$

similarly

$$\frac{dy}{d\tau} = v_2/\sqrt{1 - v^2/c^2}, \frac{dz}{d\tau} = v_3/\sqrt{1 - v^2/c^2}.$$

Hence

$$(1/\sqrt{1 - v^2/c^2}, v_1/\sqrt{1 - v^2/c^2}, v_2/\sqrt{1 - v^2/c^2}, v_3/\sqrt{1 - v^2/c^2})$$

is a four-vector; i.e. $(1/\sqrt{1 - v^2/c^2}, \vec{v}/\sqrt{1 - v^2/c^2})$ is a four-vector. Since m is an invariant: $(m/\sqrt{1 - v^2/c^2}, m\vec{v}/\sqrt{1 - v^2/c^2}) = (E/c^2, \vec{p})$ is a four-vector, where: $E = mc^2/\sqrt{1 - v^2/c^2}$ = energy, and $\vec{p} = m\vec{v}/\sqrt{1 - v^2/c^2}$ = momentum. Hence $\left(\frac{1}{c^2} \cdot \frac{dE}{d\tau}, \frac{d\vec{p}}{d\tau}\right)$ is also a four vector. But:

$$\frac{dE}{d\tau} = \frac{dE}{dt} \cdot \frac{dt}{d\tau} = \frac{\sigma}{\sqrt{1 - v^2/c^2}} \text{ and}$$

$$\frac{d\vec{p}}{d\tau} = \frac{d\vec{p}}{dt} \cdot \frac{dt}{d\tau} = \frac{\vec{f}}{\sqrt{1 - v^2/c^2}}.$$

Thus we have shown that $(\sigma/c^2\sqrt{1 - v^2/c^2}, \vec{f}/\sqrt{1 - v^2/c^2}) = (\sigma/c^2\sqrt{1 - v^2/c^2}, f_1/\sqrt{1 - v^2/c^2}, f_2/\sqrt{1 - v^2/c^2}, f_3/\sqrt{1 - v^2/c^2})$ is a four-vector, i.e. an entity which transforms like the coordinate-differentials. Therefore:

$$f'_1/\sqrt{1 - v'^2/c^2} = \beta[(f_1/\sqrt{1 - v^2/c^2}) - (u\sigma/c^2\sqrt{1 - v^2/c^2})]$$

$$f'_2/\sqrt{1 - v'^2/c^2} = f_2/\sqrt{1 - v^2/c^2}; f'_3/\sqrt{1 - v'^2/c^2} = f_3/\sqrt{1 - v^2/c^2}.$$

By (6) and (9), it follows that:

$$f'_1 = [f_1 - u\sigma/c^2]/(1 - uv_1/c^2) = \left[f_1 - u\left(\vec{f}\cdot\vec{v} + c^2\sqrt{1 - v^2/c^2}\cdot\frac{dm}{dt}\right)\Big/c^2\right]/(1 - uv_1/c^2)$$

$$= \left[f_1 - u\left(f_1v_1 + f_2v_2 + f_3v_3 + c^2\sqrt{1 - v^2/c^2}\cdot\frac{dm}{dt}\right)\Big/c^2\right]/(1 - uv_1/c^2)$$

$$= \left[f_1 - \frac{u}{c^2}\frac{f_2v_2 + f_3v_3}{(1 - uv_1/c^2)} - \frac{u\sqrt{1 - v^2/c^2}\,(dm/dt)}{(1 - uv_1/c^2)}\right]$$

Similarly, by (6) $f'_2 = f_2/\beta(1 - uv_1/c^2)$, $f'_3 = f_3/\beta(1 - uv_1/c^2)$. Hence:

(11)
$$\begin{cases} f'_1 = f_1 - \dfrac{u}{c^2}\left(\dfrac{f_2v_2 + f_3v_3}{1 - uv_1/c^2}\right) - \dfrac{u\sqrt{1 - v^2/c^2}\cdot\dfrac{dm}{dt}}{1 - uv_1/c^2}; \\ f'_2 = \dfrac{f_2}{\beta(1 - uv_1/c^2)}; f'_3 = \dfrac{f_3}{\beta(1 - uv_1/c^2)} \\ \Bigg(\text{If the rest-mass m is constant, then: } dm/dt = 0, \text{ so:} \\ f'_1 = f_1 - \dfrac{u}{c^2}\left(\dfrac{f_2v_2 + f_3v_3}{1 - uv_1/c^2}\right)\Bigg) \end{cases}$$

APPENDIX TWO

MOTION OF A PARTICLE UNDER A POTENTIAL FIELD φ (see Kilmister 1970, p. 45)

Write $x^0 = t$, $x^1 = x$, $x^2 = y$ and $x^3 = z$. Hence:

(1) $ds^2 = c^2dt^2 - dx^2 - dy^2 - dz^2 = c^2d\tau^2$

$= g_{\mu\nu}dx^\mu dx^\nu$; where:

(2) $(g_{\mu\nu}) = \begin{pmatrix} c^2 & & & O \\ & -1 & & \\ & & -1 & \\ O & & & -1 \end{pmatrix}$, hence: $(g^{\mu\nu}) = \begin{pmatrix} 1/c^2 & & & O \\ & -1 & & \\ & & -1 & \\ O & & & -1 \end{pmatrix}$

Since φ is an invariant, $\frac{\partial\varphi}{\partial x^\nu}$ is a covariant vector; therefore: $g^{\mu\nu}\frac{\partial\varphi}{\partial x^\nu}$ is a contravariant vector. By (2):

(3) $g^{\mu\nu}\frac{\partial\varphi}{\partial x^\gamma} = \left(\frac{1}{c^2}\frac{\partial\varphi}{\partial t}, -\frac{\partial\varphi}{\partial x}, -\frac{\partial\varphi}{\partial y}, -\frac{\partial\varphi}{\partial z}\right) = \left(\frac{1}{c^2}\frac{\partial\varphi}{\partial t}, -\nabla\varphi\right)$.

(It can of course be directly verified that $\left(\frac{1}{c^2}\frac{\partial\varphi}{\partial t}, -\nabla\varphi\right)$ is a contravariant vector.)

Let m be the rest mass of the particle in the absence of any potential. If the potential φ is present, the rest-mass increases by φ/c^2 (by the relation: mass = energy/c^2). Therefore:

(4) $\left(\frac{m + \varphi/c^2}{\sqrt{1 - v^2/c^2}}, \frac{(m + \varphi/c^2)\vec{v}}{\sqrt{1 - v^2/c^2}}\right) =$

(contravariant) four-momentum of the particle.

Therefore:

(5) $\left(\frac{d}{d\tau}\left(\frac{m + \varphi/c^2}{\sqrt{1 - v^2/c^2}}\right), \frac{d}{d\tau}\left(\frac{(m + \varphi/c^2)\vec{v}}{\sqrt{1 - v^2/c^2}}\right)\right) =$ four-force.

Remember that $\frac{d}{d\tau} = \frac{d}{dt} \cdot \frac{dt}{d\tau} = \frac{1}{\sqrt{1 - v^2/c^2}} \frac{d}{dt}$.

Equating (3) and (5), we obtain:

$$(6)\quad \begin{cases} \dfrac{1}{c^2}\dfrac{\partial \varphi}{\partial t} = \dfrac{d}{d\tau}\left(\dfrac{m + \varphi/c^2}{\sqrt{1 - v^2/c^2}}\right) = \dfrac{1}{\sqrt{1 - v^2/c^2}} \cdot \dfrac{d}{dt}\left(\dfrac{m + \varphi/c^2}{\sqrt{1 - v^2/c^2}}\right) \\ - \nabla\varphi = \dfrac{d}{d\tau}\left(\dfrac{(m + \varphi/c^2)\vec{v}}{\sqrt{1 - v^2/c^2}}\right) = \dfrac{1}{\sqrt{1 - v^2/c^2}} \cdot \dfrac{d}{dt}\left(\dfrac{(m + \varphi/c^2)\vec{v}}{\sqrt{1 - v^2/c^2}}\right) \end{cases}$$

APPENDIX THREE

ELECTROMAGNETIC INERTIA
(see Lorentz 1906, pp. 26–32, 334–335)

(I)
THE ELECTROMAGNETIC MOMENTUM

We consider a velocity and charge density distribution which give rise to an electromagnetic field in the ether. The field acts back on its sources, thereby altering their motion. We mark out a volume w bounded by a surface σ and we propose to calculate the net force which the field exerts on w. By Lorentz's Law, the force acting on an infinitesimal volume dw where the charge density is ρ is given by:

$$\rho dw\left(\vec{E} + \frac{\vec{v}}{c} \times \vec{H}\right).$$

Hence the total force is:

$$\vec{F} = \iiint \rho\left(\vec{E} + \frac{\vec{v}}{c} \times \vec{H}\right).dw$$

By Maxwell's equations:

$$\nabla.\vec{E} = \rho,\ \nabla.\vec{H}. = O,\ \nabla \times \vec{H} = \frac{1}{c}\left(\frac{\partial \vec{E}}{\partial t} + \rho\vec{v}\right) \text{ and } \nabla \times \vec{E} = -\frac{1}{c}\frac{\partial \vec{H}}{\partial t}.$$

Hence $\rho\frac{\vec{v}}{c} = \nabla \times \vec{H} - \frac{1}{c}\frac{\partial \vec{E}}{\partial t}$. Substituting for ρ and $\rho\frac{\vec{v}}{c}$ in the expression of $\vec{F}$:

$$(1)\ \vec{F} = \iiint \rho\left(\vec{E} + \frac{\vec{v}}{c} \times \vec{H}\right)dw =$$

$$\iiint\left[(\nabla.\vec{E})\vec{E} + (\nabla \times \vec{H}) \times \vec{H} - \frac{1}{c}\frac{\partial \vec{E}}{\partial t} \times \vec{H}\right]dw$$

(2) Define $\vec{P} = c\,(\vec{E} \times \vec{H})$.

Hence:

$$\frac{\partial \vec{P}}{\partial t} = c\left(\frac{\partial \vec{E}}{\partial t} \times \vec{H} + \vec{E} \times \frac{\partial \vec{H}}{\partial t}\right) =$$

$$c\left[\frac{\partial \vec{E}}{\partial t} \times \vec{H} - c\vec{E} \times (\nabla \times \vec{E})\right];$$

since, by Maxwell's equations $\partial \vec{H}/\partial t = -c\,\nabla \times \vec{E}$. Substituting from equation (2) into (1):

$$\textbf{(3)}\quad \vec{F} = -\frac{1}{c^2}\iiint\frac{\partial \vec{P}}{\partial t}\,dw + \iiint[(\nabla.\vec{E})\,\vec{E} +$$

$$(\nabla \times \vec{E}) \times \vec{E}]dw + \iiint[(\nabla \times \vec{H}) \times \vec{H}]dw$$

$$= \vec{F}_1' + \vec{F}_1'' + \vec{F}_2\text{ ; where:}$$

$$\textbf{(4)}\quad \vec{F}_1' = \iiint[(\nabla.\vec{E})\vec{E} + (\nabla \times \vec{E}) \times \vec{E}]dw\ ;$$

$$\vec{F}_1'' = \iiint[(\nabla \times \vec{H}) \times \vec{H}]\,dw = \iiint[(\nabla.\vec{H})\,\vec{H} +$$

$$(\nabla \times \vec{H}) \times \vec{H}]dw \text{ (since } \nabla.\vec{H} = 0\text{) and:}$$

$$\vec{F}_2 = -\frac{1}{c^2}\iiint\frac{\partial \vec{P}}{\partial t}\,dw = -\frac{1}{c^2}\frac{d}{dt}\left(\iiint\vec{P}\,dw\right) =$$

$$-\frac{1}{c}\frac{d}{dt}\left(\iiint\vec{E} \times \vec{H}\,dw\right)$$

(Since $\vec{P} \underset{\text{Def}}{=} c(E \times \vec{H})$).

We write: $\vec{E} = (E_x, E_y, E_z)$, $\vec{H} = (H_x, H_y, H_z)$, $\vec{F}_1' = (F_{1x}', F_{1y}', F_{1z}')$, $\vec{F}_1'' = (F_{1x}'', F_{1y}'', F_{1z}'')$. By the first equation in (4):

$$F_{1x}' = \iiint\left(\frac{\partial E_x}{\partial x} + \frac{\partial E_y}{\partial y} + \frac{\partial E_z}{\partial z}\right)E_x +$$

$$\left(\frac{\partial E_x}{\partial z} - \frac{\partial E_z}{\partial x}\right)E_z - \left(\frac{\partial E_y}{\partial x} - \frac{\partial E_x}{\partial y}\right)E_y\right]dw.$$

Rearranging the terms:

$$\textbf{(5)}\ F'_{1x} = \iiint\left[\left(\frac{\partial E_x^2}{\partial x} + \frac{\partial(E_xE_y)}{\partial y} + \frac{\partial(E_xE_z)}{\partial z}\right) - \frac{1}{2}\frac{\partial}{\partial x}(E_x^2 + E_y^2 + E_z^2)\right]dw$$

$$= \iiint\left[\nabla\cdot(E_x\vec{E}) - \frac{1}{2}\nabla\cdot(E^2\vec{\imath}_x)\right]dw;\ \text{where}$$

$E^2 = |\vec{E}|^2 = E_x^2 + E_y^2 + E_z^2$ and $\vec{\imath}_x = (1,0,0) =$ unit vector along x-axis.

By Gauss's theorem:

$$\textbf{(6)}\ F'_{1x} = \iint_\sigma\left[E_x\vec{E}.\vec{n} - \frac{1}{2}E^2\vec{\imath}_x.\vec{n}\right]d\sigma;$$

where $\vec{n}$ is the outward normal to the enclosing surface σ. Similarly

$$\textbf{(7)}\begin{cases} F'_{1y} = \iint\left[E_y\vec{E}.\vec{n} - \frac{1}{2}E^2\vec{\imath}_y.\vec{n}\right]d\sigma \\ F'_{1z} = \iint\left[E_z\vec{E}.\vec{n} - \frac{1}{2}E^2\vec{\imath}_z.\vec{n}\right]d\sigma \end{cases}$$

Suppose we are dealing with a system in which ρ vanishes outside σ. We want to calculate the total force acting on all the sources of the field, i.e. on all the electrons. It can be shown that, if σ recedes to infinity, the surface integrals in (6) and (7) tend to 0. i.e. $\vec{F}'_1 \to \mathbf{0}$; similarly $\vec{F}''_1 \to \mathbf{0}$. By (3), the net force $\vec{F}$ reduces to:

$$\textbf{(8)}\ \vec{F} = -\frac{1}{c^2}\frac{d}{dt}\left(\iiint\vec{P}\,dw\right) = -\frac{1}{c}\cdot\frac{d}{dt}\left(\iiint\vec{E}\times\vec{H}\,dw\right)$$

where the integration is to be extended to the whole of space.

(II)
SELF-FIELD OF THE ELECTRON

The electron is supposed to be a thin spherical shell of radius R. The charge e is supposed to be evenly distributed on the shell.

a) THE ELECTROSTATIC CASE

Suppose that the electron is at rest and that its centre coincides with the origin of coordinates. Inside the spherical shell which constitutes the electron the potential is constant and so the electric field is zero. Outside the shell the electrostatic potential φ is given by: $\varphi = -\,ke/r = -\,ke\,(x^2 + y^2 + z^2)^{-1/2}$; where k is some constant. Therefore: $\partial\varphi/\partial x = kex\,(x^2 + y^2 + z^2)^{-3/2}$. For reasons which will become clear later we propose to calculate the integral $\int\int\int[(\partial\varphi/\partial x)^2 + (\partial\varphi/\partial y)^2 + (\partial\varphi/\partial z)^2]dw$, where the domain of integration is the whole of space.

$$\left(\frac{\partial\varphi}{\partial x}\right)^2 + \left(\frac{\partial\varphi}{\partial y}\right)^2 + \left(\frac{\partial\varphi}{\partial z}\right)^2 = k^2e^2\,\frac{(x^2 + y^2 + z^2)}{(x^2 + y^2 + z^2)^3} = \frac{k^2e^2}{r^4}$$

where: $\vec{r} = (x,y,z)$, hence $r = (x^2 + y^2 + z^2)^{1/2}$. Using spherical coordinates:

$$\int\int\int\left[\left(\frac{\partial\varphi}{\partial x}\right)^2 + \left(\frac{\partial\varphi}{\partial y}\right)^2 + \left(\frac{\partial\varphi}{\partial z}\right)^2\right]dx\,dy\,dz =$$

$$\int\int\int\frac{k^2e^2}{r^4}\,.\,r^2\sin\theta.dr.d\theta.d\psi$$

$$= \int_R^\infty dr\int_o^{2\pi}d\psi\int_o^\pi\frac{k^2e^2}{r^2}\,.\,\sin\theta.d\theta = \frac{4\pi k^2e^2}{R}\left(= \frac{e^2}{4\pi R}\,,\ \text{if } k = \frac{1}{4\pi}\right)$$

We carried out the integration, keeping in mind that, inside the spherical shell, the potential φ is constant; hence $\partial\varphi/\partial x = \partial\varphi/\partial y = \partial\varphi/\partial z = 0$ for $r < R$. Note that: the electrostatic energy $= \frac{1}{2}\int\int\int E^2dw =$ $\int\int\int\frac{1}{2}(\nabla\varphi)^2.dw = \frac{e^2}{8\pi R}$ if $k = \frac{1}{4\pi}$

b) FIELD OF A MOVING ELECTRON

Consider an electron moving with velocity $\vec{v} = (v,o,o)$ along the x-axis of an inertial frame K. We propose to calculate the force with which the field created by the electron acts on the electron itself. By (I) this force is given by $-\frac{1}{c}\frac{d}{dt}\left(\int\int\int\vec{E}\times\vec{H}dw\right)$ where the integral is extended to the whole of space. Thus we have to determine the electric and the magnetic fields created by a single moving electron. Consider the inertial system K′ whose origin coincides with the centre of the electron. If (t′,x′,y′,z′) are the coordinates in K′, then: $x' = \beta(x - vt)$, $y' = y$, $z' = z$,

$t' = \beta(t - vx/c^2)$, where $\beta = (1 - v^2/c^2)^{-1/2}$. In K′ we have a simple electrostatic field given by: $H'_x = H'_y = H'_z = 0$ and $E'_x = -\partial\varphi'/\partial x'$, $E'_y = -\partial\varphi'/\partial y'$, $E'_z = -\partial\varphi'/\partial z'$ where: $\varphi' = e/4\pi r'$ and $\vec{r}' = (x',y',z')$, $r' = (x'^2 + y'^2 + z'^2)^{1/2}$. Using the transformation equations for the field which Lorentz found in 1904:

$$E_x = E'_x = -\frac{\partial\varphi'}{\partial x'}\ ;\ E_y = -\beta\frac{\partial\varphi'}{\partial y'}\ ;\ E_z = -\beta\frac{\partial\varphi'}{\partial z'}$$

$$H_x = O\ ;\ H_y = \beta\frac{v}{c}\frac{\partial\varphi'}{\partial z'}\ ;\ H_z = -\beta\frac{v}{c}\frac{\partial\varphi'}{\partial y'}\ .$$

Therefore

$$\vec{E} \times \vec{H} = \left(\frac{\beta^2 v}{c}\left[\left(\frac{\partial\varphi'}{\partial y'}\right)^2 + \left(\frac{\partial\varphi'}{\partial z'}\right)^2\right],\ -\frac{\beta v}{c}\cdot\frac{\partial\varphi'}{\partial x'}\cdot\frac{\partial\varphi'}{\partial y'},\ -\frac{\beta v}{c}\cdot\frac{\partial\varphi'}{\partial x'}\cdot\frac{\partial\varphi'}{\partial z'}\right)$$

The first component of $\iiint(\vec{E} \times \vec{H})dw$ is given by:

$$\frac{\beta^2 v}{c}\iiint\left[\left(\frac{\partial\varphi'}{\partial y'}\right)^2 + \left(\frac{\partial\varphi'}{\partial z'}\right)^2\right]dx.dy.dz.$$

Changing to the variables

$x' = \beta(x - vt),\ y' = y,\ z' = z:$

$dx' = \beta dx,\ dy' = dy,\ dz' = dz.$

Hence:

$$\frac{\beta^2 v}{c}\iiint\left[\left(\frac{\partial\varphi'}{\partial y'}\right)^2 + \left(\frac{\partial\varphi'}{\partial z'}\right)^2\right]dx.dy.dz = \frac{\beta v}{c}\iiint\left[\left(\frac{\partial\varphi'}{\partial y'}\right)^2 + \left(\frac{\partial\varphi'}{\partial z'}\right)^2\right]dx'.dy'.dz'$$

Because of the symmetry of φ' in x',y',z', we have:

$$\iiint\left(\frac{\partial\varphi'}{\partial x'}\right)^2 dw' = \iiint\left(\frac{\partial\varphi'}{\partial y'}\right)^2 dw' = \iiint\left(\frac{\partial\varphi'}{\partial z'}\right)^2 dw.$$

Using the results established in a), we conclude:

$$\frac{\beta v}{c}\iiint\left[\left(\frac{\partial\varphi'}{\partial y'}\right)^2 + \left(\frac{\partial\varphi'}{\partial z'}\right)^2\right]\cdot dw' =$$

$$\frac{\beta v}{c} \cdot \frac{2}{3} \iiint \left[\left(\frac{\partial \varphi'}{\partial x'} \right)^2 + \left(\frac{\partial \varphi'}{\partial y'} \right)^2 + \left(\frac{\partial \varphi'}{\partial z'} \right)^2 dw' = \right.$$

$$\frac{2\beta v}{3c} \cdot \frac{e^2}{4\pi R} = \frac{e^2 \beta v}{6\pi R c}$$

This is the first component of $\iiint \vec{E} \times \vec{H} dw$. The second component is:

$$-\frac{\beta v}{c} \iiint \frac{\partial \varphi'}{\partial x'} \cdot \frac{\partial \varphi'}{\partial y'} \cdot dw = -\frac{v}{c} \iiint \frac{\partial \varphi'}{\partial x'} \cdot \frac{\partial \varphi'}{\partial y'} \cdot dw' =$$

$$-\frac{v}{c} \iiint \frac{k^2 e^2 x' y'}{r'^6} dw' = 0$$

[See a) for the expressions of $\partial \varphi' / \partial x'$ and $\partial \varphi' / \partial y'$.] The last integral vanishes because it is the sum of two integrals each of which is equal and opposite to the other: the integral over the region $x' \geqslant 0$ and the integral over the region $x' \leqslant 0$. To each product $x'y'$ in the first region corresponds the product $-x'y'$ in the second region.

Thus the second component, and similarly also the third component, of $\iiint \vec{E} \times \vec{H} dw$ vanishes. Hence:

$$\iiint \vec{E} \times \vec{H} . dw = \left(\frac{e^2 \beta v}{6\pi R c}, 0, 0 \right) = \frac{e^2}{6\pi R c} \beta \vec{v}$$

$$= \frac{e^2}{6\pi R c} \cdot \frac{\vec{v}}{\sqrt{1 - v^2/c^2}}$$

Hence the force acting on the electron is:

$$-\frac{1}{c} \frac{d}{dt} \left(\iiint \vec{E} \times \vec{H} . dw \right) = -\frac{e^2}{6\pi R c^2} \frac{d}{dt} \left(\frac{\vec{v}}{\sqrt{1 - v^2/c^2}} \right)$$

If the electron moves uniformly, i.e. if $\vec{v}$ = constant, the above force vanishes. If the electron accelerates and if we assume that the last equation still holds, then the self-field of the electron brakes the electron with a force equal to:

$$-\frac{d}{dt} \left(\frac{\mu_0 \cdot \vec{v}}{\sqrt{1 - v^2/c^2}} \right), \text{ where:}$$

$$\mu_0 = \frac{e^2}{6\pi R c^2} = \frac{4}{3} \frac{e^2/8\pi R}{c^2} = \frac{4}{3} \frac{\text{electrostatic energy}}{c^2}.$$

APPENDIX FOUR

MAXWELL'S EQUATIONS
(see Adler-Bazin-Schiffer 1965, ch. 4)

(I)

We shall be mainly concerned with an N-dimensional affine space S_N. Let θ be any mapping of $\{1, 2, \ldots, N\}$ into itself. E_θ is defined as follows: $E_\theta = 0$ if θ is not one-one; $E_\theta = 1$ if θ is an even permutation; $E_\theta = -1$ if θ is an odd permutation. The following relations are presupposed: $E_{\theta\varphi} = E_\theta \cdot E_\varphi$ for all mappings θ & φ of $\{1, 2, \ldots, N\}$ into itself; $E_{\psi^{-1}} = E_\psi$ for any permutation ψ.

(II)

Let $A_{\alpha_1 \ldots \alpha_n}$ be any (covariant) tensor in S_N. The tensor $[A_{\alpha_1 \ldots \alpha_n}]$ is defined as follows:

$$[A_{\alpha_1 \ldots \alpha_n}] = \frac{1}{n!} \sum_\theta E_\theta \, A_{\alpha_{\theta(1)} \alpha_{\theta(2)} \ldots \alpha_{\theta(n)}};$$

where θ ranges over all the permutations of $\{1, 2, \ldots, N\}$ (or, equivalently, where θ ranges over all the mappings of $\{1, 2, \ldots, N\}$ into itself; because, if θ is not a permutation, then $E_\theta = 0$)

Consequence: $[A_{\alpha_1 \ldots \alpha_n} + B_{\alpha_1 \ldots \alpha_n}] = \frac{1}{n!}\sum_\theta E_\theta \, (A_{\alpha_{\theta(1)} \ldots \alpha_{\theta(n)}} +$

$$B_{\alpha_{\theta(1)} \ldots \alpha_{\theta(n)}}) = \frac{1}{n!}\sum_\theta E_\theta \, A_{\alpha_{\theta(1)} \ldots \alpha_{\theta(n)}} + \frac{1}{n!}\sum_\theta E_\theta \, B_{\alpha_{\theta(1)} \ldots \alpha_{\theta(n)}} =$$

$$[A_{\alpha_1 \ldots \alpha_n}] + [B_{\alpha_1 \ldots \alpha_n}].$$

(III)

Proposition 1: Let σ be any mapping of {1, 2, . . . , N} into itself. Then

$$[A_{\alpha_{\sigma(1)} \ldots \alpha_{\sigma(n)}}] = E_\sigma\,[A_{\alpha_1 \ldots \alpha_n}].$$

Proof: (i) Suppose σ is a permutation. Hence $E_\sigma = \pm 1$, so $(E_\sigma)^2 = 1$. Therefore:

$$E_{\sigma\theta} \cdot E_\sigma = E_\sigma \cdot E_\theta\, E_\sigma = E_\theta.$$

By definition:

$$[A_{\alpha_{\sigma(1)} \ldots \alpha_{\sigma(n)}}] = \frac{1}{n!} \sum_\theta E_\theta\, A_{\alpha_{\sigma\theta(1)} \ldots \alpha_{\sigma\theta(n)}} =$$

$$\frac{1}{n!} \sum_\theta E_\sigma\, E_{\sigma\theta}\, A_{\alpha_{\sigma\theta(1)} \ldots \alpha_{\sigma\theta(n)}} = E_\sigma \cdot \frac{1}{n!} \sum_\gamma E_\gamma\, A_{\alpha_{\gamma(1)} \ldots \alpha_{\gamma(n)}};$$

where $\gamma = \sigma\theta$; as θ ranges over the set of all permutations of {1, 2, . . . , N}, so does γ; γ assumes the value ψ when θ assumes the value $\sigma^{-1}\psi$. (ii) Suppose σ is not one-one. Hence $E_\sigma = 0$; also there exist elements i and j of {1, 2, . . . , N} such that: $i \neq j$ and $\sigma(i) = \sigma(j)$. Put $\alpha_{\sigma(m)} = \tau_m$ for all m = 1, 2, . . . , n. Hence:

$$A_{\alpha_{\sigma(1)} \ldots \alpha_{\sigma(n)}} = A_{\tau_1 \ldots \tau_n},$$

so

$$[A_{\alpha_{\sigma(1)} \ldots \alpha_{\sigma(n)}}] = [A_{\tau_1 \ldots \tau_n}],$$

where:

$$\tau_i = \alpha_{\sigma(i)} = \alpha_{\sigma(j)} = \tau_j.$$

Let μ be the following permutation (transposition): $\mu(i) = j$; $\mu(j) = i$ and $\mu(m) = m$ for all $m \in \{1, \ldots, N\} \setminus \{i,j\}$. Since $\tau_i = \tau_j$, $A_{\tau_1 \ldots \tau_n} = A_{\tau_{\mu(1)} \ldots \tau_{\mu(n)}}$, so $[A_{\tau_1 \ldots \tau_n}] = [A_{\tau_{\mu(1)} \ldots \tau_{\mu(n)}}]$. But, by (i):

$$[A_{\tau_{\mu(1)} \ldots \tau_{\mu(n)}}] = E_\mu\,[A_{\tau_1 \ldots \tau_n}] = -[A_{\tau_1 \ldots \tau_n}],$$

since, μ being a transposition, $E_\mu = -1$. Thus:

$$[A_{\tau_1 \ldots \tau_n}] = -[A_{\tau_1 \ldots \tau_n}];\ \text{i.e.}\ [A_{\tau_1 \ldots \tau_n}] = 0.$$

Therefore:

$$[A_{\alpha_{\sigma(1)}} \ldots {}_{\alpha_{\sigma(n)}}] = [A_{\tau_1} \ldots {}_{\tau_n}] = 0$$
$$= E_\sigma[A_{\alpha_1} \ldots {}_{\alpha_n}] \text{ (since } E_\sigma = 0)$$

Proposition 2: Let $i \neq j$. Then: (1) If $\alpha_i = \alpha_j$, we have $[A_{\alpha_1} \ldots {}_{\alpha_n}] = 0$. (2) If $A_{\xi_1} \ldots {}_{\xi_n}$ is symmetric in the ith & j^{th} indices, then $[A_{\alpha_1} \ldots {}_{\alpha_n}] = 0$ for any choice of indices $\alpha_1, \ldots, \alpha_n$

Proof: (1) See part (ii) of the above proof. There we showed that, if $\tau_i = \tau_j$ for $i \neq j$, then $[A_{\tau_1} \ldots, {}_{\tau_n}] = 0$. (Note that this result applies to an arbitrary tensor $A_{\xi_1} \ldots, {}_{\xi_n}$ and any choice $\tau_1, \ldots, \tau_n$ of the indices for which $\tau_i = \tau_j$ for some $i \neq j$).

(2) Suppose $A_{\xi_1} \ldots {}_{\xi_i} \ldots {}_{\xi_j} \ldots {}_{\xi_n} = A_{\xi_1} \ldots {}_{\xi_j} \ldots {}_{\xi_i} \ldots {}_{\xi_n}$ for any choice of indices $\xi_1 \ldots \xi_n$.

Let $\alpha_1 \ldots \alpha_n$ be any sequence of indices. To show that $[A_{\alpha_1} \ldots {}_{\alpha_n}] = 0$. By definition:

$$[A_{\alpha_1} \ldots {}_{\alpha_n}] = \frac{1}{n!}\sum_\theta E_\theta A_{\alpha_{\theta(1)}} \ldots {}_{\alpha_{\theta(n)}}.$$

Let μ be the transposition: $\mu(i) = j$; $\mu(j) = i$; and $\mu(m) = m$ for all $m \notin \{i, j\}$. For any permutation θ, $\theta\mu$ is also a permutation. Hence:

$$E_{\theta\mu} A_{\alpha_{\theta\mu(1)}} \ldots {}_{\alpha_{\theta\mu(n)}}$$

occurs in the sum above. But

$$E_{\theta\mu} = E_\theta \cdot E_\mu = -E_\theta$$

since $E_\mu = -1$, μ being a transposition; also, by the symmetry of $A_{\xi_1} \ldots {}_{\xi_n}$ in the ith and j^{th} indices:

$$A_{\alpha_{\theta\mu(1)}} \ldots {}_{\alpha_{\theta\mu(n)}} = A_{\alpha_{\theta(i)}} \ldots {}_{\alpha_{\theta(n)}}.$$

Therefore:

$$E_{\theta\mu} A_{\alpha_{\theta\mu(1)}} \ldots {}_{\alpha_{\theta\mu(n)}} = -E_\theta A_{\alpha_{\theta(1)}} \ldots {}_{\alpha_{\theta(n)}}.$$

In other words: to every term $E_\theta A_{\alpha_{\theta(1)}} \ldots {}_{\alpha_{\theta(n)}}$ corresponds a term $-E_\theta A_{\alpha_{\theta(1)}} \ldots {}_{\alpha_{\theta(n)}}$. Hence:

$$[A_{\alpha_1} \ldots {}_{\alpha_n}] = 0.$$

(IV)

We consider a space S_N in which symmetric affine connections Γ^i_{jm} are defined; i.e. $\Gamma^i_{jm} = \Gamma^i_{mj}$, for all i, j and m.

Theorem 1: $[A_{\alpha_1 \ldots \alpha_n \| \alpha_{n+1}}] = [A_{\alpha_1 \ldots \alpha_n | \alpha_{n+1}}]$.

(where one stroke denotes ordinary differentiation and two strokes covariant differentiation).

Proof: By definition:

$$A_{\alpha_1 \ldots \alpha_n \| \alpha_{n+1}} = (A_{\alpha_1 \ldots \alpha_n | \alpha_{n+1}} - \Sigma \Gamma^{\sigma}_{\alpha_m \alpha_{n+1}} A_{\alpha_1 \ldots \alpha_{m-1} \sigma \alpha_{m+1} \ldots \alpha_n})$$

Therefore:

$$[A_{\alpha_1 \ldots \alpha_n \| \alpha_{n+1}}] = [A_{\alpha_1 \ldots \alpha_n | \alpha_{n+1}}] -$$

$$\Sigma [\Gamma^{\sigma}_{\alpha_m \alpha_{n+1}} A_{\alpha_1 \ldots \alpha_{m-1} \sigma \alpha_{m+1} \ldots \alpha_n}]$$

But, for any choice of $\gamma_1, \ldots, \gamma_n, \gamma_{n+1}$:

$$\Gamma^{\sigma}_{\gamma_m \gamma_{n+1}} = \Gamma^{\sigma}_{\gamma_{n+1} \gamma_m},$$

hence:

$$\Gamma^{\sigma}_{\gamma_m \gamma_{n+1}} A_{\gamma_1 \ldots \gamma_{m-1} \sigma \gamma_{m+1} \ldots \gamma_n} = \Gamma^{\sigma}_{\gamma_{n+1} \gamma_m} A_{\gamma_1 \ldots \gamma_{m-1} \sigma \gamma_{m+1} \ldots \gamma_n}$$

Hence, by Proposition 2(2):

$$[\Gamma^{\sigma}_{\alpha_m \alpha_{n+1}} A_{\alpha_1 \ldots \alpha_{m-1} \sigma \alpha_{m+1} \ldots \alpha_n}] = 0$$

Therefore:

$$[A_{\alpha_1 \ldots \alpha_n \| \alpha_{n+1}}] = [A_{\alpha_1 \ldots \alpha_n | \alpha_{n+1}}].$$

Consequently, since $[A_{\alpha_1 \ldots \alpha_n \| \alpha_{n+1}}]$ is a tensor, so is $[A_{\alpha_1 \ldots \alpha_n | \alpha_{n+1}}]$.

Theorem 2:

$$[\,[A_{\alpha_1 \ldots \alpha_n | \alpha_{n+1}}]_{| \alpha_{n+2}}] = 0$$

for any choice of the indices $\alpha_1, \ldots, \alpha_n, \alpha_{n+1}, \alpha_{n+2}$.

Proof: Let $T_{\alpha_1 \ldots \alpha_n \alpha_{n+1} \alpha_{n+2}}$ stand for $[\,[A_{\alpha_1 \ldots \alpha_n | \alpha_{n+1}}]_{| \alpha_{n+2}}]$

By definition:

$$T_{\alpha_1 \ldots \alpha_{n+2}} = \frac{1}{(n+2)!} \sum_{\sigma} E_{\sigma} \, [A_{\alpha_{\sigma(1)} \ldots \alpha_{\sigma(n)} | \alpha_{\sigma(n+1)}}]_{| \alpha_{\sigma(n+2)}}$$

$$= \frac{1}{(n+2)!} \cdot \frac{1}{(n+1)!} \sum_{\sigma} \sum_{\theta} E_{\sigma} E_{\theta} A_{\alpha_{\sigma\theta(1)} \cdots \alpha_{\sigma\theta(n)}|\alpha_{\sigma\theta(n+1)}|\alpha_{\sigma(n+2)}};$$

where σ and θ range over all the permutations of {1, 2, . . . , n + 2} and {1, 2, . . . , n + 1} respectively. Given θ, let θ′ be the permutation of {1, 2, . . . , n + 1, n + 2} defined by: θ′(q) = θ(q) for all q∈ {1, 2, . . . , n + 1} and θ′(n + 2) = n + 2. Obviously $E_\theta = E_{\theta'}$, because: since n + 2 = θ′(n + 2) = largest element of {1, 2, . . . , n + 1, n + 2}, the two sequences (θ(1), . . . ,θ(n + 1)) and (θ(1), ...θ(n + 1), θ(n + 2) = (θ′(1), . . . , θ′(n + 1), θ′(n + 2)) have the same number of inversions.

By the above, we can express $T_{\alpha_1 \cdots \alpha_{n+2}}$ as follows:

$$T_{\alpha_1 \cdots \alpha_{n+2}} = \frac{1}{(n+2)!} \cdot \frac{1}{(n+1)!} \sum_{\sigma} \sum_{\theta} E_{\sigma} E_{\theta'} A_{\alpha_{\sigma\theta'(1)} \cdots \alpha_{\sigma\theta'(n)}|\alpha_{\sigma\theta'(n+1)}|\alpha_{\sigma\theta'(n+2)}}$$

$$= \frac{1}{(n+1)!} \cdot \frac{1}{(n+2)!} \sum_{\sigma} \sum_{\theta} E_{\sigma\theta'} A_{\alpha_{\sigma\theta'(1)} \cdots \alpha_{\sigma\theta'(n)}|\alpha_{\sigma\theta'(n+1)}|\alpha_{\sigma\theta'(n+2)}}$$

$$= \frac{1}{(n+1)!} \sum_{\theta} \frac{1}{(n+2)!} \sum_{\sigma} E_{\mu} A_{\alpha_{\mu(1)} \cdots \alpha_{\mu(n)}|\alpha_{\mu(n+1)}|\alpha_{\mu(n+2)}}$$

$$= \frac{1}{(n+1)!} \sum_{\theta} [A_{\alpha_1 \cdots \alpha_n|\alpha_{n+1}|\alpha_{n+2}}];$$

where μ = σθ′; as σ ranges over all the permutations of {1, 2, . . . , n + 2} so does σθ′ (for fixed θ); σθ′ assumes the value φ when σ assumes the value $\varphi\theta'^{-1}$, which is always possible because θ′ is a permutation, i.e. θ′ is one-one. Note that θ ranges over all permutations of {1, 2, . . . , n + 1} and that there are (n + 1)! such permutations. Thus:

$$T_{\alpha_1 \cdots \alpha_{n+2}} = \frac{1}{(n+1)!} \sum_{\theta} [A_{\alpha_1 \cdots \alpha_n|\alpha_{n+1}|\alpha_{n+2}}]$$

$$= \frac{(n+1)!}{(n+1)!} [A_{\alpha_1 \cdots \alpha_n|\alpha_{n+1}|\alpha_{n+2}}]$$

$$= [A_{\alpha_1 \cdots \alpha_n|\alpha_{n+1}|\alpha_{n+2}}].$$

For any choice of the indices $\gamma_1, \ldots, \gamma_{n+2}$, we have, putting $B = A_{\gamma_1 \ldots \gamma_n}$:

$$A_{\gamma_1 \ldots \gamma_n|\gamma_{n+1}|\gamma_{n+2}|} = \frac{\partial^2 B}{\partial x^{\gamma_{n+2}} \partial x^{\gamma_{n+1}}} = \frac{\partial^2 B}{\partial x^{\gamma_{n+1}} \partial x^{\gamma_{n+2}}}$$
$$= A_{\gamma_1 \ldots \gamma_n|\gamma_{n+2}|\gamma_{n+1}}$$

By Proposition 2(2):

$$[A_{\alpha_1 \ldots \alpha_n|\alpha_{n+1}|\alpha_{n+2}}] = 0$$

for any choice of the indices $\alpha_1 \ldots \alpha_{n+2}$.
Therefore:

$$[\,[A_{\alpha_1 \ldots \alpha_n|\alpha_{n+1}}]_{|\alpha_{n+2}}] = T_{\alpha_1 \ldots \alpha_{n+2}} = [A_{\alpha_1 \ldots \alpha_n|\alpha_{n+1}|\alpha_{n+2}}] = 0$$

(V)

The main aim of this section is to establish that, if t_{ij} is an antisymmetric tensor such that $[t_{ij|\ell}] = 0$ for all choices of the indices i, j and ℓ, then there exists a tensor T_i such that $t_{ij} = T_{i|j} - T_{j|i}$. Note that the converse of this proposition is true by the above theorem: for let

$$t_{ij} = T_{i|j} - T_{j|i} = 2[T_{i|j}]; \text{ then } [t_{ij|\ell}] = 2[[T_{i|j}]_{|\ell}] = 0$$

Proposition 3. Let t_{ij} be antisymmetric tensor. Then:

$$[t_{ij|\ell}] = \frac{1}{3}(t_{ij|\ell} + t_{j\ell|i} + t_{\ell i|j})$$

Proof: Since t_{ij} is anti-symmetric: $t_{pq} = -t_{qp}$ for all p & q. By definition of $[t_{ij|\ell}]$:

$$[t_{ij|\ell}] = \frac{1}{3!}(t_{ij|\ell} + t_{j\ell|i} + t_{\ell i|j} - t_{ji|\ell} - t_{\ell j|i} - t_{i\ell|j})$$

But

$$t_{ji|\ell} = \frac{\partial t_{ji}}{\partial x^\ell} = -\frac{\partial t_{ij}}{\partial x^\ell} = -t_{ij|\ell}, \text{ since } t_{ji} = -t_{ij}.$$

Similarly: $t_{\ell j|i} = -t_{j\ell|i}$ and $t_{i\ell|j} = -t_{\ell i|j}$. Substituting in the expression of $[t_{ij|\ell}]$:

$$[t_{ij|e}] = \frac{1}{3!}(2t_{ij|e} + 2t_{je|i} + 2t_{ei|j})$$

$$= \frac{1}{3}(t_{ij|e} + t_{je|i} + t_{ei|j})$$

Lemma: Let $b_1, b_2, \ldots, b_n$ be functions of the variables $x^1, \ldots, x^n$, such that: $b_{i|j} = b_{j|i}$ for all i & j in $\{1, 2, \ldots, n\}$. Then there exists a function F such that $b_i = F_{|i} = \frac{\partial F}{\partial x^i}$ for all $i \in \{1, \ldots, n\}$.

Proof: By induction on n. Suppose the lemma holds for n. Let $b_1, \ldots, b_n, b_{n+1}$ be functions of $x^1, \ldots, x^n, x^{n+1}$ such that:

$$0 = b_{i|j} - b_{j|i} \equiv \frac{\partial b_i}{\partial x^j} - \frac{\partial b_j}{\partial x^i}$$

for all i & j in $\{1, \ldots, n, n+1\}$. Since $0 = b_{p|q} - b_{q|p}$ for all p and q in $\{1, 2, \ldots, n\}$, we conclude by our induction hypothesis that there exists a function H such that $b_m = \partial H/\partial x^m$ for all $m \in \{1, \ldots, n\}$. H will of course also depend on the 'parameter' x^{n+1} which occurs in b_m. But we also have by hypothesis $\partial b_{n+1}/\partial x^i = \partial b_i/\partial x^{n+1}$ for all $i \in \{1, \ldots, n\}$. Since $b_i = \partial H/\partial x^i$, we can write:

$$\frac{\partial b_{n+1}}{\partial x^i} = \frac{\partial b_i}{\partial x^{n+1}} = \frac{\partial}{\partial x^{n+1}}\left(\frac{\partial H}{\partial x^i}\right) = \frac{\partial}{\partial x^i}\left(\frac{\partial H}{\partial x^{n+1}}\right).$$

In other words:

$$\frac{\partial}{\partial x^i}\left(b_{n+1} - \frac{\partial H}{\partial x^{n+1}}\right) = 0$$

for all $i \in \{1, \ldots, n\}$. Hence:

$$\left(b_{n+1} - \frac{\partial H}{\partial x^{n+1}}\right)$$

is independent of $x^1, \ldots, x^n$. We can therefore write:

$$b_{n+1} - \frac{\partial H}{\partial x^{n+1}} = k(x^{n+1}) = \frac{dK}{dx^{n+1}},$$

where $k(x^{n+1})$ is a function of x^{n+1} and $K = \int k(x^{n+1})dx^{n+1}$. Therefore:

$$b_{n+1} = \frac{\partial H}{\partial x^{n+1}} + \frac{dK}{dx^{n+1}} = \frac{\partial}{\partial x^{n+1}}(H + K).$$

On the other hand, for $i \in \{1, 2, \ldots, n\}$, since K is independent of x^i:

$$\frac{\partial}{\partial x^i}(H + K) = \frac{\partial H}{\partial x^i} + \frac{\partial K}{\partial x^i} = \frac{\partial H}{\partial x^i} + 0 = b_i + 0 = b_i.$$

Taking $F = H+K$, we conclude: for any $q \in \{1, 2, \ldots, n, n + 1\}$,

$$\frac{\partial F}{\partial x^q} = \frac{\partial H}{\partial x^q} + \frac{\partial K}{\partial x^q} = b_q.$$

This completes our proof by induction.

Theorem 3: Let t_{ij} be any antisymmetric tensor in an n-dimensional space such that $[t_{ij|\ell}] = 0$ for all i, j and ℓ. Then there exists a tensor T_i such that:

$$t_{ij} = T_{i|j} - T_{j|i} = 2[T_{i|j}].$$

Proof: By induction on the dimension n. Suppose Theorem 3 holds for n. Let t_{ij} be an antisymmetric tensor in an (n + 1)-dimensional space $(x^1, \ldots, x^{n+1})$. Suppose further that $[t_{ik|\ell}] = 0$; i.e. by Proposition 3:

$$\textbf{(1)}\ 0 = (t_{ik|\ell} + t_{k\ell|i} + t_{\ell i|k}) = \left(\frac{\partial t_{ik}}{\partial x^\ell} + \frac{\partial t_{k\ell}}{\partial x^i} + \frac{\partial t_{\ell i}}{\partial x^k}\right).$$

Keep x^{n+1} fixed for the time being and consider t_{ik} for all i and k in $\{1, 2, \ldots, n\}$. By our induction hypothesis, there exist functions $T_1, \ldots, T_n$ such that:

$$\textbf{(2)}\ t_{ik} = T_{i|k} - T_{k|i} = \frac{\partial T_i}{\partial x^k} - \frac{\partial T_k}{\partial x^i};\ \text{for all } i \ \&\ k \text{ in } \{1, \ldots n\}.$$

To complete the induction we have to prove that there exists a function T_{n+1} such that: $t_{i(n+1)} = T_{i|(n+1)} - T_{(n+1)|i}$ for all $i \in \{1, \ldots, n\}$. Note that, since t_{pq} is antisymmetric, then automatically: $t_{(n+1)(n+1)} = 0 = T_{(n+1)|(n+1)} - T_{(n+1)|(n+1)}$.

Taking $\ell = (n + 1)$ in 1:

$$\textbf{(3)}\ \frac{\partial t_{ik}}{\partial x^{n+1}} + \frac{\partial t_{k(n+1)}}{\partial x^i} + \frac{\partial t_{(n+1)i}}{\partial x^k} = 0$$

But: $t_{k(n+1)} = -t_{(n+1)k}$ since t_{ij} is antisymmetric and $t_{ik} = (\partial T_i/\partial x^k) - (\partial T_k/\partial x^i)$ by 2. Substituting in 3:

$$\frac{\partial^2 T_i}{\partial x^{n+1}\partial x^k} - \frac{\partial^2 T_k}{\partial x^{n+1}\partial x^i} - \frac{\partial t_{(n+1)k}}{\partial x^i} + \frac{\partial t_{(n+1)i}}{\partial x^k} = 0$$

This can be written as:

$$(4)\quad \frac{\partial}{\partial x^k}\left(\frac{\partial T_i}{\partial x^{n+1}} + t_{(n+1)i}\right) - \frac{\partial}{\partial x^i}\left(\frac{\partial T_k}{\partial x^{n+1}} + t_{(n+1)k}\right) = 0;$$

or:

$$(5)\quad \frac{\partial b_i}{\partial x^k} - \frac{\partial b_k}{\partial x^i} = 0, \text{ where } b_i = \frac{\partial T_i}{\partial x^{n+1}} + t_{(n+1)i} \text{ for all } i \in \{1, 2, \ldots, n\}$$

By the previous lemma, there exists a function T_{n+1} of $x^1, \ldots, x^n$ and of the 'parameter' x^{n+1} such that:

$$(6)\quad \frac{\partial T_i}{\partial x^{n+1}} + t_{(n+1)i} = b_i = \frac{\partial T_{n+1}}{\partial x^i}; \text{ in other words:}$$

$$t_{(n+1)i} = \frac{\partial T_{n+1}}{\partial x^i} - \frac{\partial T_i}{\partial x^{n+1}}.$$

By the antisymmetry of t_{pq}:

$$(7)\quad t_{i(n+1)} = -t_{(n+1)i} = \frac{\partial T_i}{\partial x^{n+1}} - \frac{\partial T_{n+1}}{\partial x^i}$$

Now we let $T_1, \ldots, T_n, T_{n+1}$ transform like the components of a covariant vector. Equations 2, 6 and 7 are now tensor equations which, because they hold in one coordinate system, automatically hold in all coordinate systems.

This completes our proof by induction.

(VI)
Maxwell's Equations in Generally Covariant Form

Choose an inertial frame and put $x^0 = ct$, $x^1 = x$, $x^2 = y$ and $x^3 = z$. $ds^2 = c^2dt^2 - dx^2 - dy^2 - dz^2 = (dx^0)^2 - (dx^1)^2 - (dx^2)^2 - (dx^3)^2$. For a mobile moving with velocity $\vec{v}$: $ds^2 = c^2dt^2(1 - (dx^2 + dy^2 + dz^2)/(c^2dt^2)) = c^2dt^2(1 - v^2/c^2)$ Hence:

$$(1)\quad ds = cdt\sqrt{1 - v^2/c^2}; \text{ or } ds/dt = c\sqrt{1 - v^2/c^2}$$

Maxwell's equations can be divided into 2 sets. The so-called source equations are:

$$(2)\quad \nabla \times \vec{H} = \frac{1}{c}\frac{\partial \vec{E}}{\partial t} + \frac{1}{c}\vec{j} = \frac{1}{c}\left(\frac{\partial \vec{E}}{\partial t} + \rho\vec{v}\right);$$

$$\nabla . \vec{E} = p;\ \text{where}\ \vec{j} = \rho \vec{v}.$$

The internal equations are:

(3) $\nabla \times \vec{E} = -\frac{1}{c}\frac{\partial \vec{H}}{\partial t};\ \nabla . \vec{H} = \mathbf{0}$. Note that:

(4) $\rho = \rho_0 / \sqrt{1 - v^2/c^2}$;

where p_0 is an invariant; p_0 is in fact the charge density in the rest frame of the point (ct, x, y, z). We now define a contravariant vector S^σ as follows:

$$\textbf{(5)}(S^\sigma) = \rho_0\left(\frac{dx^0}{ds}, \frac{dx^1}{ds}, \frac{dx^2}{ds}, \frac{dx^3}{ds}\right) = \rho^0\frac{dt}{ds}\left(\frac{d(ct)}{dt}, \frac{dx}{dt}, \frac{dy}{dt}, \frac{dz}{dt}\right)$$

$$= \frac{\rho_0}{c\sqrt{1 - v^2/c^2}}(c, \vec{v}) = \frac{\rho}{c}(c, \vec{v}) = (\rho, \frac{\rho}{c}\vec{v}) = (\rho, \frac{1}{c}\vec{j})$$

Define the contravariant tensor $F^{\mu\nu}$ as follows:

$$\textbf{(6)}\ (F^{\mu\nu}) = \begin{pmatrix} 0 & E_x & E_y & E_z \\ -E_x & 0 & H_z & -H_y \\ -E_y & -H_z & 0 & H_x \\ -E_z & H_y & -H_x & 0 \end{pmatrix}$$

where: $\vec{E} = (E_x, E_y, E_z)$ and $\vec{H} = (H_x, H_y, H_z)$.

Since $ds^2 = (dx^0)^2 - (dx^1)^2 - (dx^2)^2 - (dx^3)^2 = g_{\mu\nu}dx^\mu dx^\nu$, we have

$$\textbf{(7)}\ (g_{\mu\nu}) = \begin{pmatrix} 1 & 0 & 0 & 0 \\ 0 & -1 & 0 & 0 \\ 0 & 0 & -1 & 0 \\ 0 & 0 & 0 & -1 \end{pmatrix}$$

(hence $g_{\alpha\beta} = 0$ if $\alpha \neq \beta$; $g_{00} = 1$; $g_{11} = g_{22} = g_{33} = -1$).

As usual, we define $F_{\mu\nu} = g_{\mu\alpha}\, g_{\nu\beta}\, F^{\alpha\beta}$. It is easily verified that:

$$\textbf{(8)}\ (F_{\mu\nu}) = \begin{pmatrix} 0 & -E_x & -E_y & -E_z \\ E_x & 0 & H_z & -H_y \\ E_y & -H_z & 0 & H_x \\ E_z & H_y & -H_x & 0 \end{pmatrix}$$

Similarly:

$$\textbf{(9)}\ (F^\mu{}_\nu) = (F^{\mu\sigma}g_{\sigma\nu}) = \begin{pmatrix} 0 & E_x & E_y & E_z \\ -E_x & 0 & H_z & -H_y \\ -E_y & -H_z & 0 & H_x \\ -E_z & H_y & -H_x & 0 \end{pmatrix}\begin{pmatrix} 1 & 0 & 0 & 0 \\ 0 & -1 & 0 & 0 \\ 0 & 0 & -1 & 0 \\ 0 & 0 & 0 & -1 \end{pmatrix}$$

$$= \begin{pmatrix} 0 & -E_x & -E_y & -E_z \\ -E_x & 0 & -H_z & H_y \\ -E_y & H_z & 0 & -H_x \\ -E_z & -H_y & H_x & 0 \end{pmatrix}$$

It can be easily verified that the source equations are given by: $F^{\mu\nu}{}_{|\nu} = S^{\mu}$. For example, taking $\mu = 2$, we have:

$$\begin{aligned} F^{2\nu}{}_{|\nu} &= \left(\frac{\partial F^{20}}{\partial x^0} + \frac{\partial F^{21}}{\partial x^1} + \frac{\partial F^{22}}{\partial x^2} + \frac{\partial F^{23}}{\partial x^3}\right) \\ &= -\frac{\partial E_y}{c\partial t} - \frac{\partial H_z}{\partial x} + 0 + \frac{\partial H_x}{\partial z} \\ &= -\frac{1}{c}\frac{\partial E_y}{\partial t} + \left(\frac{\partial H_x}{\partial z} - \frac{\partial H_z}{\partial x}\right). \end{aligned}$$

By definition of (S^{σ}): $S^2 = \frac{1}{c}\rho\nu_y$ where $\vec{\nu} = (\nu_x, \nu_y, \nu_z)$. Hence the equation $F^{2\nu}{}_{|\nu} = S^2$ is the second component of the 3-vector equation $\nabla x \vec{H} = \frac{1}{c}\frac{\partial \vec{E}}{\partial t} + \frac{1}{c}\rho\vec{\nu}$.

Since we are dealing with flat space, the affine connections vanish, so covariant differentiation coincides with ordinary (partial) differentiation. Hence: $F^{\mu\nu}{}_{|\nu} = S^{\mu}$ is equivalent to:

(10) $F^{\mu\nu}{}_{||\nu} = S^{\mu}$. $(\mu = 0,1,2,3)$.

Equation (10) is obviously generally covariant. Similarly, it can be verified that the internal equations can be written in the form:

$$\frac{\partial F_{\mu\nu}}{\partial x^{\sigma}} + \frac{\partial F_{\nu\sigma}}{\partial x^{\mu}} + \frac{\partial F_{\sigma\mu}}{\partial x^{\nu}} = 0.$$

In view of the fact that $F_{\mu\nu}$ is antisymmetric, the last equation is equivalent to

(11) $[F_{\mu\nu|\sigma}] = 0$.

Equation (11) is generally covariant, for we know by Theorem 1 that: $[F_{\mu\nu|\sigma}] = [F_{\mu\nu||\sigma}]$ and $[F_{\mu\nu||\sigma}]$ is a tensor.

By Theorem 3, equation (11) is equivalent to the proposition that there exists a (covariant) vector k_{μ} such that:

(12) $F_{\mu\nu} = k_{\mu|\nu} - k_{\nu|\mu} = \dfrac{\partial k_{\mu}}{\partial x^{\nu}} - \dfrac{\partial k_{\nu}}{\partial x^{\mu}}$.

APPENDIX FIVE

THE GEODESIC

The notion of a geodesic can be approached from two different standpoints, one of which is more general than the other. We naturally start from the more general point of view.

(I)

Consider any N-dimensional space in which the affine connections Γ^{i}_{jk} are defined as well-behaved functions at each point.

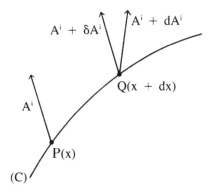

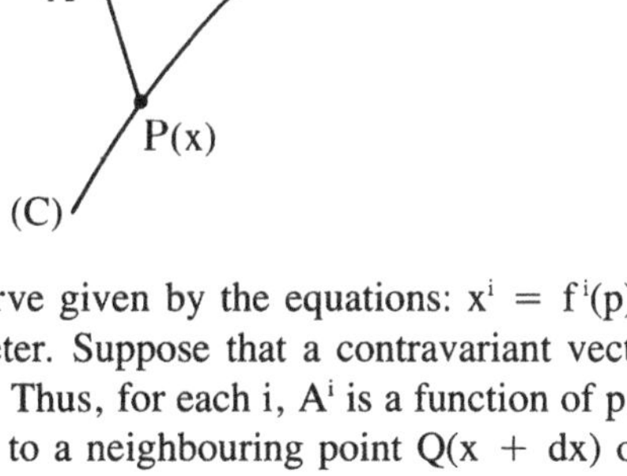

Let C be a curve given by the equations: $x^i = f^i(p)$, where p is some (invariant) parameter. Suppose that a contravariant vector A^i is defined at each point P of C. Thus, for each i, A^i is a function of p. When transported from a point P(x) to a neighbouring point Q(x + dx) on C, the vector A^i becomes:

(1) $$A^i + \delta A^i = A^i - \Gamma^i_{jk} \cdot A^j \cdot dx^k = A^i - \Gamma^i_{jk} \cdot A^j \cdot \frac{dx^k}{dp} \cdot dp$$

$$= A^i - \Gamma^i_{jk} \cdot A^j \cdot \frac{dx^k}{dp} \cdot h,$$

where we have put: dp = h.

At Q the (local) value of the vector is:

$$\textbf{(2)}\ A^i + dA^i = A^i + \frac{dA^i}{dp} \cdot dp = A^i + \frac{dA^i}{dp} \cdot h$$

[1] Let us now require that the transported vector ($A^i + \delta A^i$) be parallel to the vector ($A^i + dA^i$) as determined at Q. (Note: both vectors are compared at Q.) This means that there exists a function $\psi(p,h)$ such that:

$$\textbf{(3)}\ \frac{A^i + dA^i}{A^i + \delta A^i} = \psi(p,h), \text{ for every } i.$$

Let $dp = h \rightarrow 0$. Thus: $dA^i = \frac{dA^i}{dp} h \rightarrow 0$ and $\delta A^i = -\Gamma^i_{jk} \cdot A^j \cdot \frac{dx^k}{dp} \cdot h \rightarrow 0$. From (3) it follows that: $\frac{A^i}{A^i} = \psi(p,0)$. That is:

(4) $\psi(p,0) = 1$. This holds for all p.

Now go back to (3), which can be rewritten as:

$$A^i + dA^i = \psi(p,h)[A^i + \delta A^i].$$

In conjunction with (1) and (2), we obtain:

$$A^i + \frac{dA^i}{dp} \cdot h = \psi(p,h)\left[A^i - \Gamma^i_{jk} \cdot A^j \cdot \frac{dx^k}{dp} \cdot h\right].$$

Thus, rearranging, then dividing by h:

$$\textbf{(5)}\ \frac{dA^i}{dp} + \psi(p,h) \cdot \Gamma^i_{jk} \cdot A^j \cdot \frac{dx^k}{dp} = \left(\frac{\psi(p,h) - 1}{h}\right) A^i = \left(\frac{\psi(p,h) - \psi(p,0)}{h}\right) \cdot A^i,$$

since, by (4): $\psi(p,0) \equiv 1$.

In (5), let $h \rightarrow 0$. $\psi(p,h) \rightarrow \psi(p,0) = 1$. By definition of the partial derivatives of $\psi(p,h)$,

$$\left(\frac{\psi(p,h) - \psi(p,0)}{h}\right) \rightarrow \left(\frac{\partial\psi}{\partial h}\right)_{h=o} = \psi'_h(p,0)$$

Putting $\varphi(p) \underset{\text{Def}}{=} \psi'_h(p,o)$, we infer from (5):

$$(6)\quad \frac{dA^i}{dp} + \Gamma^i_{jk} \cdot A^j \cdot \frac{dx^k}{dp} = \varphi(p) \cdot A^i$$

[2] Thus, equation (6) expresses the condition that, at the point Q 'near' P on C, the transported vector is parallel to the (local) vector at Q. Had we required that these two vectors, namely $A^i + dA^i$ and $A^i + \delta A^i$, should actually coincide, then we should have obtained the equations: $\psi(p,h) \equiv 1$, hence:

$$(7)\quad \frac{dA^i}{dp} + \Gamma^i_{jk} \cdot A^j \frac{dx^k}{dp} = 0, \text{ since } \varphi(p) = \psi'_h(p,o) = 0.$$

[3] Now consider the special case where:

$$A^i = \frac{dx^i}{dp} = \frac{d}{dp}(f^i(p)).$$

A^i is thus tangent, at P, to the curve C. In this case, equations (6) and (7) respectively assume the forms:

$$(6')\quad \frac{d^2x^i}{dp^2} + \Gamma^i_{jk} \cdot \frac{dx^j}{dp} \cdot \frac{dx^k}{dp} = \varphi(p) \cdot \frac{dx^i}{dp}\text{; and}$$

$$(7')\quad \frac{d^2x^i}{dp^2} + \Gamma^i_{jk} \cdot \frac{dx^i}{dp} \cdot \frac{dx^k}{dp} = 0$$

Let us anyway show that, given (6′), a change of the parameter p will reduce (6′) to the form (7′).

Let s be any new parameter, i.e. let s be a function of p. Writing $\frac{d}{dp} = \frac{ds}{dp} \cdot \frac{d}{ds}$, we transform (6′) into:

$$\frac{ds}{dp} \cdot \frac{d}{ds}\left(\frac{ds}{dp}\frac{dx^i}{ds}\right) + \left(\frac{ds}{dp}\right)^2 \Gamma^i_{jk} \frac{dx^j}{ds} \cdot \frac{dx^k}{ds} = \varphi(p)\frac{ds}{dp} \cdot \frac{dx^i}{ds}$$

Dividing by $\frac{ds}{dp}$, then expanding, we obtain:

$$(8)\quad \left(\frac{ds}{dp}\right)\left[\frac{d^2x^i}{ds^2} + \Gamma^i_{jk}\frac{dx^j}{ds}\frac{dx^k}{ds}\right] = \left[-\frac{d}{ds}\left(\frac{ds}{dp}\right) + \varphi(p)\right]\frac{dx^i}{ds}$$

Thus, it suffices for the right-hand side of (8) to vanish in order for the equation of the geodesic to assume the form of (7′). So we require:

$$\varphi(p) = \frac{d}{ds}\left(\frac{ds}{dp}\right) = \frac{d}{dp}\left(\frac{ds}{dp}\right) \cdot \frac{dp}{ds} = \left(\frac{d^2s}{dp^2}\right)\Big/\left(\frac{ds}{dp}\right).$$

Integrating:

$$\varphi_1(p) + \alpha = \log_e\left(\frac{ds}{dp}\right),$$

where $\varphi_1(p)$ is an indefinite integral $\int\varphi(p) \cdot dp$, and α is an arbitrary constant. Therefore:

$$\frac{ds}{dp} = e^{\varphi_1(p) + \alpha} = ae^{\varphi_1(p)}, \text{ where } a \underset{\text{Def}}{=} e^{\alpha}.$$

Integrating again:

(9) $s = a \cdot F(p) + b$; where: $F(p)$ is an indefinite integral $\int e^{\varphi_1(p)} \cdot dp$ and b is an arbitrary constant.

Since a and b are arbitrary constants, s is determined to within an arbitrary linear transformation. For any such s, the equation of the geodesic is:

(10) $$\frac{d^2x^i}{ds^2} + \Gamma^i_{jk}\frac{dx^j}{ds} \cdot \frac{dx^k}{ds} = 0.$$

(Note: there is a sense in which, even in a purely affine space, s represents a length along the geodesic; see Schrödinger 1950, ch. VII.)

(II)

Now consider the case of a Riemannian space in which a 'distance' ds is defined by

(11) $$\begin{cases} ds^2 = g_{ij}\,dx^i \cdot dx^j \text{ and } \Gamma^i_{jk} = \left\{{i \atop jk}\right\} = g^{iu}[jk,u] \underset{\text{Def}}{=} \\ \qquad \tfrac{1}{2}\, g^{iu}\left(\dfrac{\partial g_{ju}}{\partial x^k} + \dfrac{\partial g_{ku}}{\partial x^j} - \dfrac{\partial g_{jk}}{\partial x^u}\right) \end{cases}$$

We can define the geodesic by means of equation (10), or of equation (7′) where we take dp to be the infinitesimal distance ds. This is tantamount to defining the geodesic as a curve C along which the unit tangent vector $\frac{dx^i}{ds}$, when transported from P to a neighbouring point Q, coincides with the unit tangent vector at Q.

In a Riemannian space, there is however an alternative way of defining the geodesic. Consider any curve C given in terms of some parameter p; and let P and Q be any two points on C (Q need not be 'close' to P). The distance between P and Q along C is given by the integral:

$$\textbf{(12)}\quad I = \int_{P}^{Q} (g_{\mu\nu} \cdot dx^{\mu} \cdot dx^{\nu})^{1/2} = \int (g_{\mu\nu} \cdot \dot{x}^{\mu} \cdot \dot{x}^{\nu})^{1/2} \cdot dp$$

where we put:

$$\dot{x}^{\mu} = \frac{dx^{\mu}}{dp},\ \dot{x}^{\nu} = \frac{dx^{\nu}}{dp}.$$

The integral I will thus generally depend on the path C.

Let us now define the geodesic as a curve which extremises the value of I. Let $T = (g_{\mu\nu}\dot{x}^{\mu}\dot{x}^{\nu})^{1/2} = H^{1/2}$, where:

$$H = g_{\mu\nu}\frac{dx^{\mu}}{dp}\frac{dx^{\nu}}{dp} = g_{\mu\nu}\dot{x}^{\mu}\dot{x}^{\nu}.$$

By the Euler-Lagrange equations:

$$\frac{d}{dp}\left(\frac{\partial T}{\partial \dot{x}^{\mu}}\right) - \frac{\partial T}{\partial x^{\mu}} = 0.$$

In other words:

$$\textbf{(13)}\quad \frac{d}{dp}\left(\tfrac{1}{2} H^{-1/2}\frac{\partial H}{\partial \dot{x}^{\mu}}\right) - \tfrac{1}{2} H^{-1/2}\frac{\partial H}{\partial x^{\mu}} = 0;$$

where:

$$H = g_{\mu\nu}\dot{x}^{\mu}\dot{x}^{\nu}.$$

Therefore:

$$\frac{\partial H}{\partial \dot{x}^{\mu}} = g_{\mu\nu}\dot{x}^{\nu} + g_{\nu\mu}\dot{x}^{\nu} = 2g_{\mu\nu}\dot{x}^{\nu}$$

$$\frac{\partial H}{\partial x^{\mu}} = \frac{\partial}{\partial x^{\mu}}(g_{\alpha\beta}\dot{x}^{\alpha}\dot{x}^{\beta}) = \frac{\partial g_{\alpha\beta}}{\partial x^{\mu}}\dot{x}^{\alpha}\dot{x}^{\beta} = g_{\alpha\beta|\mu}\dot{x}^{\alpha}\dot{x}^{\beta}.$$

(where $g_{\alpha\beta|\mu} = \partial g_{\alpha\beta}/\partial x^{\mu}$)

Substituting in (8):

$$(14)\quad \begin{cases} \dfrac{d}{dp}(H^{-1/2}\, g_{\mu\nu}\, \dot{x}^{\nu}) - \tfrac{1}{2} H^{-1/2}\, g_{\alpha\beta|\mu}\, \dot{x}^{\alpha}\, \dot{x}^{\beta} = 0;\ \text{i.e.} \\[2ex] \dfrac{d}{dp}\left(H^{-1/2}\, g_{\mu\nu}\, \dfrac{dx^{\nu}}{dp}\right) - \tfrac{1}{2} H^{-1/2}\, g_{\alpha\beta|\mu} \cdot \dfrac{dx^{\alpha}}{dp} \cdot \dfrac{dx^{\beta}}{dp} = 0 \end{cases}$$

Now take the parameter p to be s, where s denotes the length along the geodesic. Then:

$$H = g_{\mu\nu} \cdot \frac{dx^{\mu}}{dp} \cdot \frac{dx^{\nu}}{dp} = g_{\mu\nu} \frac{dx^{\mu}}{ds} \frac{dx^{\nu}}{ds} = 1, \text{ by definition of } ds.$$

Equation (14) becomes:

$$(15)\quad \frac{d}{ds}\left(g_{\mu\nu} \frac{dx^{\nu}}{ds}\right) - \tfrac{1}{2}\, g_{\alpha\beta|\mu} \frac{dx^{\alpha}}{ds} \cdot \frac{dx^{\beta}}{ds} = 0.$$

Expanding the first term:

$$\frac{d}{ds}\left(g_{\mu\nu} \frac{dx^{\nu}}{ds}\right) = \left(\frac{dg_{\mu\nu}}{ds} \cdot \frac{dx^{\nu}}{ds} + g_{\mu\nu} \frac{d^2x^{\nu}}{ds^2}\right) = \left(\frac{\partial g_{\mu\nu}}{\partial x^{\beta}} \frac{dx^{\beta}}{ds} \frac{dx^{\nu}}{ds} + g_{\mu\nu} \frac{d^2x^{\nu}}{ds^2}\right)$$

In other words, substituting dummy α for dummy ν:

$$\frac{d}{ds}\left(g_{\mu\nu} \frac{dx^{\nu}}{ds}\right) = \frac{\partial g_{\mu\alpha}}{\partial x^{\beta}} \frac{dx^{\beta}}{ds} \frac{dx^{\alpha}}{ds} + g_{\mu\alpha} \frac{d^2x^{\alpha}}{ds^2}.$$

Substituting in (15):

$$(16)\quad g_{\mu\alpha} \frac{d^2x^{\alpha}}{ds^2} + g_{\mu\alpha|\beta} \frac{dx^{\alpha}}{ds} \frac{dx^{\beta}}{ds} - \tfrac{1}{2}\, g_{\alpha\beta|\mu} \frac{dx^{\alpha}}{ds} \frac{dx^{\beta}}{ds} = 0$$

(Remember: $g_{\mu\alpha|\beta} \underset{\text{Def}}{=} \partial g_{\mu\alpha} / \partial x^{\beta}$)

Interchanging dummies α and β:

$$g_{\mu\alpha|\beta} \frac{dx^{\alpha}}{ds} \frac{dx^{\beta}}{ds} = g_{\mu\beta|\alpha} \frac{dx^{\beta}}{ds} \frac{dx^{\alpha}}{ds} = g_{\mu\beta|\alpha} \frac{dx^{\alpha}}{ds} \frac{dx^{\beta}}{ds}.$$

Hence:

$$g_{\mu\alpha|\beta} \frac{dx^{\alpha}}{ds} \frac{dx^{\beta}}{ds} = \tfrac{1}{2}\,(g_{\mu\alpha|\beta} + g_{\mu\beta|\alpha}) \frac{dx^{\alpha}}{ds} \frac{dx^{\beta}}{ds}.$$

Substituting in (16):

$$g_{\mu\alpha} \frac{d^2x^\alpha}{ds^2} + \tfrac{1}{2}\,(g_{\mu\alpha|\beta} + g_{\mu\beta|\alpha} - g_{\alpha\beta|\mu}) \frac{dx^\alpha}{ds} \cdot \frac{dx^\beta}{ds} = 0$$

i.e.

$$g_{\mu\alpha} \frac{d^2x^\alpha}{ds^2} + [\alpha\beta,\mu] \frac{dx^\alpha}{ds} \cdot \frac{dx^\beta}{ds} = 0.$$

Multiplying by $g^{\sigma\mu}$:

$$\delta^\sigma_\alpha \frac{d^2x^\alpha}{ds^2} + \left\{{\sigma \atop \alpha\beta}\right\} \frac{dx^\alpha}{ds} \cdot \frac{dx^\beta}{ds} = 0.$$

i.e.:

(17) $$\frac{d^2x^\sigma}{ds^2} + \left\{{\sigma \atop \alpha\beta}\right\} \frac{dx^\alpha}{ds} \cdot \frac{dx^\beta}{ds} = 0, \text{ or } \frac{d^2x^\sigma}{ds^2} + \Gamma^\sigma_{\alpha\beta} \frac{dx^\alpha}{ds} \frac{dx^\beta}{ds} = 0$$

Thus, the two approaches yield the same result in the case of Riemannian geometry.

APPENDIX SIX

THE CURVATURE TENSOR

(I)

We consider an N-dimensional space S_N in which affine connections Γ^i_{mn} are defined. For the time being we do not assume that Γ^i_{mn} is symmetric in m and n.

The easiest way of introducing the curvature tensor is to consider the difference $A_{i\|jk} - A_{i\|kj}$, where A_i is an arbitrary covariant vector. We know that, if A_i is a well-behaved function of the coordinates, then:

$$A_{i|jk} - A_{i|kj} = -\partial^2 A_i/\partial x^j \partial x^k + \partial^2 A_i/\partial x^k \partial x^j = 0.$$

It will turn out that in general $A_{i\|jk} - A_{i\|kj} \neq 0$.

(i) To compute $A_{i\|jk} - A_{i\|kj}$.

$$A_{i\|j} = \partial A_i/\partial x^j - \Gamma^\sigma_{ij} A_\sigma.$$

$$A_{i\|jk} = \frac{\partial A_{i\|j}}{\partial x^k} - \Gamma^\sigma_{ik} A_{\sigma\|j} - \Gamma^\sigma_{jk} A_{i\|\sigma}$$

$$= \left[\frac{\partial^2 A_i}{\partial x^k \partial x^j} - \frac{\partial \Gamma^\sigma_{ij}}{\partial x^k} A_\sigma - \Gamma^\sigma_{ij}\frac{\partial A_\sigma}{\partial x^k} - \Gamma^\sigma_{ik}\left(\frac{\partial A_\sigma}{\partial x^j} - \Gamma^\upsilon_{\sigma j} A_\upsilon\right) - \Gamma^\sigma_{jk} A_{i\|\sigma}\right]$$

Interchanging j and k:

$$A_{i\|kj} = \left[\frac{\partial^2 A_i}{\partial x^j \partial x^k} - \frac{\partial \Gamma^\sigma_{ik}}{\partial x^j} A_\sigma - \Gamma^\sigma_{ik}\frac{\partial A_\sigma}{\partial x^j} - \right.$$

$$\left.\Gamma^\sigma_{ij}\left(\frac{\partial A_\sigma}{\partial x^k} - \Gamma^\alpha_{\sigma k} A_\alpha\right) - \Gamma^\sigma_{kj} A_{i\|\sigma}\right].$$

Therefore:

$$\textbf{(1)}\quad A_{i\|jk} - A_{i\|kj} = \left[\left(-\frac{\partial \Gamma^\sigma_{ij}}{\partial x^k} + \Gamma^\alpha_{ik}\Gamma^\sigma_{\alpha j} + \frac{\partial \Gamma^\sigma_{ik}}{\partial x^j} - \Gamma^\alpha_{ij}\Gamma^\sigma_{\alpha k}\right) A_\sigma - \right.$$

$$\left.(\Gamma^\sigma_{jk} - \Gamma^\sigma_{kj}) A_{i\|\sigma}\right]$$

[we interchanged the dummies α and σ in the second and fourth terms]

This can be written as:

$$(2)\quad \begin{cases} A_{i\|jk} - A_{i\|kj} = B^{\sigma}_{ijk}A_{\sigma} - (\Gamma^{\sigma}_{jk} - \Gamma^{\sigma}_{kj})A_{i\|\sigma}; \text{ where:} \\ B^{\sigma}_{ijk} = \Gamma^{\sigma}_{\upsilon j}\Gamma^{\upsilon}_{ik} - \dfrac{\partial \Gamma^{\sigma}_{ij}}{\partial x^{k}} - \Gamma^{\sigma}_{\upsilon k}\Gamma^{\upsilon}_{ij} + \dfrac{\partial \Gamma^{\sigma}_{ik}}{\partial x^{j}} \end{cases}$$

The difference $(\Gamma^{\sigma}_{jk} - \Gamma^{\sigma}_{kj})$ of two affine connections is a tensor. Hence $(\Gamma^{\sigma}_{jk} - \Gamma^{\sigma}_{kj})A_{i\|\sigma}$ is a tensor. By (2):

$$(3)\quad B^{\sigma}_{ijk}A_{\sigma} = (A_{i\|jk} - A_{i\|kj}) + (\Gamma^{\sigma}_{jk} - \Gamma^{\sigma}_{kj})A_{i\|\sigma}$$

Since the right-hand side is a tensor, so is the left-hand side. In view of the fact that A_{σ} is an arbitrary covariant vector, it follows by the quotient theorem that B^{σ}_{ijk} is a tensor. B^{σ}_{ijk} is called the Riemann-Christoffel curvature tensor.

Note that, if the affine connections are symmetric, i.e. if $\Gamma^{\sigma}_{jk} = \Gamma^{\sigma}_{kj}$, then, by (3):

$$(3')\quad B^{\sigma}_{ijk}\,A_{\sigma} = A_{i\|jk} - A_{i\|kj}.$$

(ii) C U(x) A^{σ} Q(x + ξ + dξ) P(x + ξ)

Fix a point $U(x^1, \ldots, x^N)$ in S_N and consider an infinitesimal closed curve C about U. Transplant the contravariant vector A^{σ} first from $U(x^1, \ldots, x^N)$ to $P(x^1 + \xi^1, \ldots, x^N + \xi^N)$ and then from $P(x^1 + \xi^1, \ldots, x^N + \xi^N)$ to $Q(x^1 + \xi^1 + d\xi^1, \ldots, x^N + \xi^N + d\xi^N)$

At P the transplanted vector is $A^{\sigma} - \Gamma^{\sigma}_{\alpha\beta}A^{\alpha}\xi^{\beta}$

At P the value of the affine connection is $\Gamma^{\sigma}_{\alpha\beta} + \dfrac{\partial \Gamma^{\sigma}_{\alpha\beta}}{\partial x^{\tau}}\xi^{\tau}$.

As we go from P to Q the transplanted vector is subject to a variation:

$$\delta^{*}A^{\sigma} = -\left(\Gamma^{\sigma}_{\alpha\beta} + \frac{\partial \Gamma^{\sigma}_{\alpha\beta}}{\partial x^{\tau}}\xi^{\tau}\right)(A^{\alpha} - \Gamma^{\alpha}_{\mu\nu}A^{\mu}\xi^{\nu})d\xi^{\beta}$$

$$(4)\quad \begin{aligned} &= \left(-\Gamma^{\sigma}_{\alpha\beta}A^{\alpha}d\xi^{\beta} + \Gamma^{\sigma}_{\alpha\beta}\Gamma^{\alpha}_{\mu\nu}A^{\mu}\xi^{\nu}d\xi^{\beta} - \frac{\partial \Gamma^{\sigma}_{\alpha\beta}}{\partial x^{\tau}}A^{\alpha}\xi^{\tau}d\xi^{\beta}\right) \\ &= \left[-\Gamma^{\sigma}_{\alpha\beta}A^{\alpha}d\xi^{\beta} + \left(\Gamma^{\sigma}_{\alpha\beta}\Gamma^{\alpha}_{\mu\nu} - \frac{\partial \Gamma^{\sigma}_{\mu\beta}}{\partial x^{\nu}}\right)A^{\mu}\xi^{\nu}d\xi^{\beta}\right] \end{aligned}$$

to within third order quantities in ξ^{ν} and $d\xi^{\mu}$.

Remembering that the values $\Gamma^{\sigma}_{\alpha\beta}$, $\partial\Gamma^{\sigma}_{\alpha\beta}/\partial x^{\tau}$ are taken at the fixed point U, let us integrate (4) around C:

(5) $$\Delta A^\sigma = \oint \delta^* A^\sigma = -\Gamma^\sigma_{\alpha\beta} A^\alpha \oint d\xi^\beta + \left(\Gamma^\sigma_{\alpha\beta}\Gamma^\alpha_{\mu\nu} - \frac{\partial \Gamma^\sigma_{\mu\beta}}{\partial x^\nu}\right) A^\mu \oint \xi^\nu d\xi^\beta$$

$$= \left(\Gamma^\sigma_{\alpha\beta}\Gamma^\alpha_{\mu\nu} - \frac{\partial \Gamma^\sigma_{\mu\beta}}{\partial x^\nu}\right) A^\mu \oint \xi^\nu d\xi^\beta,$$

since $\oint d\xi^\beta = 0$, $d\xi^\beta$ being a perfect differential. For the same reason: $0 = \oint d(\xi^\nu \xi^\beta) = \oint \xi^\nu d\xi^\beta + \oint \xi^\beta d\xi^\nu$. Hence:

6. $$\oint \xi^\beta d\xi^\nu = -\oint \xi^\nu d\xi^\beta;\ \text{so: } \oint \xi^\beta d\xi^\nu = \frac{1}{2}\oint(\xi^\beta d\xi^\nu - \xi^\nu d\xi^\beta).$$

Interchanging dummies ν and β in (5):

(7) $$\Delta A^\sigma = \left(\Gamma^\sigma_{\alpha\nu}\Gamma^\alpha_{\mu\beta} - \frac{\partial \Gamma^\sigma_{\mu\nu}}{\partial x^\beta}\right) A^\mu \oint \xi^\beta d\xi^\nu =$$

$$- \left(\Gamma^\sigma_{\alpha\nu}\Gamma^\alpha_{\mu\beta} - \frac{\partial \Gamma^\sigma_{\mu\nu}}{\partial x^\beta}\right) A^\mu \oint \xi^\nu d\xi^\beta \text{ (by (6))}$$

By (5), (6), and (7):

$$\Delta A^\sigma = \frac{1}{2}\left[\Gamma^\sigma_{\alpha\beta}\Gamma^\alpha_{\mu\nu} - \frac{\partial \Gamma^\sigma_{\mu\beta}}{\partial x^\nu} - \Gamma^\sigma_{\alpha\nu}\Gamma^\alpha_{\mu\beta} + \frac{\partial \Gamma^\sigma_{\mu\nu}}{\partial x^\beta}\right] A^\mu \oint \xi^\nu d\xi^\beta$$

$$= \frac{1}{2} B^\sigma_{\mu\beta\nu} A^\mu \oint \xi^\nu d\xi^\beta = -\frac{1}{2} B^\sigma_{\mu\beta\nu} A^\mu \oint \xi^\beta d\xi^\nu$$

$$= \frac{1}{4} B^\sigma_{\mu\beta\nu} A^\mu \oint (\xi^\nu d\xi^\beta - \xi^\beta d\xi^\nu) \text{ (see Lawden 1962, ch. 5, sec. 37)}$$

(II)

In the case of a Riemannian space where $ds^2 = g_{ij} dx^i dx^j$, $\Gamma^i_{jk} = \{^{\ i}_{jk}\} = \{^{\ i}_{kj}\} = \Gamma^i_{kj} = \frac{1}{2} g^{im}(g_{jm|k} + g_{km|j} - g_{jk|m})$. Hence: $B^\sigma{}_{ijk} = \{^{\ \sigma}_{kj}\}\{^{\ \nu}_{ik}\} - \frac{\partial}{\partial x^k}\{^{\ \sigma}_{ij}\} - \{^{\ \sigma}_{\alpha k}\}\{^{\ \alpha}_{ij}\} + \frac{\partial}{\partial x^j}\{^{\ \sigma}_{ik}\}$.

APPENDIX SEVEN

SYMMETRY PROPERTIES OF THE CURVATURE TENSOR

Remember that in the very general case of an affinely connected space:

$$\text{(1)}\quad B^i_{jk\ell} = \left(\Gamma^i_{\sigma k}\,\Gamma^\sigma_{j\ell} - \frac{\partial \Gamma^i_{jk}}{\partial x^\ell}\right) - \left(\Gamma^i_{\sigma\ell}\,\Gamma^\sigma_{jk} - \frac{\partial \Gamma^i_{j\ell}}{\partial x^k}\right)$$

(1) Since the second bracket in the expression of $B^i_{jk\ell}$ is obtained by interchanging k and ℓ in the first bracket, $B^i_{jk\ell}$ is antisymmetric in k and ℓ; i.e.

$$\text{(2)}\quad B^i_{jk\ell} = -B^i_{j\ell k}.$$

(2) Suppose that the affine connections are symmetric; i.e. $\Gamma^i_{jk} = \Gamma^i_{kj}$ for all i,j,k. Through circularly permuting j,k,ℓ in equation (1), we obtain:

$$\text{(3)}\quad B^i_{jk\ell} + B^i_{k\ell j} + B^i_{j\ell k} = 0.$$

(3) Assume now that we have a (symmetric) fundamental tensor g_{ij} and an infinitesimal distance ds defined by:

(4) $ds^2 = g_{ij}\,dx^i dx^j$. We also suppose that:

$$\text{(5)}\quad \Gamma^i_{jk} = \left\{{i \atop jk}\right\} = g^{im}[jk,\, m] = \frac{1}{2}\, g^{im}(g_{jm|k} + g_{km|j} - g_{jk|m})$$

where $g_{jm|k}$ denotes $\partial g_{jm}/\partial x^k$.

We define: $B_{ijk\ell} = g_{im}\,B^m{}_{jk\ell}$ and we propose to express $B_{ijk\ell}$ in such a form that its symmetry properties become apparent. By (1) and (5):

$$B_{ijk\ell} = g_{is}\,B^s{}_{jk\ell} = g_{is}\left(\left\{{s \atop rk}\right\}\left\{{r \atop j\ell}\right\} - \left\{{s \atop r\ell}\right\}\left\{{r \atop jk}\right\} - \frac{\partial}{\partial x^\ell}\left\{{s \atop jk}\right\} + \frac{\partial}{\partial x^k}\left\{{s \atop j\ell}\right\}\right)$$

Note that: $g_{is}\left\{{s \atop rk}\right\} = g_{is}\,g^{sm}[rk,m] = \delta^m_i[rk,m] = [rk,i].$

Substituting in the last expression of $B_{ijk\ell}$:

$$B_{ijk\ell} = [rk,i]\begin{Bmatrix} r \\ j\ell \end{Bmatrix} - [r\ell,i]\begin{Bmatrix} r \\ jk \end{Bmatrix} + g_{is}\frac{\partial}{\partial x^k}\begin{Bmatrix} s \\ j\ell \end{Bmatrix} - g_{is}\frac{\partial}{\partial x^\ell}\begin{Bmatrix} s \\ jk \end{Bmatrix}$$

$$= \left([rk,i]\begin{Bmatrix} r \\ j\ell \end{Bmatrix} - [r\ell,i]\begin{Bmatrix} r \\ jk \end{Bmatrix} + \frac{\partial}{\partial x^k}\left(g_{is}\begin{Bmatrix} s \\ j\ell \end{Bmatrix}\right)\right.$$

$$\left. - \frac{\partial}{\partial x^\ell}\left(g_{is}\begin{Bmatrix} s \\ jk \end{Bmatrix}\right) - \begin{Bmatrix} s \\ j\ell \end{Bmatrix}\frac{\partial g_{is}}{\partial x^k} + \begin{Bmatrix} s \\ jk \end{Bmatrix}\frac{\partial g_{is}}{\partial x^\ell}\right)$$

Note that:

$$[ik,s] + [sk,i] = \frac{1}{2}(g_{is|k} + g_{ks|i} - g_{ik|s} + g_{si|k} + g_{ki|s} - g_{sk|i})$$

$$= g_{is|k} \underset{\text{Def}}{=} \partial g_{is}/\partial x^k, \text{ since } g_{ij} = g_{ji} \text{ for all } i,j.$$

By the last expression of $B_{ijk\ell}$:

$$B_{ijk\ell} = \left([rk,i]\begin{Bmatrix} r \\ j\ell \end{Bmatrix} - [r\ell,i]\begin{Bmatrix} r \\ jk \end{Bmatrix} + \frac{\partial}{\partial x^k}[j\ell,i] - \frac{\partial}{\partial x^\ell}[jk,i]\right.$$

$$\left. - \begin{Bmatrix} s \\ j\ell \end{Bmatrix}([ik,s] + [sk,i]) + \begin{Bmatrix} s \\ jk \end{Bmatrix}([i\ell,s] + [s\ell,i])\right)$$

$$= \left(\frac{\partial}{\partial x^k}[j\ell,i] - \frac{\partial}{\partial x^\ell}[jk,i] - \begin{Bmatrix} s \\ j\ell \end{Bmatrix}[ik,s] + \begin{Bmatrix} s \\ jk \end{Bmatrix}[i\ell,s]\right).$$

We write:

$$\frac{\partial}{\partial x^k}[j\ell,i] = \frac{1}{2}\left(\frac{\partial^2 g_{ji}}{\partial x^k \partial x^\ell} + \frac{\partial^2 g_{\ell i}}{\partial x^k \partial x^j} - \frac{\partial^2 g_{j\ell}}{\partial x^k \partial x^i}\right)$$

$$\frac{\partial}{\partial x^\ell}[jk,i] = \frac{1}{2}\left(\frac{\partial^2 g_{ji}}{\partial x^\ell \partial x^k} + \frac{\partial^2 g_{ki}}{\partial x^\ell \partial x^j} - \frac{\partial^2 g_{jk}}{\partial x^\ell \partial x^i}\right)$$

$$[ik,s] = g_{sr}\begin{Bmatrix} r \\ ik \end{Bmatrix} \text{ and } [i\ell,s] = g_{sr}\begin{Bmatrix} r \\ i\ell \end{Bmatrix}.$$

Substituting in the last expression of $B_{ijk\ell}$:

$$\textbf{(6)}\ B_{ijk\ell} = \left(\frac{1}{2}\left(\frac{\partial^2 g_{\ell i}}{\partial x^k \partial x^j} + \frac{\partial^2 g_{jk}}{\partial x^\ell \partial x^i} - \frac{\partial^2 g_{j\ell}}{\partial x^k \partial x^i} - \frac{\partial^2 g_{ki}}{\partial x^\ell \partial x^j}\right)\right.$$

$$\left. - g_{sr}\begin{Bmatrix} s \\ j\ell \end{Bmatrix}\begin{Bmatrix} r \\ ik \end{Bmatrix} + g_{sr}\begin{Bmatrix} s \\ jk \end{Bmatrix}\begin{Bmatrix} r \\ i\ell \end{Bmatrix}\right)$$

By interchanging i and j in (6), we obtain: $B_{jik\ell} = -B_{ijk\ell}$. By interchanging i and k, j and ℓ, we obtain: $B_{k\ell ij} = B_{ijk\ell}$. By lowering the index i in (2) and (3): $B_{ijk\ell} = -B_{ij\ell k}$ and $B_{ijk\ell} + B_{ik\ell j} + B_{i\ell jk} = 0$. Summing up:

$$(7)\begin{cases} B_{ijk\ell} = -B_{jik\ell} = -B_{ij\ell k} = B_{k\ell ij}; \text{ and:} \\ B_{ijk\ell} + B_{ik\ell j} + B_{i\ell jk} = 0 \end{cases}$$

APPENDIX EIGHT

EINSTEIN'S TENSOR

(I)
THE BIANCHI IDENTITIES

We consider an affinely connected space S_N in which the affine connections $\Gamma^{\,i}_{jm}$ are symmetric. (i.e. $\Gamma^{\,i}_{jm} = \Gamma^{\,i}_{mj}$ for all i, j and m).

Theorem 1: For any point A of S_N, there exists a coordinate system such that the affine connections vanish at A.

Proof: Let A have coordinates $(a^1, \ldots, a^N)$ in some coordinate system $K(x^1, \ldots, x^N)$. We propose to determine a coordinate system $\overline{K}(\overline{x}^1, \ldots, \overline{x}^N)$ such that $(\overline{\Gamma}^i_{mj})_A = 0$ for all i, m, j; where $(\overline{\Gamma}^i_{mj})_A$ denotes the value of $\overline{\Gamma}^i_{mj}$ at A.

The transformation $(x^1, \ldots, x^N) \rightarrow (\overline{x}^1, \ldots, \overline{x}^N)$ will be of the form:

(1) $\overline{x}^i = (x^i - a^i) + b^i_{mj}(x^m - a^m)(x^j - a^j)$; where:

the $b^{\,i}_{mj}$'s are constant and symmetric in m and j (in other words $b^i_{mj} = b^i_{jm}$ for all i, m and j). We shall determine b^i_{mj} in such a way that $\overline{\Gamma}^i_{mj} = 0$ (at A).

By (1): $\overline{a}^i = 0$ for all i. Hence A has coordinates $(0, \ldots, 0)$ in the system $\overline{K}(\overline{x}^1, \ldots, \overline{x}^N)$. Also by (1):

$$\textbf{(2)}\quad \frac{\partial \overline{x}^i}{\partial x^n} = \delta^i_n + b^i_{mn}(x^m - a^m) + b^i_{nj}(x^j - a^j)$$

$$= \delta^i_n + b^i_{mn}(x^m - a^m) + b^i_{nm}(x^m - a^m)$$

$$= \delta^i_n + 2b^i_{mn}(x^m - a^m).$$

Differentiating once more:

$$\textbf{(3)}\quad \frac{\partial^2 \overline{x}^i}{\partial x^m \partial x^n} = 2b^i_{mn} = \text{constant (for all points P of } S_N).$$

Taking equation (2) at the point A (i.e. $x^m = a^m$ for all m):

$$\text{(4)}\quad \left(\frac{\partial \bar{x}^i}{\partial x^n}\right)_A = \delta^i_n.$$

We have, in general:

$$\frac{\partial \bar{x}^i}{\partial x^n} \cdot \frac{\partial x^n}{\partial \bar{x}^m} = \delta^i_m.$$

In particular:

$$\left(\frac{\partial \bar{x}^i}{\partial x^n}\right)_A \left(\frac{\partial x^n}{\partial \bar{x}^m}\right) = \delta^i_m.$$

By (4), it follows that:

$$\delta^i_n \left(\frac{\partial x^n}{\partial \bar{x}^m}\right)_A = \delta^i_m.$$

In other words:

$$\text{(5)}\quad \left(\frac{\partial x^i}{\partial \bar{x}^m}\right)_A = \delta^i_m.$$

Now we turn to the affine connections, by the transformation law for Γ^i_{mn}:

$$\text{(6)}\quad \bar{\Gamma}^i_{mn} = \frac{\partial \bar{x}^i}{\partial x^\alpha}\frac{\partial x^\mu}{\partial \bar{x}^m}\frac{\partial x^\nu}{\partial \bar{x}^n}\Gamma^\alpha_{\mu\nu} - \frac{\partial x^\mu}{\partial \bar{x}^m}\cdot\frac{\partial x^\nu}{\partial \bar{x}^n}\frac{\partial^2 \bar{x}^i}{\partial x^\mu \partial x^\nu}$$

$$= \frac{\partial \bar{x}^i}{\partial x^\alpha}\frac{\partial x^\mu}{\partial \bar{x}^m}\frac{\partial x^\nu}{\partial \bar{x}^n}\Gamma^\alpha_{\mu\nu} - 2b^i_{\mu\nu}\frac{\partial x^\mu}{\partial \bar{x}^m}\frac{\partial x^\nu}{\partial \bar{x}^n}\ \textit{(by (3))}$$

By (4) and (5), we have, at the point A:

$$\text{(7)}\quad (\bar{\Gamma}^i_{mn})_A = \delta^i_\alpha\,\delta^\mu_m\,\delta^\nu_n(\Gamma^\alpha_{\mu\nu})_A - 2b^i_{\mu\nu}\,\delta^\mu_m\,\delta^\nu_n = (\Gamma^i_{mn})_A - 2b^i_{mn}$$

Remember that we want $(\bar{\Gamma}^i_{mn})_A$ to vanish for all i, m and n. By (7), it suffices to choose $b^i_{mn} = \frac{1}{2}(\Gamma^i_{mn})_A$.Q.E.D.

Note: If the space is a Riemannian space in which $ds^2 = g_{\mu\nu}dx^\mu dx^\nu$, then: we first carry out a transformation which makes $\left\{{i \atop mn}\right\}$ vanish at A for all i, m and n; we then carry out a transformation which is linear and which reduces the matrix $(g_{\mu\nu})_A$ to diagonal form. By (6) above, this second linear transformation does not alter the fact that the affinities vanish at A, since: $\partial^2\bar{x}^i/\partial x^\mu \partial x^\nu = 0$ at all points P of the manifold.

Theorem 2 (Bianchi Identities). For any choice of the indices α, μ, ν, σ and τ:

$$B^{\alpha}_{\mu\nu\sigma\|\tau} + B^{\alpha}_{\mu\sigma\tau\|\nu} + B^{\alpha}_{\mu\tau\nu\|\sigma} = 0;$$

where $B^{\alpha}_{\mu\nu\sigma}$ denotes the Curvature tensor. (Remember that, throughout this section, the affine connection $\Gamma^{\alpha}_{\beta\gamma}$ are supposed to be symmetric.)

Proof: Fix an arbitrary point P and choose a coordinate system in which the affine connections vanish at P. i.e. $\Gamma^{\alpha}_{\beta\gamma} = 0$ for all α, β, γ (at P). Hence, at P, the covariant derivative of any tensor coincides with its ordinary (partial) derivative; e.g. $B^{\mu}_{\nu\sigma\tau\|\gamma} = B^{\mu}_{\nu\sigma\tau|\gamma}$. Thus:

(1) $B^{\alpha}_{\mu\nu\sigma\|\tau} = B^{\alpha}_{\mu\nu\sigma|\tau} = \partial B^{\alpha}_{\mu\nu\sigma}/\partial x^{\tau}$; where

(2) $B^{\alpha}_{\mu\nu\sigma} = \Gamma^{\alpha}_{\gamma\nu}\,\Gamma^{\gamma}_{\mu\sigma} - \partial\Gamma^{\alpha}_{\mu\nu}/\partial x^{\sigma} - \Gamma^{\alpha}_{\gamma\sigma}\,\Gamma^{\gamma}_{\mu\nu} + \partial\Gamma^{\alpha}_{\mu\sigma}/\partial x^{\nu}$

Hence

$(\Gamma^{\alpha}_{\gamma\nu}\,\Gamma^{\gamma}_{\mu\sigma})_{|\tau} = \Gamma^{\alpha}_{\gamma\nu|\tau}\,\Gamma^{\gamma}_{\mu\sigma} + \Gamma^{\alpha}_{\gamma\nu}\,\Gamma^{\gamma}_{\mu\sigma|\tau} = 0$ (at P: since all affine connections vanish at P).

Similarly:

$(\Gamma^{\alpha}_{\gamma\sigma}\,\Gamma^{\gamma}_{\mu\nu})_{|\tau} = 0$ (at P).

Therefore, by (2):

(3) $$B^{\alpha}_{\mu\nu\sigma|\tau} = [-(\partial\Gamma^{\alpha}_{\mu\nu}/\partial x^{\sigma}) + (\partial\Gamma^{\alpha}_{\mu\sigma}/\partial x^{\nu})]_{|\tau} = -\frac{\partial^2\Gamma^{\alpha}_{\mu\nu}}{\partial x^{\tau}\partial x^{\sigma}} + \frac{\partial^2\Gamma^{\alpha}_{\mu\sigma}}{\partial x^{\tau}\partial x^{\nu}}$$

Through circularly permuting ν, σ, τ, we obtain:

(4) $$B^{\alpha}_{\mu\sigma\tau|\nu} = -\frac{\partial^2\Gamma^{\alpha}_{\mu\sigma}}{\partial x^{\nu}\partial x^{\tau}} + \frac{\partial^2\Gamma^{\alpha}_{\mu\tau}}{\partial x^{\nu}\partial x^{\sigma}}\text{; and}$$

(5) $$B^{\alpha}_{\mu\tau\nu|\sigma} = -\frac{\partial^2\Gamma^{\alpha}_{\mu\tau}}{\partial x^{\sigma}\partial x^{\nu}} + \frac{\partial^2\Gamma^{\alpha}_{\mu\nu}}{\partial x^{\sigma}\partial x^{\tau}}$$

Remembering that $\Gamma^{\alpha}_{\beta\gamma}$ is symmetric in β and γ, we obtain by adding the last 3 equations:

(6) $B^{\alpha}_{\mu\nu\sigma|\tau} + B^{\alpha}_{\mu\sigma\tau|\nu} + B^{\alpha}_{\mu\tau\nu|\sigma} = 0$ (at P).

Therefore:

(7) $B^{\alpha}_{\mu\nu\sigma\|\tau} + B^{\alpha}_{\mu\sigma\tau\|\nu} + B^{\alpha}_{\mu\tau\nu\|\sigma} = 0;$

because the covariant and ordinary derivatives coincide at P. This last equation is a tensor equation, which must consequently hold in all frames of reference because it holds in one of them (namely in the system in which the affine connections vanish at P). But P is an arbitrary point of our space. Therefore: equation (7) is generally valid.

(II)
EINSTEIN'S TENSOR IN RIEMANNIAN SPACE

We now consider a Riemannian space in which $ds^2 = g_{\mu\nu}dx^\mu dx^\nu$, where $g_{\mu\nu}$ is a symmetric covariant tensor. We define $B_{\alpha\mu\nu\sigma} = g_{\alpha\beta}B^\beta_{\mu\nu\sigma}$. We shall need the following (symmetry) properties of the covariant curvature tensor:

(8) $B_{\alpha\mu\nu\sigma} = -B_{\mu\alpha\nu\sigma} = -B_{\alpha\mu\sigma\nu} = B_{\nu\sigma\alpha\mu}$.

(These relations were proved in a previous appendix.)
Through lowering the index α in the Bianchi identity (7), i.e. through multiplying by $g_{\beta\alpha}$ and then substituting α for β:

(9) $B_{\alpha\mu\nu\sigma\|\tau} + B_{\alpha\mu\sigma\tau\|\nu} + B_{\alpha\mu\tau\nu\|\sigma} = 0$.

By the last equation in (8), equation (9) is equivalent to:

(10) $B_{\nu\sigma\alpha\mu\|\tau} + B_{\sigma\tau\alpha\mu\|\nu} + B_{\tau\nu\alpha\mu\|\sigma} = 0$.

Remember that $R_{\mu\nu} \underset{\text{Def}}{=} B^\sigma_{\mu\nu\sigma}$. Let us show that $R_{\mu\nu} = R_{\nu\mu}$. We have:

$$\begin{aligned} R_{\mu\nu} &= B^\sigma_{\mu\nu\sigma} = g^{\sigma\alpha}B_{\alpha\mu\nu\sigma} = +g^{\sigma\alpha}B_{\nu\sigma\alpha\mu} \\ &= -g^{\sigma\alpha}B_{\sigma\nu\alpha\mu} = +g^{\sigma\alpha}B_{\sigma\nu\mu\alpha} = g^{\alpha\sigma}B_{\sigma\nu\mu\alpha} \\ &= B^\alpha_{\nu\mu\alpha} \underset{\text{Def}}{=} R_{\nu\mu}. \end{aligned}$$

(We have used (8) throughout) Thus:

(11) $R_{\mu\nu} = R_{\nu\mu}$; i.e. $R_{\mu\nu}$ is symmetric.

Define: $G_{\mu\nu} = R_{\mu\nu} - \frac{1}{2} g_{\mu\nu} R = R_{\mu\nu} - \frac{1}{2} g_{\mu\nu}g^{\alpha\beta}R_{\alpha\beta}$.

It follows that:

$$G^{\mu\nu} = R^{\mu\nu} - \frac{1}{2} g^{\mu\nu}R \text{ and } G_\mu{}^\nu = R_\mu{}^\nu - \frac{1}{2}\delta^\nu_\mu R.$$

$G_{\mu\nu}$ is called Einstein's tensor.

Theorem 3: $G^{\mu\nu}{}_{\|\nu} = 0$; i.e. the divergence of Einstein's tensor vanishes identically.

Proof: Lowering the index μ, we see that it suffices to show that: $G_{\mu}{}^{\nu}{}_{\|\nu} = 0$; i.e. to show that: $(R_{\mu}{}^{\nu}{}_{\|\nu} - \frac{1}{2}(\delta^{\nu}_{\mu}R)_{\|\nu}) = 0$; i.e. to show that: $R_{\mu}{}^{\nu}{}_{\|\nu} - \frac{1}{2}\delta^{\nu}_{\mu}R_{\|\nu} = 0$, since δ^{ν}_{μ} behaves like a constant with respect to covariant differentiation. But R is an invariant; hence $R_{\|\nu} = R_{|\nu} = \partial R/\partial x^{\nu}$. Hence it suffices to show that $R_{\mu}{}^{\nu}{}_{\|\nu} - \frac{1}{2}\delta^{\nu}_{\mu}\,\partial R/\partial x^{\nu} = 0$ i.e. to show that

$$R_{\mu}{}^{\nu}{}_{\|\nu} = \frac{1}{2}\frac{\partial R}{\partial x^{\mu}}.$$

By definition of $R_{\mu\nu}$:

(12) $R_{\mu}{}^{\nu}{}_{\|\nu} = R_{\mu}{}^{\gamma}{}_{\|\gamma} = g^{\gamma\nu}R_{\mu\nu\|\gamma} = g^{\gamma\nu}B^{\sigma}_{\mu\nu\sigma\|\gamma} = g^{\gamma\nu}g^{\sigma\alpha}B_{\alpha\mu\nu\sigma\|\gamma}$.

By equation (10):

(13) $B_{\alpha\mu\nu\sigma\|\gamma} = -B_{\mu\gamma\nu\sigma\|\alpha} - B_{\gamma\alpha\nu\sigma\|\mu}$.

By (12) & (13):

$$R_{\mu}{}^{\nu}{}_{\|\nu} = g^{\gamma\nu}g^{\sigma\alpha}B_{\alpha\mu\nu\sigma\|\gamma} = [-g^{\gamma\nu}g^{\sigma\alpha}B_{\mu\gamma\nu\sigma\|\alpha} - g^{\gamma\nu}g^{\sigma\alpha}B_{\gamma\alpha\nu\sigma\|\mu}] = [g^{\gamma\nu}g^{\sigma\alpha}B_{\gamma\mu\nu\sigma\|\alpha} + g^{\gamma\nu}g^{\sigma\alpha}B_{\alpha\gamma\nu\sigma\|\mu}]$$

$$= [-g^{\alpha\sigma}g^{\nu\gamma}B_{\gamma\mu\sigma\nu\|\alpha} + g^{\gamma\nu}g^{\sigma\alpha}B_{\alpha\gamma\nu\sigma\|\mu}]$$

$$= [-g^{\alpha\sigma}B^{\nu}_{\mu\sigma\nu\|\alpha} + g^{\gamma\nu}B^{\sigma}_{\gamma\nu\sigma\|\mu}]$$

i.e.

$$\begin{aligned} R_{\mu}{}^{\nu}{}_{\|\nu} &= -g^{\alpha\sigma}R_{\mu\sigma\|\alpha} + g^{\gamma\nu}R_{\gamma\nu\|\mu} \\ &= -(g^{\alpha\sigma}R_{\mu\sigma})_{\|\alpha} + (g^{\gamma\nu}R_{\gamma\nu})_{\|\mu} \\ &= -R_{\mu}{}^{\alpha}{}_{\|\alpha} + R_{\|\mu} \\ &= -R_{\mu}{}^{\nu}{}_{\|\nu} + \frac{\partial R}{\partial x^{\mu}} \end{aligned}$$

since $g_{\alpha\beta}$ and $g^{\alpha\beta}$ behave like constants in covariant differentiation. By this last equation:

(14) $2R_{\mu}{}^{\nu}{}_{\|\nu} = \dfrac{\partial R}{\partial x^{\mu}}$; i.e. $R_{\mu}{}^{\nu}{}_{\|\nu} = \dfrac{1}{2}\dfrac{\partial R}{\partial x^{\mu}}$

We have already shown that equation (14) implies:

$G^{\mu\nu}{}_{\|\nu} = 0.$

Q.E.D.

APPENDIX NINE

CONSERVATION LAWS IN GTR (see Einstein, Lorentz, Weyl & Minkowski 1923, pp. 143-150)

In the course of obtaining the conservation laws in GTR we shall need the following simple relations.

(I)

Let $g = \det(g_{ij})$ and let Q^{ij} be the cofactor of g_{ij} in g. Since $g_{ij} = g_{ji}$, it follows that $Q^{ij} = Q^{ji}$; i.e. Q^{ij} is symmetric in i and j.

By the definition of a determinant: $g_{ij}Q^{jm} = g_{ij}Q^{mj} = \delta^m_i g$. Hence $g_{ij}\,(\frac{1}{g}\,Q^{jm}) = \delta^m_i$. By definition of the matrix (g^{ij}) as $(g_{ij})^{-1}$, it follows that: $g^{ij} = \frac{1}{g}\,Q^{ij} = \frac{1}{g}\,Q^{ji}$. Hence $Q^{ij} = g \cdot g^{ij}$

Again, by the definition of a determinant: $g = \sum_j g_{ij}Q^{ij}$ (no summation over i). Hence:

$$\text{(1)}\quad \begin{cases} \partial g / \partial g_{ij} = Q^{ij} = g \cdot g^{ij}. \text{ In other words:} \\ g^{ij} = \dfrac{1}{g}\dfrac{\partial g}{\partial g_{ij}} = \dfrac{1}{(-g)}\dfrac{\partial(-g)}{\partial g_{ij}} = \dfrac{\partial \log(-g)}{\partial g_{ij}}. \end{cases}$$

(II)

It follows from 1 that:

$$\text{(2)}\quad \frac{\partial g}{\partial x^\sigma} = \frac{\partial g}{\partial g_{ij}} \cdot \frac{\partial g_{ij}}{\partial x^\sigma} = g \cdot g^{ij}\,\frac{\partial g_{ij}}{\partial x^\sigma}$$

(III)

Remember that $g_{ij}g^{jm} = \delta^m_i$. Differentiating with respect to x^σ: $(\partial g_{ij}/\partial x^\sigma)\cdot g^{jm} + g_{ij}\cdot \partial g^{jm}/\partial x^\sigma = 0$. Multiplying by $g^{\alpha i}$: $g^{\alpha i}\cdot g^{jm}\cdot(\partial g_{ij}/\partial x^\sigma) + \delta^\alpha_j(\partial g^{jm}/\partial x^\sigma) = 0$. In other words:

$$\textbf{(3)}\quad \frac{\partial g^{\alpha m}}{\partial x^\sigma} = -\, g^{\alpha i}\cdot g^{mj}\frac{\partial g_{ij}}{\partial x^\sigma}\cdot \text{(since } g^{mj} = g^{jm}\text{)}$$

Multiplying (3) by $g_{\tau\alpha}g_{\beta m}$: $g_{\tau\alpha}g_{\beta m}\dfrac{\partial g^{\alpha m}}{\partial x^\sigma} = -\,\delta^i_\tau\,\delta^j_\beta\dfrac{\partial g_{ij}}{\partial x^\sigma}$

In other words:

$$\textbf{(4)}\quad \frac{\partial g_{\tau\beta}}{\partial x^\sigma} = -\, g_{\tau\alpha}\cdot g_{\beta m}\frac{\partial g^{\alpha m}}{\partial x^\sigma}.$$

Definitions: $[\alpha\beta,\gamma] \underset{\text{def.}}{=} \dfrac{1}{2}\left(\dfrac{\partial g_{\alpha\gamma}}{\partial x^\beta} + \dfrac{\partial g_{\beta\gamma}}{\partial x^\alpha} - \dfrac{\partial g_{\alpha\beta}}{\partial x^\gamma}\right)$

and: $\Gamma^\sigma_{\alpha\beta} = \left\{{\sigma \atop \alpha\beta}\right\} = g^{\sigma\gamma}\,[\alpha\beta,\gamma] = \dfrac{1}{2}g^{\sigma\gamma}\left(\dfrac{\partial g_{\alpha\gamma}}{\partial x^\beta} + \dfrac{\partial g_{\beta\gamma}}{\partial x^\alpha} - \dfrac{\partial g_{\alpha\beta}}{\partial x^\gamma}\right)$

(IV)

To express $\partial g_{\beta\gamma}/\partial x^\alpha$ and $\partial g^{\beta\gamma}/\partial x^\alpha$ in terms of the $\left\{{\alpha \atop \beta\gamma}\right\}$'s and the $[\beta\gamma,\alpha]$'s, write $g_{\alpha\beta|\gamma}$ for $\partial g_{\alpha\beta}/\partial x^\gamma$.

$$[\alpha\beta,\gamma] + [\alpha\gamma,\beta] = \frac{1}{2}\,(g_{\alpha\gamma|\beta} + g_{\beta\gamma|\alpha} - g_{\alpha\beta|\gamma} +$$

$$g_{\alpha\beta|\gamma} + g_{\gamma\beta|\alpha} - g_{\alpha\gamma|\beta}) = \frac{1}{2}\cdot 2g_{\beta\gamma|\alpha} = g_{\beta\gamma|\alpha}.$$

[Remember that $g_{\gamma\beta} = g_{\beta\gamma}$; hence $\partial g_{\gamma\beta}/\partial x^\alpha = \partial g_{\beta\gamma}/\partial x^\alpha$, i.e. $g_{\gamma\beta|\alpha} = g_{\beta\gamma|\alpha}$]

(5) We have shown that: $\dfrac{\partial g_{\beta\gamma}}{\partial x^\alpha} = [\alpha\beta,\gamma] + [\alpha\gamma,\beta]$.

By equations (3) and (5):

$$\textbf{(6)}\quad \frac{\partial g^{\alpha m}}{\partial x^\sigma} = -\, g^{\alpha i}\cdot g^{mj}\cdot\frac{\partial g_{ij}}{\partial x^\sigma} = -g^{\alpha i}\cdot g^{mj}([\sigma i, j] + [\sigma j, i])$$

By the definitions of $\Gamma^{\alpha}_{\beta\gamma}$ and $\left\{ {\alpha \atop \beta\gamma} \right\}$ and by (6)

$$(7)\ \frac{\partial g^{\alpha m}}{\partial x^{\sigma}} = (- g^{\alpha i} \cdot g^{mj}[\sigma i, j] - g^{\alpha i} \cdot g^{mj}[\sigma j, i]) =$$

$$\left(- g^{\alpha i}\left\{ {m \atop \sigma i} \right\} - g^{mj}\left\{ {\alpha \atop \sigma j} \right\}\right) = \left(- g^{\alpha i}\,\Gamma^{m}_{\sigma i} - g^{mj}\Gamma^{\alpha}_{\sigma j}\right)$$

Note that $[\alpha\beta, \gamma]$ and $\left\{ {\gamma \atop \alpha\beta} \right\}$ are interchangeable in the following sense. By definition:

$$\left\{ {\gamma \atop \alpha\beta} \right\} = \Gamma^{\gamma}_{\alpha\beta} = + g^{\gamma\sigma}[\alpha\beta, \sigma].\ \text{Multiplying by } g_{\tau\gamma}:$$

$$g_{\tau\gamma} \cdot \left\{ {\gamma \atop \alpha\beta} \right\} = g_{\tau\gamma}\Gamma^{\gamma}_{\alpha\beta} = \delta^{\sigma}_{\tau}[\alpha\beta, \sigma] = [\alpha\beta, \tau]$$

(V)

To show that $\Gamma^{\alpha}_{i\alpha} = \Gamma^{\alpha}_{\alpha i} = \left\{ {\alpha \atop i\alpha} \right\} = \left\{ {\alpha \atop \alpha i} \right\} = \dfrac{\partial \log\sqrt{-g}}{\partial x^{i}}$

$\left\{ {\alpha \atop \alpha i} \right\} = g^{\alpha\sigma}[\alpha i, \sigma] = \dfrac{1}{2}g^{\alpha\sigma}\left(\dfrac{\partial g_{\alpha\sigma}}{\partial x^{i}} + \dfrac{\partial g_{i\sigma}}{\partial x^{\alpha}} - \dfrac{\partial g_{\alpha i}}{\partial x^{\sigma}}\right)$. But $g^{\alpha\sigma}\dfrac{\partial g_{i\sigma}}{\partial x^{\alpha}} = g^{\sigma\alpha}\dfrac{\partial g_{i\alpha}}{\partial x^{\sigma}}$ [by interchange of dummies α & σ] $= g^{\alpha\sigma}\dfrac{\partial g_{\alpha i}}{\partial x^{\sigma}}$ [by symmetry of $g^{\alpha\sigma}$ & $g_{\alpha i}$].

Thus, by the last two equations, and by (1):

$$\left\{ {\alpha \atop \alpha i} \right\} = \frac{1}{2} g^{\alpha\sigma} \cdot \frac{\partial g_{\alpha\sigma}}{\partial x^{i}}$$

$$= \frac{1}{2} \cdot \frac{1}{(-g)} \cdot \frac{\partial(-g)}{\partial g_{\alpha\sigma}} \cdot \frac{\partial g_{\alpha\sigma}}{\partial x^{i}}$$

$$= \frac{1}{2} \cdot \frac{1}{(-g)} \cdot \frac{\partial(-g)}{\partial x^{i}}$$

$$= \frac{1}{2}\frac{\partial \log(-g)}{\partial x^{i}}$$

$$= \frac{\partial\left(\frac{1}{2}\log(-g)\right)}{\partial x^{i}}$$

$$= \frac{\partial \log\sqrt{-g}}{\partial x^i}$$

$$= \frac{1}{\sqrt{-g}} \cdot \frac{\partial\sqrt{-g}}{\partial x^i}.$$ We have thus proved:

(8) $\Gamma^{\alpha}_{\alpha i} = \Gamma^{\alpha}_{i\alpha} = \begin{Bmatrix} \alpha \\ \alpha i \end{Bmatrix} = \begin{Bmatrix} \alpha \\ i\alpha \end{Bmatrix} = \frac{1}{\sqrt{-g}} \frac{\partial\sqrt{-g}}{\partial x^i} = \frac{\partial \log\sqrt{-g}}{\partial x^i}$

(VI)

Einstein's field equations for free space are as follows. Let:

(9) $B^{\alpha}_{\cdot\ \beta\gamma\sigma}$ = curvature tensor = $\left(\Gamma^{\alpha}_{i\gamma}\, \Gamma^{i}_{\beta\sigma} - \frac{\partial\Gamma^{\alpha}_{\beta\gamma}}{\partial x^{\sigma}} - \Gamma^{\alpha}_{i\sigma}\, \Gamma^{i}_{\beta\gamma} + \frac{\partial\Gamma^{\alpha}_{\beta\sigma}}{\partial x^{\gamma}}\right)$

Contracting α and σ:

(10) $R_{\beta\gamma} = B^{\alpha}_{\cdot\ \beta\gamma\alpha} = \left(\Gamma^{\alpha}_{i\gamma}\, \Gamma^{i}_{\beta\alpha} - \frac{\partial\Gamma^{\alpha}_{\beta\gamma}}{\partial x^{\alpha}} - \Gamma^{\alpha}_{i\alpha}\, \Gamma^{i}_{\beta\gamma} + \frac{\partial\Gamma^{\alpha}_{\beta\alpha}}{\partial x^{\gamma}}\right)$

Remember that, by (8′): $\Gamma^{\alpha}_{i\alpha} = \Gamma^{\alpha}_{\alpha i} = \frac{1}{\sqrt{-g}} \cdot \frac{\partial\sqrt{-g}}{\partial x_i}$.

The field equations for free space are $R_{\beta\gamma} = 0$. We can always choose a coordinate system throughout which $\sqrt{-g} = 1$ or $g = -1$. In such a coordinate system: $\Gamma^{\alpha}_{i\alpha} = \frac{1}{\sqrt{-g}} \cdot \frac{\partial\sqrt{-g}}{\partial x^i} = 1 \cdot \frac{\partial 1}{\partial x^i} = 0$. Hence, by (2), $R_{\beta\gamma}$ reduces to $(\Gamma^{\alpha}_{i\gamma}\, \Gamma^{i}_{\beta\alpha} - \partial\Gamma^{\alpha}_{\beta\gamma}/\partial x^{\alpha})$. Therefore the field equations can be written in the form:

(11) $\begin{cases} R_{\mu\nu} = \frac{\partial\Gamma^{\alpha}_{\mu\nu}}{\partial x^{\alpha}} + \Gamma^{\alpha}_{i\nu}\Gamma^{i}_{\mu\alpha} = \frac{\partial\Gamma^{\alpha}_{\mu\nu}}{\partial x^{\alpha}} + \Gamma^{\alpha}_{\mu;}\, \Gamma^{\beta}_{\nu\alpha} = 0 \\ \sqrt{-g} = 1 \text{ or } g = -1. \end{cases}$

Einstein proposed to construct a Hamiltonian function H and write the field equations in the form $\delta \int H\, dw = 0$, where: $dw = \sqrt{-g} \cdot dx^0 \cdot dx^1 \cdot dx^2 \cdot dx^3 = dx^0 \cdot dx^1 \cdot dx^2 \cdot dx^3$ (in view of $\sqrt{-g} = 1$). He took:

(12) $H = g^{\mu\nu}\, \Gamma^{\alpha}_{\mu\beta}\, \Gamma^{\beta}_{\gamma\alpha};$

where H is taken to be a function of the quantities $g^{\mu\nu}$ and $g^{\mu\upsilon}_{\sigma} \underset{\text{def}}{=} \partial g^{\mu\upsilon}/\partial x^{\sigma}$.

Einstein proceeded as follows: By (12):

$$\delta H = [\delta(g^{\mu\nu})\,\Gamma^{\alpha}_{\mu\beta}\,\Gamma^{\beta}_{\nu\alpha} + g^{\mu\nu}\cdot\delta\,\Gamma^{\alpha}_{\mu\beta}\cdot\Gamma^{\beta}_{\nu\alpha} + g^{\mu\nu}\cdot\Gamma^{\alpha}_{\mu\beta}\cdot\delta\,\Gamma^{\beta}_{\nu\alpha}].$$

By interchanging dummies α & β, μ & ν:

$$g^{\mu\nu}\cdot\delta(\Gamma^{\alpha}_{\mu\beta})\cdot\Gamma^{\beta}_{\nu\alpha} = g^{\nu\mu}\cdot\delta(\Gamma^{\beta}_{\nu\alpha})\cdot\Gamma^{\alpha}_{\mu\beta} = g^{\mu\nu}\,\Gamma^{\alpha}_{\mu\beta}\,\delta(\Gamma^{\beta}_{\nu\alpha}).$$

Hence, by the last expression for δH:

(13) $\delta H = \delta g^{\mu\nu}\cdot\Gamma^{\alpha}_{\mu\beta}\cdot\Gamma^{\beta}_{\nu\alpha} + 2g^{\mu\nu}\cdot\Gamma^{\alpha}_{\mu\beta}\cdot\delta\Gamma^{\beta}_{\nu\alpha}$. We can write:

$g^{\mu\nu}\cdot\delta\Gamma^{\beta}_{\nu\alpha} = \delta(g^{\mu\nu}\,\Gamma^{\beta}_{\nu\alpha}) - \delta g^{\mu\nu}\cdot\Gamma^{\beta}_{\nu\alpha}$. Hence, by (13):

(14) $$\delta H = \delta g^{\mu\nu}\cdot\Gamma^{\alpha}_{\mu\beta}\,\Gamma^{\beta}_{\nu\alpha} + 2\,\Gamma^{\alpha}_{\mu\beta}\cdot\delta(g^{\mu\nu}\,\Gamma^{\beta}_{\nu\alpha}) - 2\delta g^{\mu\nu}\cdot\Gamma^{\alpha}_{\mu\beta}\,\Gamma^{\beta}_{\nu\alpha} = - \delta g^{\mu\nu}\cdot\Gamma^{\alpha}_{\mu\beta}\,\Gamma^{\beta}_{\nu\alpha} + 2\Gamma^{\alpha}_{\mu\beta}\,\delta(g^{\mu\nu}\Gamma^{\beta}_{\nu\alpha}).$$

By definition of $\Gamma^{\alpha}_{\beta\gamma,}$ we have

(15) $$\delta(g^{\mu\nu}\,\Gamma^{\beta}_{\nu\alpha}) = \frac{1}{2}\,\delta\left(g^{\mu\nu}g^{\beta\lambda}\left(\frac{\partial g_{\nu\lambda}}{\partial x^{\alpha}} + \frac{\partial g_{\alpha\lambda}}{\partial x^{\nu}} - \frac{\partial g_{\nu\alpha}}{\partial x^{\lambda}}\right)\right)\cdot$$

$$= \frac{1}{2}\,\delta\left(g^{\mu\nu}g^{\beta\lambda}\,\frac{\partial g_{\nu\lambda}}{\partial x^{\alpha}}\right) + \frac{1}{2}\,\delta\left(g^{\mu\nu}g^{\beta\lambda}\cdot\frac{\partial g_{\alpha\lambda}}{\partial x^{\nu}}\right) - \frac{1}{2}\,\delta\left(g^{\mu\nu}g^{\beta\lambda}\,\frac{\partial g_{\nu\alpha}}{\partial x^{\lambda}}\right)$$

$$= -\frac{1}{2}\,\delta\left(\frac{\partial g^{\mu\beta}}{\partial x^{\alpha}}\right) + \frac{1}{2}\,\delta\left(g^{\mu\nu}g^{\beta\lambda}\frac{\partial g_{\alpha\lambda}}{\partial x^{\nu}}\right) - \frac{1}{2}\,\delta\left(g^{\mu\nu}g^{\beta\lambda}\frac{\partial g_{\nu\alpha}}{\partial x^{\lambda}}\right)$$

[by (3)]. Remember that, by (14), we have to multiply $\delta(g^{\mu\nu}\,\Gamma^{\beta}_{\nu\alpha})$ by $\Gamma^{\alpha}_{\mu\beta}$ to show that the last 2 terms in (15) give no contribution.

Interchanging dummies μ and β, λ and ν:

(16) $$\Gamma^{\alpha}_{\mu\beta}\cdot\delta\left(g^{\mu\nu}\,g^{\beta\lambda}\,\frac{\partial g_{\alpha\lambda}}{\partial x^{\nu}}\right) = \Gamma^{\alpha}_{\beta\mu}\,\delta\left(g^{\beta\lambda}g^{\mu\nu}\frac{\partial g_{\alpha\nu}}{\partial x^{\lambda}}\right) = \Gamma^{\alpha}_{\mu\beta}\,\delta\left(g^{\mu\nu}\,g^{\beta\lambda}\,\frac{\partial g_{\nu\alpha}}{\partial x^{\lambda}}\right)$$

By equations (14), (15) and (16)

(17) $$\delta H = -\delta g^{\mu\nu}\,\Gamma^{\alpha}_{\mu\beta}\,\Gamma^{\beta}_{\nu\alpha} + 2\Gamma^{\alpha}_{\mu\beta}\cdot\left(-\frac{1}{2}\right)\cdot\delta\left(\frac{\partial g^{\mu\beta}}{\partial x^{\alpha}}\right) = \left(-\Gamma^{\alpha}_{\mu\beta}\cdot\Gamma^{\beta}_{\nu\alpha}\cdot\delta g^{\mu\nu} - \Gamma^{\alpha}_{\mu\beta}\cdot\delta g^{\mu\beta}_{\alpha}\right)$$

[Remember: $g^{\mu\beta}_{\alpha} = \partial g^{\mu\beta}/\partial x^{\alpha}$] Since, trivially:
$\delta H = (\partial H/\partial g^{\mu\nu})\,\delta g^{\mu\nu} + (\partial H/\partial g^{\mu\beta}_{\alpha})\cdot\delta g^{\mu\beta}_{\alpha}$, we obtain from (17)

$$(18)\ \frac{\partial H}{\partial g^{\mu\nu}} = -\Gamma^{\alpha}_{\mu\beta}\,\Gamma^{\beta}_{\nu\alpha} \text{ and } \frac{\partial H}{\partial g^{\mu\beta}_{\alpha}} = -\Gamma^{\alpha}_{\mu\beta}.$$

Let us consider the condition $\delta(\int H dw) = 0$. By the Euler-Lagrange equations:

$$(19)\ \begin{cases} \dfrac{\partial}{\partial x^{\alpha}}\left(\dfrac{\partial H}{\partial g^{\mu\nu}_{\alpha}}\right) - \dfrac{\partial H}{\partial g^{\mu\nu}} = 0; \text{ i.e. by (18)} \\[2ex] -\dfrac{\partial \Gamma^{\alpha}_{\mu\nu}}{\partial x^{\alpha}} + \Gamma^{\alpha}_{\mu\beta}\,\Gamma^{\beta}_{\nu\alpha} = 0. \end{cases}$$

These are the free-space field equations $R_{\mu\nu} = 0$. (Condition: $\sqrt{-g} = 1$.) We now propose to determine a matrix t^{α}_{σ} such that the ordinary divergence $\partial t^{\alpha}_{\sigma}/\partial x^{\alpha}$ vanishes. Multiplying (19) by $g^{\mu\nu}_{\sigma} \underset{\text{Def}}{=} \partial g^{\mu\nu}/\partial x^{\sigma}$:

$$(20)\ g^{\mu\nu}_{\sigma} \cdot \frac{\partial}{\partial x^{\alpha}}\left(\frac{\partial H}{\partial g^{\mu\nu}_{\alpha}}\right) - \frac{\partial H}{\partial g^{\mu\nu}} \cdot \frac{\partial g^{\mu\nu}}{\partial x^{\sigma}} = 0.$$

We can write:

$$(21)\ g^{\mu\nu}_{\sigma}\frac{\partial}{\partial x^{\alpha}}\left(\frac{\partial H}{\partial g^{\mu\nu}_{\alpha}}\right) = \frac{\partial}{\partial x^{\alpha}}\left(g^{\mu\nu}_{\sigma}\frac{\partial H}{\partial g^{\mu\nu}_{\alpha}}\right) - \frac{\partial H}{\partial g^{\mu\nu}_{\alpha}} \cdot \frac{\partial g^{\mu\nu}_{\sigma}}{\partial x^{\alpha}}$$

$$= \frac{\partial}{\partial x^{\alpha}}\left(g^{\mu\nu}_{\sigma}\frac{\partial H}{\partial g^{\mu\nu}_{\alpha}}\right) - \frac{\partial H}{\partial g^{\mu\nu}_{\alpha}} \cdot \frac{\partial g^{\mu\nu}_{\alpha}}{\partial x^{\sigma}};$$

$$\text{since: } \frac{\partial g^{\mu\nu}_{\sigma}}{\partial x^{\alpha}} = \frac{\partial}{\partial x^{\alpha}}\left(\frac{\partial g^{\mu\nu}}{\partial x^{\sigma}}\right) = \frac{\partial}{\partial x^{\sigma}}\left(\frac{\partial g^{\mu\nu}}{\partial x^{\alpha}}\right) = \frac{\partial g^{\mu\nu}_{\alpha}}{\partial x^{\sigma}}.$$

It follows from (20) and (21) that:

$$\frac{\partial}{\partial x^{\alpha}}\left(g^{\mu\nu}_{\sigma}\frac{\partial H}{\partial g^{\mu\nu}_{\alpha}}\right) - \left(\frac{\partial H}{\partial g^{\mu\nu}_{\alpha}} \cdot \frac{\partial g^{\mu\nu}_{\alpha}}{\partial x^{\sigma}} + \frac{\partial H}{\partial g^{\mu\nu}} \cdot \frac{\partial g^{\mu\nu}}{\partial x^{\sigma}}\right) = 0$$

Since H is a function of the $g^{\mu\nu}_{\alpha}$'s and the $g^{\mu\nu}$'s, the last equation is identical with the following:

$$\frac{\partial}{\partial x^{\alpha}}\left(g^{\mu\nu}_{\sigma} \cdot \frac{\partial H}{\partial g^{\mu\nu}}\right) - \frac{\partial H}{\partial x^{\sigma}} = 0; \text{ i.e.}$$

$$\frac{\partial}{\partial x^{\alpha}}\left(g^{\mu\nu}_{\sigma}\frac{\partial H}{\partial g^{\mu\nu}_{\alpha}}\right) - \frac{\partial}{\partial x^{\alpha}}(\delta^{\alpha}_{\sigma}H) = 0; \text{ i.e.}$$

$$(22)\begin{cases}\dfrac{\partial}{\partial x^\alpha}\left(g_\sigma^{\mu\nu}\dfrac{\partial H}{\partial g_\alpha^{\mu\nu}} - \delta_\sigma^\alpha H\right) = 0;\text{ or }\dfrac{\partial t_\sigma^\alpha}{\partial x^\alpha} = 0;\text{ where we put}\\[2ex] t_\sigma^\alpha = -\dfrac{1}{2k}\left(g_\sigma^{\mu\nu}\cdot\dfrac{\partial H}{\partial g_\alpha^{\mu\nu}} - \delta_\sigma^\alpha H\right),\text{ k being some constant.}\end{cases}$$

Let us express t_σ^α in terms of the fundamental tensor. By definition of H: $H = g^{\mu\nu}\Gamma_{\mu\beta}^{\gamma}\,\Gamma_{\nu\gamma}^{\beta}$. By (18) and (7) (i.e. by the expression of $\partial g^{\mu\nu}/\partial x^\sigma$ in terms of the $\Gamma_{\beta\lambda}^{\alpha}$'s):

$$g_\sigma^{\mu\nu}\frac{\partial H}{\partial g_\alpha^{\mu\nu}} = \frac{\partial g^{\mu\nu}}{\partial x^\sigma}\cdot\frac{\partial H}{\partial g_\sigma^{\mu\nu}} = (g^{\mu\tau}\Gamma_{\tau\sigma}^{\nu} + g^{\nu\tau}\Gamma_{\tau\sigma}^{\mu})\cdot\Gamma_{\mu\nu}^{\alpha} =$$

$$(g^{\tau\mu}\Gamma_{\tau\sigma}^{\nu}\,\Gamma_{\mu\nu}^{\alpha} + g^{\tau\nu}\,\Gamma_{\tau\sigma}^{\mu}\,\Gamma_{\nu\mu}^{\alpha}) = 2g^{\mu\nu}\,\Gamma_{\mu\tau}^{\alpha}\,\Gamma_{\nu\sigma}^{\tau}$$

Hence, by (22):

$$-\,2k\,t_\sigma^\alpha = \left(g_\sigma^{\mu\nu}\cdot\frac{\partial H}{\partial g_\alpha^{\mu\nu}} - \delta_\sigma^\alpha H\right) = \left(2g^{\mu\nu}\,\Gamma_{\mu\tau}^{\alpha}\,\Gamma_{\nu\sigma}^{\tau} - \delta_\sigma^\alpha g^{\mu\nu}\,\Gamma_{\mu\beta}^{\gamma}\,\Gamma_{\nu\gamma}^{\beta}\right);\text{ i.e.}$$

$$(23)\quad k t_\sigma^\alpha = \frac{1}{2}\,\delta_\sigma^\alpha g^{\mu\nu}\,\Gamma_{\mu\beta}^{\gamma}\,\Gamma_{\nu\gamma}^{\beta} - g^{\mu\nu}\,\Gamma_{\mu\tau}^{\alpha}\,\Gamma_{\nu\sigma}^{\tau}$$

$$= \frac{1}{2}\,\delta_\sigma^\alpha g^{\mu\nu}\,\Gamma_{\mu\beta}^{\gamma}\,\Gamma_{\nu\gamma}^{\beta} - g^{\mu\nu}\,\Gamma_{\mu\sigma}^{\tau}\,\Gamma_{\nu\tau}^{\alpha}$$

[for, interchanging dummies μ & ν and remembering that $g^{\mu\nu}$ is symmetric:

$$g^{\mu\nu}\,\Gamma_{\mu\tau}^{\alpha}\,\Gamma_{\nu\sigma}^{\tau} = g^{\nu\mu}\,\Gamma_{\nu\tau}^{\alpha}\,\Gamma_{\mu\sigma}^{\tau} = g^{\mu\nu}\Gamma_{\mu\sigma}^{\tau}\,\Gamma_{\nu\tau}^{\alpha}]$$

Let us briefly show why the vanishing of an ordinary divergence, e.g. $\partial t_\sigma^\alpha/\partial x^\alpha = 0$, represents a conservation law. Take x^0 to be the time and x^1, x^2, x^3 to be the space coordinates.

$$0 = \frac{\partial t_\sigma^\alpha}{\partial x^\alpha} = \frac{\partial t_\sigma^0}{\partial x^0} + \frac{\partial t_\sigma^1}{\partial x^1} + \frac{\partial t_\sigma^2}{\partial x^2} + \frac{\partial t_\sigma^3}{\partial x^3}.$$

Integrating with respect to x^1, x^2, x^3 and using Gauss's theorem:

$$0 = \iiint\left(\frac{\partial t_\sigma^0}{dx^0} + \frac{\partial t_\sigma^1}{\partial x^1} + \frac{\partial t_\sigma^2}{\partial x^2} + \frac{\partial t_\sigma^3}{\partial x^3}\right)dx^1dx^2dx^3$$

$$= \iiint\frac{\partial t_\sigma^0}{\partial x^0}\,dw + \iiint\nabla\cdot(t_\sigma^1,\,t_\sigma^2,\,t_\sigma^3)\,dw$$

$$= \frac{d}{dx^0}\left(\iiint t_\sigma^0\,dw\right) + \iint(t_\sigma^1,\,t_\sigma^2,\,t_\sigma^3)\cdot\vec{n}\,dS;$$

where: $dw = dx^1 \cdot dx^2 \cdot dx^3$, S is the surface enclosing the volume w of integration and $\vec{n}$ is the outward normal to S. The last equation can be written as:

$$\frac{d}{dx^0}\left(\iiint t^0_\sigma \, dw\right) = -\iint (t^1_\sigma, t^2_\sigma, t^3_\sigma) \cdot \vec{n} \, dS.$$

(VII)

We now propose to rewrite the field equations for free space in a form which exhibits their dependence on the energy matrix t^α_σ. We shall then extend the field equations to the case where the space is non-empty. We shall proceed as follows: to begin with, when we speak of free space, we mean a region void of *non-gravitational* energy. Suppose now that non-gravitational energy, as represented by a divergenceless tensor $T^{\mu\nu}$, is also present. In the field equations for free space, which depend on t^α_σ, we add T^α_σ to t^α_σ and T to t, thus generalising the equations to the case of non-empty space. Let us go back to the field equations:

$$\textbf{(11)} \quad -\frac{\partial \Gamma^\alpha_{\mu\nu}}{\partial x^\alpha} + \Gamma^\alpha_{\mu\beta}\Gamma^\beta_{\nu\alpha} \equiv R_{\mu\nu} = 0; \sqrt{-g} = 1; \text{ i.e. } g = -1.$$

Since we want to exhibit the dependence of these equations on the matrix t^α_σ, we should rewrite them in mixed form. Multiplying by $g^{\sigma\nu}$:

$$-g^{\sigma\nu}\frac{\partial \Gamma^\alpha_{\mu\nu}}{\partial x^\alpha} + g^{\sigma\nu}\,\Gamma^\alpha_{\mu\beta}\,\Gamma^\beta_{\nu\alpha} = 0.$$

As usual we write:

$$g^{\sigma\nu}\frac{\partial \Gamma^\alpha_{\mu\nu}}{\partial x^\alpha} = \frac{\partial}{\partial x^\alpha}(g^{\sigma\nu}\,\Gamma^\alpha_{\mu\nu}) - \Gamma^\alpha_{\mu\nu} \cdot \frac{\partial g^{\sigma\nu}}{\partial x^\alpha}. \text{ Hence:}$$

$$\textbf{(24)} \quad -\frac{\partial}{\partial x^\alpha}(g^{\sigma\nu}\,\Gamma^\alpha_{\mu\nu}) + \Gamma^\alpha_{\mu\nu}\frac{\partial g^{\sigma\nu}}{\partial x^\alpha} + g^{\sigma\nu}\,\Gamma^\alpha_{\mu\beta}\,\Gamma^\beta_{\nu\alpha} = 0.$$

By (7):

$$-\frac{\partial g^{\sigma\nu}}{\partial x^\alpha} = g^{\sigma\tau}\,\Gamma^\nu_{\alpha\tau} + g^{\nu\tau}\,\Gamma^\sigma_{\alpha\tau} = g^{\sigma\tau}\,\Gamma^\nu_{\tau\alpha} + g^{\nu\tau}\,\Gamma^\sigma_{\tau\alpha}$$

Hence, by (24):

(25) $\frac{\partial}{\partial x^\alpha}(g^{\sigma\nu}\,\Gamma^\alpha_{\mu\nu}) = -\ g^{\sigma\tau}\,\Gamma^\nu_{\tau\alpha}\,\Gamma^\alpha_{\mu\nu} - g^{\nu\tau}\,\Gamma^\sigma_{\tau\alpha}\,\Gamma^\alpha_{\mu\nu} + g^{\sigma\nu}\,\Gamma^\alpha_{\mu\beta}\,\Gamma^\beta_{\nu\alpha}$

$$= -\ g^{\nu\tau}\,\Gamma^\sigma_{\tau\alpha}\,\Gamma^\alpha_{\mu\nu} = -\ g^{\nu\tau}\,\Gamma^\sigma_{\tau\alpha}\,\Gamma^\alpha_{\nu\mu}$$

[since $g^{\sigma\tau}\,\Gamma^\alpha_{\mu\nu}\,\Gamma^\nu_{\tau\alpha} = g^{\sigma\nu}\,\Gamma^\alpha_{\mu\beta}\,\Gamma^\beta_{\nu\alpha}$ by interchange of dummies τ and ν, then τ and β]

We now propose to express the right-hand side of (25) in terms of $kt^\sigma_\mu = \frac{1}{2}\,\delta^\sigma_\mu g^{\tau\nu}\Gamma^\lambda_{\tau\beta}\Gamma^\beta_{\nu\lambda} - g^{\tau\nu}\Gamma^\sigma_{\tau\beta}\Gamma^\beta_{\nu\mu}$. We see that the last term of this last equation is the right-hand side of (25). Contracting μ with σ:

(26) $kt = kt^\gamma_\gamma = \frac{1}{2}\,\delta^\gamma_\gamma g^{\tau\nu}\Gamma^\lambda_{\tau\beta}\Gamma^\beta_{\nu\lambda} - g^{\tau\nu}\Gamma^\gamma_{\tau\beta}\Gamma^\beta_{\nu\gamma}$

$= g^{\tau\nu}\Gamma^\lambda_{\tau\beta}\Gamma^\beta_{\nu\lambda}$.[since $\delta^\gamma_\gamma = 4$]. Hence:

(27) $k(t^\sigma_\mu - \frac{1}{2}\,\delta^\sigma_\mu t) = -g^{\tau\nu}\Gamma^\sigma_{\tau\beta}\Gamma^\beta_{\nu\mu} = -g^{\nu\tau}\Gamma^\sigma_{\tau\beta}\Gamma^\beta_{\nu\mu}$.

By (25) and (27), the field equations for free space assume the form:

(28) $\frac{\partial}{\partial x^\alpha}(g^{\sigma\nu}\Gamma^\alpha_{\mu\nu}) = k(t^\sigma_\mu - \frac{1}{2}\,\delta^\sigma_\mu t)$.

This last equation shows the dependence of the field on its own energy, which is represented on the right-hand side by both t^σ_μ and t. The equation lends itself to immediate generalisation. Given a region of space in which other, i.e. non-gravitational, forms of energy described by a symmetric tensor $T^{\mu\nu}$ are present, we simply add T^σ_μ to t^σ_μ and T to t in eq. (28), thus obtaining:

(29) $\frac{\partial}{\partial x^\alpha}\,(g^{\sigma\nu}\Gamma^\alpha_{\mu\nu}) = k[(t^\sigma_\mu + T^\sigma_\mu) - \frac{1}{2}\,\delta^\sigma_\mu(t + T)]$.

Hence

$$\frac{\partial}{\partial x^\alpha}\,(g^{\sigma\nu}\Gamma^\alpha_{\mu\nu}) - k(t^\sigma_\mu - \tfrac{1}{2}\,\delta^\sigma_\mu t) = +k(T^\sigma_\mu - \tfrac{1}{2}\,\delta^\sigma_\mu T).$$

Retracing the steps which took us from (11) to (28):

(30) $-\frac{\partial}{\partial x^\alpha}(\Gamma^\alpha_{\mu\nu}) + \Gamma^\alpha_{\mu\beta}\Gamma^\beta_{\nu\alpha} = -k(T_{\mu\nu} - \frac{1}{2}\,g_{\mu\nu}T)$; i.e.

(31) $R_{\mu\nu} = -k(T_{\mu\nu} - \frac{1}{2}\,g_{\mu\nu}T)$ [since $\sqrt{-g} = 1 =$ constant]

This last equation is a tensor equation, which must therefore hold good in all coordinate systems. We can express (31) in a slightly different form. Multiplying both sides by $g^{\mu\nu}$:

(32) $R = -k(T - \frac{1}{2}\, \delta^{\nu}_{\nu} T) = -k(T - \frac{4}{2}\, T) = kT$

Substituting for kT in (31):

(33) $R_{\mu\nu} - \frac{1}{2}\, g_{\mu\nu} R = -k\, T_{\mu\nu}.$

(VIII)
Conservation Law in the General Case

We propose to show that, in the general case where non-gravitational energy ($T_{\mu\nu}$) may be present, $\frac{\partial}{\partial x^{\sigma}} (T^{\sigma}_{\mu} + t^{\sigma}_{\mu}) = 0$, i.e. the total energy is conserved.

Again we choose a coordinate system in which $\sqrt{-g} = 1$. The field equations are given by (29). Contracting σ with μ:

$$\textbf{(34)}\quad \frac{\partial}{\partial x^{\alpha}} (g^{\sigma\beta}\Gamma^{\alpha}_{\sigma\beta}) = k[(t + T) - \tfrac{1}{2}\, \delta^{\sigma}_{\sigma}\, (t + T)] = -k(t + T).$$

Substituting back in (29):

$$\textbf{(35)}\quad \frac{\partial}{\partial x^{\alpha}} (g^{\sigma\beta}\Gamma^{\alpha}_{\mu\beta} - \tfrac{1}{2}\, \delta^{\sigma}_{\mu}\, g^{\lambda\beta}\Gamma^{\upsilon}_{\lambda\beta}) = k(t^{\sigma}_{\mu} + T^{\sigma}_{\mu}).$$

We form the (ordinary) divergence by differentiating both sides with respect to x^{σ}:

$$\textbf{(36)}\quad \frac{\partial^2}{\partial x^{\sigma}\partial x^{\alpha}} [g^{\sigma\beta}\Gamma^{\alpha}_{\mu\beta} - \tfrac{1}{2}\, \delta^{\sigma}_{\mu} g^{\lambda\beta}\Gamma^{\alpha}_{\lambda\beta}] = k\, \frac{\partial(t^{\sigma}_{\mu} + T^{\sigma}_{\mu})}{\partial x^{\sigma}}$$

It suffices to show that the left-hand side of this equation vanishes. By definition of $\Gamma^{\alpha}_{\beta\gamma}$:

$$\textbf{(37)}\quad \frac{\partial^2}{\partial x^{\sigma}\partial x^{\alpha}} (g^{\sigma\beta}\Gamma^{\alpha}_{\mu\beta}) = \tfrac{1}{2}\, \frac{\partial^2}{\partial x^{\sigma}\partial x^{\alpha}} \left[g^{\sigma\beta} g^{\alpha\lambda} \left(\frac{\partial g_{\mu\lambda}}{\partial x^{\beta}} + \frac{\partial g_{\beta\lambda}}{\partial x^{\mu}} - \frac{\partial g_{\mu\beta}}{\partial x^{\lambda}} \right) \right]$$

By interchanging α and σ, λ and β:

$$\textbf{(38)}\quad \frac{\partial^2}{\partial x^{\sigma}\partial x^{\alpha}} \left(g^{\sigma\beta} g^{\alpha\lambda} \quad \frac{\partial g_{\mu\lambda}}{\partial x^{\beta}} \right) = \frac{\partial^2}{\partial x^{\alpha}\partial x^{\sigma}} \left(g^{\alpha\lambda} g^{\sigma\beta} \quad \frac{\partial g_{\mu\beta}}{\partial x^{\lambda}} \right) = \frac{\partial^2}{\partial x^{\sigma}\partial x^{\alpha}} \left(g^{\sigma\beta} g^{\alpha\lambda} \quad \frac{\partial g_{\mu\beta}}{\partial x^{\lambda}} \right)$$

It follows from (37) and (38) that:

$$\textbf{(39)}\quad \frac{\partial^2}{\partial x^\sigma \partial x^\alpha}(g^{\sigma\beta}\Gamma^\alpha_{\mu\beta}) = \tfrac{1}{2}\frac{\partial^2}{\partial x^\sigma \partial x^\alpha}\left(g^{\sigma\beta}g^{\alpha\lambda}\frac{\partial g_{\beta\lambda}}{\partial x^\mu}\right) =$$

$$-\tfrac{1}{2}\frac{\partial^2}{\partial x^\sigma \partial x^\alpha}\left(\frac{\partial g^{\sigma\alpha}}{\partial x^\mu}\right) = -\tfrac{1}{2}\frac{\partial^3 g^{\sigma\alpha}}{\partial x^\sigma \partial x^\alpha \partial x^\mu} \text{ [by (3)]}$$

Going back to (36), we have to compute $\partial^2(\delta^\sigma_\mu g^{\lambda\beta}\Gamma^\alpha_{\lambda\beta})/\partial x^\sigma \partial x^\alpha$

$$\textbf{(40)}\quad g^{\lambda\beta}\Gamma^\alpha_{\lambda\beta} = \tfrac{1}{2}g^{\lambda\beta}g^{\alpha\tau}\left(\frac{\partial g_{\lambda\tau}}{\partial x^\beta} + \frac{\partial g_{\beta\tau}}{\partial x^\lambda} - \frac{\partial g_{\lambda\beta}}{\partial x^\tau}\right) = \tfrac{1}{2}\left[g^{\beta\lambda}g^{\alpha\tau}\frac{\partial g_{\lambda\tau}}{\partial x^\beta} +\right.$$

$$\left. g^{\lambda\beta}g^{\alpha\tau}\frac{\partial g_{\beta\tau}}{\partial x^\lambda} - g^{\alpha\tau}g^{\lambda\beta}\frac{\partial g_{\lambda\beta}}{\partial x^\tau}\right] = \tfrac{1}{2}\left[-\frac{\partial g^{\beta\alpha}}{\partial x^\beta} - \frac{\partial g^{\lambda\alpha}}{\partial x^\lambda} - g^{\alpha\tau}g^{\lambda\beta}\frac{\partial g_{\lambda\beta}}{\partial x^\tau}\right] =$$

$$\left[-\frac{\partial g^{\beta\alpha}}{\partial x^\beta} - \tfrac{1}{2}g^{\alpha\tau}g^{\lambda\beta}\frac{\partial g_{\lambda\beta}}{\partial x^\tau}\right] \text{ [by (3)].}$$

By (2):

$$g^{\lambda\beta}\frac{\partial g_{\lambda\beta}}{\partial x^\tau} = \frac{1}{g}\frac{\partial g}{\partial x^\tau} = 0, \text{ since } g = -1 = \text{constant.}$$

Hence, by (40):

$$g^{\lambda\beta}\Gamma^\alpha_{\lambda\beta} = -\frac{\partial g^{\beta\alpha}}{\partial x^\beta} = -\frac{\partial g^{\sigma\alpha}}{\partial x^\sigma}.$$

Therefore:

$$\textbf{(41)}\quad \frac{\partial^2}{\partial x^\sigma \partial x^\alpha}(\delta^\sigma_\mu g^{\lambda\beta}\Gamma^\alpha_{\lambda\beta}) = \frac{\partial^2}{\partial x^\mu \partial x^\alpha}(g^{\lambda\beta}\Gamma^\alpha_{\lambda\beta}) = -\frac{\partial^3 g^{\sigma\alpha}}{\partial x^\sigma \partial x^\mu \partial x^\alpha}$$

By (36), (39) and (41), we finally obtain:

$$\textbf{(42)}\quad k\frac{\partial}{\partial x^\sigma}(T^\sigma_\mu + t^\sigma_\mu) = \frac{\partial^2}{\partial x^\sigma \partial x^\alpha}(g^{\sigma\beta}\Gamma^\alpha_{\mu\beta} - \tfrac{1}{2}\delta^\sigma_\mu g^{\lambda\beta}\Gamma^\alpha_{\lambda\beta}) =$$

$$\left(-\tfrac{1}{2}\frac{\partial^3 g^{\sigma\alpha}}{\partial x^\sigma \partial x^\alpha \partial x^\mu} + \tfrac{1}{2}\frac{\partial^3 g^{\sigma\alpha}}{\partial x^\sigma \partial x^\mu \partial x^\alpha}\right) = 0.$$

Q.E.D.

ABBREVIATIONS

EMB	Einstein, "On the Electrodynamics of Moving Bodies" (Einstein 1905)
GTR	General theory of relativity
LCP	Lorentz's *Collected Papers*
LE	Poincaré, *Last Essays* (Poincaré 1913)
LFC	Lorentz-Fitzgerald contraction hypothesis
MFH	Molecular forces hypothesis
MSRP	Methodology of scientific research programmes
RMEE	Lorentz, "The Relative Motion of the Earth and the Ether" (Lorentz 1892b)
SH	Poincaré, "Science and Hypothesis" (Poincaré 1902a)
SM	Poincaré, "Science et méthode" (Poincaré 1908)
STR	Special theory of relativity
TEM	Lorentz, "Théorie électromagnétique de Maxwell" (Lorentz 1892a)
Versuch	Lorentz, *Versuch einer Theorie der electrischen und optischen Erscheinungen in bewegter Körpern* (Lorentz 1895)
VS	Poincaré, *La valeur de la science* (Poincaré 1906a)

BIBLIOGRAPHY

Adler, R., Bazin, M. and Schiffer, M. [1965]: *Introduction to General Relativity,* McGraw-Hill, New York.

Becker, O. [1963]: *Grundlagen der Mathematik,* Karl Alber, Freiburg/München.
Bell, E. T. [1937]: *Men of Mathematics,* Simon & Schuster, New York.
Boltzmann, L. [1905]: *Populäre Schriften,* J. A. Barth, Leipzig.
Born, M. [1956]: "Physics and Relativity", *Helvetica Physica Acta,* Supp. IX.
Born, M. [1962]: *Einstein's Theory of Relativity,* Dover, New York.
Born, M. [1964]: *Natural Philosophy of Cause and Chance,* Dover, New York.
Bouveresse, R. [1981]: *Karl Popper, ou le rationalisme critique,* Vrin, Paris.
Bridgman, P. W. [1927]: *The Logic of Modern Physics,* Macmillan, New York.
Bridgman, P. W. [1936]: *The Nature of Physical Theory,* Wiley, New York.
Bucherer, A. H. [1909]: "Die experimentelle Bestätigung des Relativitätsprinzips", *Annalen der Physik,* 28.

Cajori, F. [1934]: *Sir Isaac Newton's Mathematical Principles,* University of California Press, Berkeley.

de Haas-Lorentz, G. L. (ed.) [1957]: *H. A. Lorentz: Impressions of his Life and Work,* North-Holland, Amsterdam.
Dingler, H. [1926]: *Die Grundgedanken der Machschen Philosophie,* J. A. Barth, Leipzig.
Dreyer, J. L. E. [1953]: *A History of Astronomy from Thales to Kepler,* Dover, New York.
Duhem, P. [1906]: *La Théorie Physique: Son Objet, Sa Structure,* Marcel Rivière, Paris; English translation by P. P. Wiener [1954]: *The Aim and Structure of Physical Theory,* Princeton University Press, Princeton.

Eddington, A. S. [1920]: *Space, Time and Gravitation,* Cambridge University Press, Cambridge.
Eddington, A. S. [1923]: *The Mathematical Theory of Relativity,* Cambridge University Press, Cambridge.
Eddington, A. S. [1939]: *The Philosophy of Physical Science,* University of Michigan, Ann Arbor.
Ehrenfest, P. [1913]: *Zur Krise der Lichtäther-Hypothese.* Reprinted in M. J. Klein (ed.): *Paul Ehrenfest, Collected Scientific Papers,* North-Holland, Amsterdam.
Ehrenfest, P. [1923]: *Professor H. A. Lorentz as Researcher,* in M. J. Klein (ed.): *Paul Ehrenfest, Collected Scientific Papers,* North-Holland, Amsterdam.
Einstein, A. [1905]: "Zur Elektrodynamik bewegter Körper", *Annalen der Physik,* 17; English translation, 'On the Electrodynamics of Moving Bodies', in A. Einstein and others [1923].
Einstein, A. [1907]: "Uber die vom Relativitätsprinzip geforderte Trägheit der Energie", *Annalen der Physik,* 23.

Einstein, A. [1912]: "Relativität und Gravitation. Erwiderung auf eine Bemerkung von M. Abraham", *Annalen der Physik,* 38.

Einstein, A. [1914]: "Die formale Grundlage der allgemeinen Relativitätstheorie", *Sitzungsberichte der preussichen Akademie der Wissenschaften,* Part 2.

Einstein, A. [1915*a*]: "Zur allgemeinen Relativitätstheorie", *Sitzungsberichte der preussichen Akademie der Wissenschaften,* Part 2.

Einstein, A. [1915*b*]: 'Zur allgemeinen Relativitätstheorie (Nachtrag)', *Sitzungsberichte der preussichen Akademie der Wissenschaften,* Part 2.

Einstein, A. [1915*c*]: "Erklärung der Perihelbewegung des Merkurs aus der allgemeinen Relativitätstheorie", *Sitzungsberichte der preussichen Akademie der Wissenschaften,* Part 2.

Einstein, A. [1915*d*]: "Die Feldgleichungen der Gravitation", *Sitzungsberichte der preussichen Akademie der Wissenschaften,* Part 2.

Einstein, A. [1915*e*]: Letter to Sommerfeld, 28.11.1915; in A. Hermann (ed.) [1968]: *Albert Einstein und Arnold Sommerfeld: Briefwechsel,* Schwabe und Co., Basel/Stuttgart.

Einstein, A. [1916]: "Grundlage der allgemeinen Relativitätstheorie", *Annalen der Physik,* 49; English translation, 'The Foundation of the General Theory of Relativity' in A. Einstein and others [1923].

Einstein, A. [1918]: "Prinzipielles zur allgemeinen Relativitätstheorie", *Annalen der Physik,* 55.

Einstein, A. [1920]: *Relativity, the Special and the General Theory,* Methuen, London.

Einstein, A. [1922]:; *The Meaning of Relativity,* Princeton University Press, Princeton.

Einstein, A. [1934]: *Mein Weltbild,* Ullstein, Frankfurt am Main.

Einstein, A. [1936]: "Physics and Reality", *Journal of the Franklin Institute,* 221.

Einstein, A. [1949]: "Autobiographical Notes" in P. A. Schilpp (ed.): *Albert Einstein: Philosopher-Scientist,* Tudor, New York.

Einstein, A. [1950]: *Out of My Later Years,* Littlefields, Adams and Co., New Jersey.

Einstein, A. and others [1923]: *The Principle of Relativity,* Dover, New York.

Einstein, A., Born, M. and Born H. [1969]: *Briefwechsel 1916–1955,* Nymphenburger Verlagshandlung, München.

Einstein, A. and Grossmann, M. [1913]: 'Entwurf einer verallgemeinerten Relativitätstheorie und einer Theorie der Gravitation', *Zeitschrift für Mathematik and Physik,* 62.

Feyerabend, P. K. [1972]: "Von der beschränkten Gültigkeit methodologischer Regeln" in Feyerabend, P. K. [1972]: *Der wissenschaftstheoretische Realismus und die Autorität der Wissenschaften,* Kapitel 10, Vieweg, Braunschweig.

Feyerabend, P. K. [1974]: "Zahar on Einstein", *The British Journal for the Philosophy of Science.*

Feyerabend, P. K. [1975]: *Against Method,* Humanities Press, London.

Feyerabend, P. K. [1980]: "Zahar on Mach, Einstein and Modern Science", *British Journal for the Philosophy of Science',* 31.

Feynman, R. [1965]: *The Character of Physical Law,* MIT Press, Cambridge, Mass.

Giedymin, J. [1982]: *Science and Convention,* Pergamon, Oxford.

Gillies, D. A. [1980]: "Brouwer's Philosophy of Mathematics", *Erkenntnis,* 15.

Grünbaum, A. [1954]: "Operationism and Relativity" in Philipp G. Frank (ed.): *The Validation of Scientific Theories,* Collier, New York.

Grünbaum, A. [1959]: "The Falsifiability of the Lorentz-Fitzgerald Contraction Hypothesis", *The British Journal for the Philosophy of Science,* 10.

Grünbaum, A. [1961]: "The Genesis of the Special Theory of Relativity" in H. Feigl and G. Maxwell (eds.): *Current Issues in the Philosophy of Science,* Holt, Rinehart and Winston, New York.

Grünbaum, A. [1963]: "The Bearing of Philosophy on the History of Science", *Science,* 143.

Grünbaum, A. [1973]: *Philosophical Problems of Space and Time,* Reidel, Dordrecht, Holland.

Hanson, N. R. [1958]: *Patterns of Discovery,* Cambridge University Press, Cambridge.
Heller, G. [1960]: *Ernst Mach,* Springer, Berlin.
Hesse, M. B. [1961]: *Forces and Fields,* Littlefields, Adams and Co., New Jersey.
Hesse, M. B. [1963]: *Models and Analogies in Science,* Sheed and Ward, London and New York.
Hilbert, D. and Courant, R. [1924]: *Methoden der mathematischen Physik,* Springer, Berlin.
Holton, G. [1969]: "Einstein, Michelson, and the 'Crucial' Experiment", *Isis,* 60.
Husserl, E. [1969]: *Cartesianische Meditationen,* Felix Meiner, Hamburg.

Kaufmann, W. [1905]: "Uber die Konstitution des Elektrons", *Sitzungsberichte der preussichen Akademie der Wissenschaften,* 1915, Part 2.
Kaufmann, W. [1906]: "Uber die Konstitution des Elektrons", *Annalen der Physik,* 19.
Kennedy, R. J. and Thorndike E. M. [1932]: "Experimental Establishment of the Relativity of Time", *Physical Review,* 42.
Kilmister, C. W. [1970]: *Special Theory of Relativity,* Pergamon, Oxford.
Kilmister, C. W. [1973]: *General Theory of Relativity,* Pergamon, Oxford.
Kompanayets, A. S. [1962]: *Theoretical Physics,* Dover, New York.
Kretschmann, E. [1917]: "Uber den physikalischen Sinn der Relativitätspostulate, A. Einstein's neue und seine ursprüngliche Relativitätstheorie", *Annalen der Physik,* 53.
Kuhn, T. S. [1962]: *The Structure of Scientific Revolutions,* University of Chicago Press, Chicago.
Kuhn, T. S. [1970]: "Logic of Discovery or Psychology of Research?" in Lakatos and Musgrave (eds.) *Criticism and the Growth of Knowledge,* Cambridge University Press, Cambridge.

Lakatos, I. [1963–4]: "Proofs and Refutations", *The British Journal for the Philosophy of Science,* 14; reprinted as Chapter 1 of Lakatos I. [1976]: *Proofs and Refutations,* Cambridge University Press, Cambridge.
Lakatos, I. [1968*a*]: "Changes in the Problem of Inductive Logic" in I. Lakatos (ed). [1968]: *The Problem of Inductive Logic,* North-Holland, Amsterdam.
Lakatos, I. [1968*b*]: "Criticism and the Methodology of Scientific Research Programmes", *Proceedings of the Aristotelian Society,* 69.
Lakatos, I. [1970]: "Falsification and the Methodology of Scientific Research Programmes" in I. Lakatos and A. Musgrave (eds.) [1970].
Lakatos, I. [1971*a*]: "History of Science and its Rational Reconstructions" in R. C. Buck and R. S. Cohen (eds.): *Boston Studies in the Philosophy of Science,* 8, reprinted in C. Howson (ed.) [1976]: *Method and Appraisal in the Physical Sciences,* Cambridge University Press, Cambridge.
Lakatos, I. [1971*b*]: "Popper zum Abgrenzungs- und Induktionsproblem" in H. Lenk (ed.): *Neue Aspekte der Wissenschaftstheorie;* translated into English as "Popper on Demarcation and Induction" in P. A. Schilpp (ed.) [1973]: *The Philosophy of Sir Karl Popper,* Tudor, New York.
Lakatos, I. [1978]: *The Methodology of Scientific Research Programmes. Philosophical Papers, Volume 1,* edited by J. Worrall and G. Currie, Cambridge University Press, Cambridge.
Lakatos, I. and Musgrave, A. (eds.) [1970]: *Criticism and the Growth of Knowledge,* Cambridge University Press, Cambridge.
Lakatos, I. and Zahar, E. G. [1976]: "Why did Copernicus's Programme Supersede Ptolemy's?" in R. Westman (ed.) [1976]: *The Copernican Achievement,* University of California Press, Berkeley.
Lanczos, C. [1972]: "Einstein's Path from Special to General Relativity" in L. O'Raifeartaigh (ed). [1972]: *General Relativity,* Oxford University Press, Oxford.
Laue, M. von [1911]: *Das Relativitätsprinzip,* Vieweg, Braunschweig.
Laue, M. von [1952]: *Die Relativitätstheorie,* Vieweg, Braunschweig.
Lawden, D. [1962]: *Introduction to Tensor Calculus and Relativity,* Methuen, London.
Lewis, G. N. [1908]: "A Revision of the Fundamental Laws of Matter and Energy", *Philosophical Magazine and Journal of Science,* Series 5, 16.

Lewis, G. N. and Tolman, R. C. [1909]: "The Principle of Relativity and Non-Newtonian Mechanics", *Philosophical Magazine,* Series 5, 18.

Lorentz, H. A. [1886]: "De l'Influence du Mouvement de la Terre sur les Phénomènes Lumineux", *Versl. Kon. Akad. Wetensch. Amsterdam,* 2; reprinted in Lorentz H. A. [1934]: *Collected Papers,* 4, Nijhoff, The Hague.

Lorentz, H. A. [1892*a*]: "La Théorie electromagnétique de Maxwell et son application aux corps mouvants", Arch Néerl., 25; reprinted in *Collected Papers,* 2.

Lorentz, H. A. [1892*b*]: "The Relative Motion of the Earth and the Ether", *Versl. Kon. Wetensch. Amsterdam,* 1; reprinted in *Collected Papers,* 4.

Lorentz, H. A. [1895]: *Versuch einer Theorie der electrischen und optischen Erscheinungen in bewegten Körpern;* reprinted in *Collected Papers,* 5.

Lorentz, H. A. [1899]: "Théorie simplifiée des phénomènes electriques et optiques dans des corps en mouvement", *Versl. Kom. Akad. Wetensch. Amsterdam,* 7; reprinted in *Collected Papers,* 5.

Lorentz, H. A. [1904]: "Electromagnetic Phenomena in a System Moving with any Velocity Less than that of Light", *Proceedings of the Royal Academy of Amsterdam,* 6; reprinted in *Collected Papers,* 5.

Lorentz, H. A. [1906]: *The Theory of Electrons and its Application to the Phenomena of Light and Radiant Heat: A Course of Lectures Delivered in Columbia University, New York, in March and April 1906,* Dover, New York.

Lorentz, H. A. [1909]: *The Theory of Electrons,* Dover, New York.

Lorentz, H. A. [1914]: "La Gravitation", *Scientia,* 16; reprinted in *Collected Papers,* 7.

Mach, E. [1883]: *Die Mechanik in ihrer Entwicklung, historisch-kritisch dargestellt,* Wissenschaftliche Buchgesellschaft, Darmstadt; English translation *The Science of Mechanics,* 1893.

Mach, E. [1893]: *The Science of Mechanics,* Open Court, La Salle, Illinois.

Mach, E. [1905]: *Erkenntnis und Irrtum,* J. A. Barth, Leipzig.

Mach, E. [1906]: *The Analysis of Sensations,* Dover, New York.

Mach, E. [1909]: *The History and the Root of the Principle of the Conservation of Energy,* Open Court, Chicago.

Mach, E. [1913]: *The Principles of Physical Optics,* Dover, New York.

Malament, D. [1977]: "Causal Theory of Time and the Conventionality of Simultaneity", *Noûs,* 11.

McCormmach, R. [1970]: "Einstein, Lorentz and the Electron Theory" in R. McCormmach (ed.): *Historical Studies in the Physical Sciences,* 2, University of Pennsylvania Press, Philadelphia.

Meyerson, E. [1908]: *Identité et réalité,* Vrin, Paris.

Meyerson, E. [1921]: *De l'explication dans les Sciences,* Payot, Paris.

Meyerson, E. [1924]: *La Déduction relativiste,* Payot, Paris.

Meyerson, E. [1933]: *Réel et Déterminism dans la Physique Qauntique,* Hermann, Paris.

Miller, A. I. [1974]: "On Lorentz's Methodology", *The British Journal for the Philosophy of Science,* 25.

Miller, A. [1975]: "A Study of Henri Poincaré's 'Sur la Dynamique de l'Electron' ", *Archs. Hist. Exact Sciences,* 10.

Newton, I. [1686]: *Principia Mathematica,* University of California Press, Berkeley.

O'Rahilly, A. [1965]: *Electromagnetic Theory,* 2, Dover, New York.

Ostwald, W. [1895]: "Emancipation from Scientific Materialism", *Science Progress,* 4.

Peierls, R. E. [1963]: 'Field Theory since Maxwell' in C. Domb (ed.): *Clerk Maxwell and Modern Science,* Athlone Press, London.

Planck, M. [1897]: *Vorlesungen über Thermodynamik,* Veit, Leipzig; English translation by A. Ogg [1945]: *Thermodynamics,* Dover, New York.

Planck, M. [1906*a*]: "Das Prinzip der Relativität und die Grundgleichungen der Mechanik", *Verhandlungen der Deutschen Physikalischen Gesellschaft*, 8.

Planck, M. [1906*b*]: "Die Kaufmannschen Messungen der Ablenkbarkeit der β-Strahlen in ihrer Bedeutung für die Dynamik der Electronen", *Verhandlungen der Deutschen Physikalischen Gesellschaft*, 8.

Planck, M. [1907]: "Nachtrag zu der Besprechung der Kaufmannschen Ablenkungsmessungen", *Verhandlungen der Deutschen Physikalischen Gesellschaft*, 9.

Planck, G. M. [1908]: "Die Einheit des physikalischen Weltbildes"; reprinted in Planck [1973]: *Vortäge und Erinnerungen*, Wissenschaftliche Buchgesellschaft, Darmstradt.

Podlaha, M. F. and Sjödin, T. [1979]: "La Formule relativiste de l'effet Doppler dans une théorie d'ether", *Acta Physica Academiae Scientarum Hungaricae*, Tomus 48(1).

Poincaré, H. [1895]: "A propos de la théorie de M. Larmor", *Eclairage Electrique*.

Poincaré, H. [1900]: "La Théorie de Lorentz et le principe de réaction", *Arch. Neerland. Sci. Nat. Exact.*, V.

Poincaré, H. [1901]: *Electricité et Optique*, Gauthier, Villars, Paris.

Poincaré, H. [1902*a*]: *La Science et l'hypothèse*, Flammarion, Paris; English translation [1952]: *Science and Hypothesis*, Dover, New York.

Poincaré, H. [1902*b*]: "Sur la valeur objective des théories physiques", *Revue de Métaphysique et de Morale*, 10ᵉ année.

Poincaré, H. [1905]: "Sur la dynamique de l'electron", *Comptes Rendus de l'Académie des Sciences*, 140.

Poincaré, H. [1906*a*]: *La Valeur de la Science*, Flammarion, Paris; English translation [1958]: *The Value of Science*, Dover, New York.

Poincaré, H. [1906*b*]: 'Sur la dynamique de l'electron', *Rendiconti del Circolo Matematico di Palermo*, 21.

Poincaré, H. [1908]: *Science et méthode*, Flammarion, Paris; English translation [1958]: *Science and Method*, Dover, New York.

Poincaré, H. [1913]: *Dernières pensées*, Flammarion, Paris; English translation [1963]: *Last Essays*, Dover, New York.

Polanyi, M. [1958]: *Personal Knowledge*, University of Chicago Press, Chicago.

Popper, K. R. [1934]: *Logik der Forschung*, Springer, Wien.

Popper, K. R. [1945]: *The Open Society and Its Enemies*, Volumes 1 and 2, Routledge and Kegan Paul, London.

Popper, K. R. [1957]: "The Aim of Science", *Ratio*, 1; reprinted in Popper [1972].

Popper, K. R. [1958]: *Logik der Forschung*, Mohr-Siebeck, Tübingen.

Popper, K. R. [1959]: *The Logic of Scientific Discovery*, Hutchison, London.

Popper, K. R. [1963]: *Conjectures and Refutations*, Routledge & Kegan Paul, London.

Popper, K. R. [1969]: *Logik der Forschung (Dritte, vermehrte Auflage)*. Mohr-Siebeck, Tübingen.

Popper, K. R. [1972]: *Objective Knowledge*, Oxford University Press, Oxford.

Popper, K. R. [1979]: *Die beiden Grundprobleme der Erkenntnistheorie*, Mohr-Siebeck, Tübingen.

Prokhovnik, S. [1967]: *The Logic of Special Relativity*. Cambridge University Press, Cambridge.

Redhead, M. L. G. [1975]: 'Symmetry in Intertheory Relations', *Synthese*, 32.

Redhead, M. L. G. [1983]: "Nonlocality and Peaceful Coexistence:" in Swinburne [1983].

Reichenbach, H. [1958]: *The Philosophy of Space and Time*, Dover, New York.

Reichenbach, H. [1965]: *Axiomatik der relativistischen Raum-Zeit-Lehre*, Vieweg, Braunschweig.

Riemann, B. [1953]: *Collected Works*, Dover, New York.

Ritz, W. [1911]: *Gesammelte Werke*, Gauthier-Villars, Paris.

Russell, B. [1910]: *Philosophical Essays, III*, George Allen and Unwin, London.

Schaffner, K. [1969]: "The Lorentz Electron Theory and Relativity", *American Journal of Physics*, 37.

Schaffner, K. [1974]: ''Einstein versus Lorentz: Research Programmes and the Logic of Comparative Theory Evaluation'', *British Journal for the Philosophy of Science*, 25.

Schopenhauer, A. [1818]: *The World as Will and Representation*, Vol. I, Dover, New York.

Schrödinger, E. [1950]: *Space-Time Structure*, Cambridge University Press, Cambridge.

Shankland, R. S. [1963]: ''Conversations with Albert Einstein'', *American Journal of Physics*, 31.

Sjödin, T. [1979]: ''Synchronization in Special Relativity and Related Theories'', *Il Nuovo Cimento*, 51B, N2.

Stachel, J. [1979]: ''Einstein and the Rigidly Rotating Disc'' in A. Held and others (eds.) [1980]: *General Relativity and Gravitation One Hundred Years After Albert Einstein*, Vol. 1, Plenum Press, New York.

Stachel, J. [1982]: ''Einstein and Michelson, the Context of Discovery and the Context of Justification'', *Astronomische Nachrichten*, Vol. 303, pp. 47–53.

Stachel, J. and Torretti, R. [1982]: ''Einstein's First Derivation of Mass-Energy Equivalence'', *American Journal of Physics*, 50.

Stephenson, G. and Kilmister, C. W. [1958]: *Special Relativity for Physicists*, Longmans, London.

Swinburne, R. [1983]: *Space, Time and Causality*, Reidel, Dordrecht.

Trautmann, A., Pirani F. A. E. and Bondi, H. [1965]: *Lectures on General Relativity*, Prentice-Hall, Englewood Cliffs, New Jersey.

Watkins, J. W. N. [1958]: ''Confirmable and Influential Metaphysics'', *Mind*, 67.

Westfall, R. S. [1971]: *Force in Newton's Physics*, Macdonald, London.

Weyl, H. [1923]: *Raum, Zeit, Materie*, Springer, Berlin.

Weyl, H. [1949]: *Philosophy of Mathematics and Natural Science*, Princeton University Press, Princeton.

Whittaker, E. T. [1910]: *A History of the Theories of the Aether and Electricity, from the Age of Descartes to the Close of the Nineteenth Century*, Thomas Nelson and Sons, London.

Whittaker, E. T. [1951]: *A History of the Theories of Aether and Electricity: The Classical Theories*, Thomas Nelson and Sons, London.

Whittaker, E. T. [1953]: *A History of the Theories of Aether and Electricity, The Modern Theories 1920–1926*, Thomas Nelson and Sons, London.

Williams, L. P. (ed). [1968]: *Relativity Theory: Its Origins and Impact on Modern Thought*, Wiley, New York.

Wittgenstein, L. [1953]: *Philosophical Investigations*, Blackwell, Oxford.

Wittgenstein, L. [1956]: *Remarks on the Foundations of Mathematics*, Blackwell, Oxford.

Worrall, J. [1975]: *The 19th Century Revolution in Optics: A Case Study in the Interaction between Philosophy of Science and History and Sociology of Science*, University of London PhD Thesis.

Worrall, J. [1976]: ''Thomas Young and the 'Refutation' of Newtonian Optics: A Case Study in the Interaction of Philosophy of Science and History of Science'' in C. Howson (ed.) [1976]: *Method and Appraisal in the Physical Sciences*, Cambridge University Press, Cambridge.

Worrall, J. [1978]: ''The Ways in Which the Methodology of Scientific Research Programmes Improves on Popper's Methodology'' in Radnitzky and Andersson (eds.) [1978]: *Progress and Rationality in Science*, Reidel, Dordrecht.

Zahar, E. G. [1973*a*]: *The Development of Relativity Theory: A Case Study in the Methodology of Scientific Research Programmes*, University of London PhD Thesis.

Zahar, E. G. [1973*b*]: ''Why Did Einstein's Programme Supersede Lorentz's?'', *The British Journal for the Philosophy of Science*, 24.

Zahar, E. G. [1977]: ''Mach, Einstein and the Rise of Modern Science'', *The British Journal for the Philosophy of Science*, 28.

Zahar, E. G. [1978]: ''Crucial Experiments: A Case Study" in Radnitzky and Andersson (eds.) [1978]: *Progress and Rationality in Science,* Reidel, Dordrecht.
Zahar, E. G. [1980]: ''Einstein, Meyerson and the Role of Mathematics in Physical Discovery'', *British Journal for the Philosophy of Science,* 31.
Zahar, E. G. [1983]: ''Logic of Discovery or Psychology of Invention?'', *British Journal for the Philosophy of Science,* 34.

INDEX

Printed in the USA
CPSIA information can be obtained
at www.ICGtesting.com
JSHW021435221024
72172JS00002B/8

9 780812 690675